THE GREAT DESIGN

The Great Design

Particles, Fields, and Creation

ROBERT K. ADAIR

New York • Oxford
OXFORD UNIVERSITY PRESS
1987

Oxford University Press

Oxford   New York   Toronto
Delhi   Bombay   Calcutta   Madras   Karachi
Petaling Jaya   Singapore   Hong Kong   Tokyo
Nairobi   Dar es Salaam   Cape Town
Melbourne   Auckland
and associated companies in
Beirut   Berlin   Ibadan   Nicosia

Published by Oxford University Press, Inc.,
200 Madison Avenue, New York, New York 10016

Oxford is a registered trademark of Oxford University Press

Library of Congress Cataloging-in-Publication Data
Adair, Robert Kemp.
The great design.
Includes index.
1. Physics. 2. Particles (Nuclear physics)
3. Field theory (Physics) I. Title.
QC21.2.A28 1987   539   87-5729
ISBN 0-19-504380-4

To the memory of my father
Robert Cleland Adair
from whom I learned about science
and to the memory of my mother
Margaret Wiegman Adair
who taught me to care for the humanities

9 8 7 6 5 4 3 2 1
Printed in the United States of America

Preface

About 60 B.C. one of the world's great poets, Titus Lucretius Carus, wrote a long poem in Latin hexameters preaching the Epicurian gospel of salvation through common sense. The atomic model of Demokritos and Leukippos, proposed five centuries before, provided the bases of the description of the universe attributed to this "common sense." Lucretius gave to his poem the title, *De rerum natura,* which can be translated as, "On the Nature of the Universe." Directed nominally to his friend the Roman statesman Gaius Memmius, the poem was written for all educated laymen of the time.

Neither Epicurean nor Poet, but with the advantage of two millenia of scientific progress, I am attempting to address the same kind of audience with a similar exposition. Writing in a less heroic age, I apologize for foresaking the hexameter form, Latin or English. But, prose though it be rather than verse, I follow Lucretius in requesting the reader to "lay aside your cares and lend undistracted ears and an attentive mind to true reason."

In this book I have tried to present those basic concepts of particles and fields and of space and time, as illuminated by modern physics, very much as a professional physicist understands them. I believe that these concepts are accessible to the nonprofessional—that which I can't explain to an interested layman, I must not understand properly myself. Which is not to say the ideas are so trivial that they can be understood by physicist or layman without the "attentive mind" to which Lucretius refers.

By and large, there is little controversy in physics and hence there is little that is controversial in the book. The exceptions, largely concerned with the interpretation of quantum mechanics, are addressed in a spirit of "naive realism" and operationalism, which represent my own views and which I believe are not very different from those held by most physicists.

The text is nonmathematical, though on occasion simple relations are expressed in algebraic forms that should be known to anyone with a high-school education. Some more complex relations that seem to be especially interesting are presented in the extensive set of footnotes. Though few of these require mathematical sophistication beyond that taught in the first few weeks of a high-school algebra course, mathematical simplicity does not translate into conceptual simplicity, and these presentations will often require careful and time-consuming thought. Once written, a book has a life of its own independent of the author's control; however, I suggest that the mathematical footnotes be sampled rather than consumed. There are those who can "read" mathematics like a light novel, but for most of us so compact an information transfer cannot be assimilated easily and the time required to penetrate the arguments interrupts the narrative flow excessively.

It has been said that all expository books are simply forms of selected plagiarism. So be it for this book. I claim originality only for the errors, and even original errors are hard to come by. Thus, I am indebted to many colleagues, over many years, for most of the forms of presentation I have used. Only rarely have the pedagogical contributions of others been so marked—and remembered by me—that I have been able to name sources.

Perhaps one-half of the book concerns physics developed during my professional lifetime (the first paper under my name was published in 1947). Hence, I am often discussing work conducted by my friends and colleagues. Physics is a human endeavor and to describe developments without using names leads to a bloodless picture of physics. On the other hand, there is little gained by burying the reader under a load of names of strangers. Originally, I tried to solve the problem of selection by using only the names of Nobel and Wolf prize winners, but this plan broke down. To those physicists I have not mentioned when I might have, I apologize.

In 1930, with money he had earned from his book, *The Bridge of San Luis Rey,* Thornton Wilder built a home in Hamden, Connecticut, for his sister Isabel, his parents, and himself. Much of his later work was written in his fine study on the second floor of that house. This book was written in that same study and I like to think viewed tolerantly by a kindly ghost.

May 1987   R. K. A.
Hamden, Connecticut

Contents

THE GREAT DESIGN

1

Concepts in Physics

Nothing troubles me more than time and
space; and yet nothing troubles me less, as I
never think about them.

Charles Lamb

The Nature of Understanding in Physics

Even upon emerging from the cradle, it is obvious to us that there is order in
the universe: the world we live in, our life itself, follows rhythms and exhibits
patterns. Day always follows night; objects fall ever down, never up. It is evident that relations exist among our observations of the universe. The study
of relations among observations we can make of our universe constitutes the
body of science, and relations among observations of the physical universe
constitutes the subject matter of physics. It is not important to define the
extent of the physical universe or the extent of physics precisely; it is enough
to state that the primary concerns of physics are the relations between particles and forces, which lead naturally to inquiries into the nature of space and
time.

The study of observations of the physical universe implicitly entails asking questions. As a practical matter these observations must be considered in
some context. We must consider the character and limitations of the questions as well as the content of the answers. A story is told of the ichthyologist
who examined fish collected after the seining of a lake. He arrived at the interesting conclusion that all fish living in the lake were more than four inches
long. The moral, of course, is that one must examine the seine as well as the
fish—the questions as well as the answers.

The fundamental questions we ask in physics—and in other sciences—
cannot address *how* something works or *why* something happens. One can
only ask how one observation relates to another. This limitation on the character of questions derives from an implicitly accepted philosophy of science,
a kind of logical positivism. Although the personal philosophies of scientists
vary, as scientists they usually proceed according to the rule that in science
the only questions that are admissible are questions that can be answered *in
principle* by observation or by controlled observations and operations called

3

experiments. Again, as a matter of practice, such a positivism (or operationalism) does not preclude the use of intermediate constructs that may not be in themselves fully operational. For example, atoms (or even quarks, particles physicists now think of as fundamental building blocks) certainly exist as concepts that tie together very large numbers of observations. However, we may also ask (nonscientifically) whether atoms or quarks are real. To answer such a question requires an inquiry not so much into the nature of atoms or quarks than into the nature or definition of reality. We can say that atoms and quarks are real in the same sense that cats are real. It is not inconceivable that the concept of the atom (and perhaps of a cat) may change somewhat with time; it is less unlikely that the concept of quark will change with time. But atoms, cats, and even quarks are similar inasmuch as each is a linguistic construct representing the results of a large set of observations. Moreover, all physicists consider atoms real, or as real as anything else. Most consider quarks real. And who would dare to question the reality of cats to a cat?

Our condemnation of "how" pertains to the deep, fundamental, questions of physics. As a matter of common speech, physicists use "how" and "why" like everyone else—and without apology—in questions that refer uncertain inferences or observations to better established constructs or observations.

We must also establish an economy of description by the construction of generalizations. We might collect a large number of observations of falling bodies. We find, for example, that a large stone falls 64 feet in 2 seconds. In the course of another observation, we notice that an apple falls 16 feet from a tree in one second. We could accumulate an enormous number of individual data. Each datum in the collection would be true and the whole would certainly constitute a body of knowledge, albeit most unwieldy. We are now prepared to make a considerable intellectual step; we can make generalizations of all of this data. A simple generalization is that all bodies fall down. A much more subtle generalization is that different bodies fall from rest the same distance in the same time. An even deeper statement encompasses bodies thrown upwards or downwards with an arbitrary initial velocity, the acceleration of all heavy bodies is 16 ft/sec^2. You notice that we have said nothing causal: we have not said *how* bodies fall or *why* bodies fall. Our contribution is to relate the observations of the fall of different bodies, to find similarities among different observations.

Generalizations consolidate a considerable body of knowledge. However, if we observe our similar falling bodies more closely, we will see differences in their rates of acceleration that we learn to ascribe to the resistance of the air. William Butler Yeats, a mystic and a most unscientific man, tells of a wise man "busy with subtle thought, distinctions between things long held the same, resemblance of things long held different." Such wise consolidation and discrimination is the essence of science as well as of poetry. To physicists, consolidation and discrimination have an aesthetic quality: the logic of nature is beautiful, and a sense of this beauty is an important guide in the construction of physical theories.

The methods used to arrive at the generalizations that constitute physics are not qualitatively different from the methods used in solving other problems in human experience. Is there a special scientific method? There is little evidence that any qualitative difference exists between the reasoning of a physicist concerning the properties of elementary particles and fundamental forces, and the reasoning of a theologian, a lawyer, a businessman, or a farmer concerning problems of good and evil, legality and illegality, profit and loss, or famine and plenty. Indeed, Albert Einstein said, "The whole of science is nothing more than a refinement of everyday thinking." There is, however, probably some quantitative difference that follows from a special feature of the logic of the physical sciences—that is, the practicality of long, complex logical constructions. Physics is narrow, precise, and simple in a manner such that history, sociology, psychology, or the study of literature are broad, hazy, and difficult. Because of this simplicity, the premises used as the foundations of logical construction in physics are few and well defined, and often their validity is well established. With so strong a foundation, extensive logical structures leading reliably to far-reaching conclusions and generalizations can be constructed. In contrast, it is very difficult to establish any small but complete set of reliable premises in other areas of scholarship. The great generalizations of historians such as Toynbee or psychologists such as Freud are simply not reliable or useful in the sense of established physical laws. Their premises cannot support extensive logical extrapolations.

As we have noted, physics is the study of the relations among observations of the physical universe. But these relations must be a part of all possible relations among the objects of any set, and the latter relations constitute the body of mathematics. The logical structure of physics is therefore a part of the logical structure of mathematics, and the language of physics is mathematics. Physics benefits from the precise and compact notation of mathematics.

We can consider that mathematics is essentially formal logic. One must not, however, confuse physics with mathematics, nor should one consider that the relation of physics and mathematics is other than the relation of physics and logic. It is a faith of physicists that the universe they study is logical—and, remarkably, it appears that this is the case. Perhaps we are too complacent about the logic of nature. Eugene Wigner, a theoretical physicist and Nobel Laureate has written an essay entitled, "The Unreasonable Effectiveness of Mathematics in the Natural Sciences."

The fundamental simplicity of physics, which allows for extensive logical structures, leads naturally to considerable specialization among physicists. Much of humanistic scholarship concerns relationships of broad categories of human experience, and humanistic research is largely a study of these relationships. In contrast, physics is devoted to analysis and testing of extensive logical structures that can be (and usually are) built on narrow bases of experience, rather than to a study of relations among broad categories of observations. Depth of insight is thus more important than breadth of knowledge, and the limitation of most physicists' ability itself limits the number of areas in which they can be sufficiently expert to contribute. Humanists must be wise; physicists only clever. Humanists must be broad; physicists must be deep, and perforce few can afford breadth.

Perhaps the most striking division in physics is the division between experimentalists, who specialize in the conduct of experimental measurements, and theorists, who are occupied primarily with the logical (and mathematical) constructions derived from observations. Einstein is an example of a theorist; Albert Michelson, who conducted some of the experiments that led Einstein to construct the Special Theory of Relativity, is an example of an experimentalist. This division of labor does not signify a division of the science of physics, which is a unified discipline. Moreover, the brilliance of men such as Einstein and Niels Bohr should not obscure the fact that physics is based on experimentally obtained knowledge of the universe. Although many physicists believe, like Leibniz, that there can be only one logically consistent universe (which Leibniz, as a theologian, considered the best of all possible worlds), we are far from being able to deduce the structure of the universe from logical principles alone. Physics, then, is founded on observations and on the controlled observations we call experiments.

Idealizations in Physics

As a part of good reasoning and certainly as a part of the construction of physical theory, the construction of reasonable idealizations is important. Generally, we attempt to build our description of the universe from a set of constructs, each of which concerns some aspect of nature. Because, to some degree, all aspects of nature influence each other, any description of one part can apply only to the extent that the influence of other sectors can be ignored.

As an example, consider idealizations made with regard to falling bodies. Previously we stated (emphasizing similarities) that all bodies fall at the same rate of acceleration, and that any two bodies that fall the same distance from rest will take the same time to traverse a distance. But such statements ignore differences and are not correct for real bodies falling over a real course. One only has to hold up a feather and a coin to see that the coin falls quickly to the floor while the feather drifts slowly down. This might seem to be an unfair demonstration because the effect of air resistance is much greater on a feather than on a coin. Yet, for a real coin and a real feather in a real experiment, the resistance of the air (if we wish to call it that) exists as part of the real world, and in that world the coin and feather do not fall at the same rate. In our observations of falling bodies, we are able to extract that which is universal from the mass of particular results only by constructing an idealization: under conditions such that air resistance is not important, all falling bodies travel a given distance in the same time.

This particular idealization or abstraction is by no means trivial. It is clear that Galileo understood the idealization very well. Other, earlier, investigators probably did not.

But how do we extract the ideal result—in this case, the independence of the rate of a body's fall from its composition—if, like Galileo, we have no

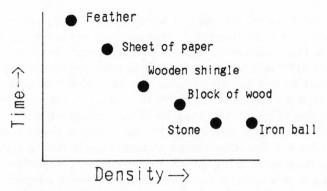

Figure 1.1 The time required for a body to fall a certain distance is plotted against an ordering of the sectional density.

way to produce a vacuum so that we can neglect air resistance? We can do this by performing a rather subtle experiment. We can drop a large variety of objects from the Leaning Tower of Pisa (the apocryphal story of Galileo's experiment) or some other location and measure the time the bodies take to reach the ground. We might then construct a graph in which the ordinate is the time of fall and the abscissa is the sectional density of the object—the mass divided by the cross-sectional area. We need not do this accurately or even assign a value to the quantity. We need only order the dropped objects, and we could probably do this using only our common sense. We might thus order a feather, a piece of paper, a small wooden shingle, a larger wooden block, a stone about the same size as the block, and an iron cannon ball the size of the stone, and rank the objects 1, 2, 3, 4, 5, and 6 in the order in which they feel more dense to us.

The graph that would result might look something like Figure 1.1. Very light objects take a long time to fall, whereas heavier, more dense objects fall more quickly. But there is a limit to this trend; there comes a point at which very dense objects all take about the same time to fall. Our general conclusion, couched in the language of modern physics, is that all objects fall at the same rate in the limit of great—or infinite—density. This is an important generalization that implicitly suggests that there is an idealization (the removal of all the air?) in which all bodies fall at the same rate.

The Character of Experiments and Theories in Physics

Almost all experiments designed to obtain information of interest in physics can be described as measurements. If the results are to be useful, both the procedures used in the measurements and the measured quantity itself must be well defined. Although measurements may be difficult in practice, the experimental methods and the interpretations of the results are almost always simple in principle. Compared to the behavioral sciences and even much of the biological sciences, physics is simple inasmuch as the physicist can nearly always provide a complete and precise description of the goals, procedures, and conclusions of an experiment. Thus, experiments in physics are reproducible in a comparatively straightforward and noncontroversial way. This reproducibility provides experimental physics with considerable rigor.

Theories must follow conventions that have been established through experience, and a reasonable discipline is necessary in the construction and dissemination of hypotheses. It is difficult to construct a precise set of rules that is not, on occasion, usefully violated, but in general a new theory should be well defined so that its precise consequences can be deduced. Those consequences must not be in conflict with well-established observations. The theory should be new in that it differs from some previously accepted view and predicts some new and different result, or in that it reformulates and extends some previously accepted theory.

Furthermore, there is an unwritten law in the physics community that the creator of a new idea must show that the idea is new, complete, and correct such that its consequences do not violate what is already known to be true. As a matter of practice and of economics, the community has no obligation to prove or disprove a carelessly formulated hypothesis. It is easy to propose a radical idea; skill and effort are required to establish the consequences of the idea and to determine the relation of these consequences to the results of experiments. Most of the effort of theoretical physicists is concerned with this examination of ideas' consequences and not with the random generation of untested hypotheses.

Some confusion is sometimes generated by different uses of the word "theory." The term is used both for well-founded descriptions of nature and for imaginative conjectures. Newton's Theory of Gravitation is certainly a correct description of nature (within well-known limitations), whereas the $N = 8$ Supersymmetry Grand Unified Field Theory is a bold (and fascinating) hypothesis that does not now seem to be a correct description of nature. This semantic problem with the word "theory" probably contributes to a misunderstanding of the degree of change in our understanding with time. Outside the profession, it may sometimes seem that all is quicksand, that everything physicists knew yesterday has been overturned by today's ideas, that everything they know today may be gone tomorrow. Actually, basic concepts are almost never—perhaps never—overturned, although grand designs take different courses. It is rather as if physicists are laborers building a cathedral, who have constructed the foundation and many rows of stone. As common workers, masons, and carpenters, they never see the architect's design, but during lunch hour they talk about their conception of the design on the basis of the work they have done. As time goes on and a buttress grows here, a nave takes form there, they change our idea of the final building. But the foundation they have dug and the stones they have placed remain unchanged.

It is useful in this context to consider gravitation. Sir Isaac Newton described gravitation more than three centuries ago in terms of action-at-a-distance, and that description is still the basis for our extraordinarily accurate predictions of celestial mechanics. In 1918 Einstein published his General

Theory of Relativity—a theory of gravitation very different in concept from that of Newton, in that action-at-a-distance was replaced by a description of the effects of curved space. We know that Einstein's theory will, and must, be replaced by a quantum theory of gravitation, and we suspect that his geometric interpretation will be replaced by quite a different picture. But for velocities that are small compared with the velocity of light, and for the weak gravitational forces we find almost everywhere in nature (outside of black holes), the results of Einstein's description of gravity are, and must be, the same as Newton's. Einstein's model must reduce to Newton's over the region of applicability of Newton's model. Similarly, the quantum theory of gravity to come will, and must, lead to the same conclusions as the General Theory over almost all facets of nature. The differences will be important only under conditions physicists believe existed only at the birth of the universe—but then the differences will be crucial. Then, the quantum theory of gravity must reduce to the General Theory over most regions of experience, and both must reduce to Newton's model over a more limited area.

Einstein's General Theory swept away the conceptual action-at-a-distance substructure of Newton but left the operational or practical consequences almost intact. Physicists now suspect that quantum gravity will sweep away the geometric conceptual basis of the General Theory, but it must leave the operational consequences of Einstein's model almost untouched. So, was Newton wrong?—yes and no. And is Einstein wrong?—yes and no. The stones Newton laid are still there and will always be there, although the cathedral is a little different than he might have thought. Honor Newton. The stones Einstein laid are also still there and will always be there, although the Cathedral is not what Einstein envisioned. Honor Einstein.

Language, Thought, and Physics

In physics we are concerned with dimensions that are very much smaller than those we can perceive directly through our senses, and with very large dimensions. We are concerned with very small and very large masses, with very short and very long times, and with very large velocities. Both the microcosm and the macrocosm are far from our direct experience. The entire structure of that experience and the entire structure of our languages (and, perhaps, the very structure of our thought) has been developed by considering the magnitudes of things that are within our immediate sensory grasp.

The English language, as well as most other European languages (that is, the Indo-European language group), seems to have developed largely from the speech of a primitive tribe that may have lived somewhere between the Elbe River and the Don in what is now East Germany, Poland, or Russia. This language was developed for the purpose of discussing cows, men and women, children, the flight of game, the growth of grain, the variation of weather, and—perhaps most abstractly—the wrath of the gods. Both the vocabulary and the structure of this language limit us to things within the

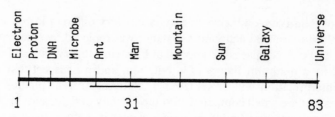

Figure 1.2 A logarithmic scale spans the dimensions of the universe, showing masses of various objects. One unit represents a factor of 10 in mass. The double lined region suggests the range of objects directly accessible to our senses, hence addressed directly by our language.

range of our direct perception if we wish to discuss matters in any precise way. The universe, which is the subject of inquiry in physics, extends over an enormous scale of magnitude. Most of that universe is far from our direct experience, and the vocabulary we use to discuss the universe cannot be easily adapted to describe it. Moreover, the very way we think is probably influenced strongly by the vocabulary and structure of our language, which may interfere with our understanding of matters involving scales of magnitude far from those of our primitive experience.

The extrapolation to the macrocosm of the galaxy and to the universe as a whole is very great. It should not be expected, then, that the General Theory of Relativity, which is basic to any description of the entire universe, could be easily understood on elementary bases. It is possible that a precise statement of the General Theory must be beyond the limits of any elementary understanding. It is then eminently reasonable that we extend our ideas by using metaphors. In discussing or considering the theory, laymen and expert alike must use analogies to link such abstract concepts as four-dimensional space-time with the accessible geometry of three-dimensional space. Of course, these analogies are at best imperfect shadows of the accurate map of reality that derives—it is hoped—from a full mathematical exposition.

For much the same reason physicists cannot properly describe the microcosm in terms of direct experience. We speak of elementary subdivisions of matter as particles and waves; however, they are neither particles or waves, but in a limited way they behave somewhat like primitive particles and somewhat like primitive waves—a little like stones and a little like ripples on the surface of a pond. We have no better words and no better language to describe these phenomena than the language and words we use in everyday life. Yet we must be careful to recognize the limitations of our language.

Figure 1.2 shows the relative masses of typical objects in our universe on a logarithmic scale where a unit length represents a factor of 10 in mass. The scale begins with the mass of an electron, the lightest discrete object we know with a nonzero mass. The proton, with a mass about 2000 times that of an electron, is a little more than 3 units to the right; a DNA molecule is about 10^5 times as massive as a proton, or 8 units to the right. We find a very small ant at 20 units, and the mass of a man is about 10^{31} times that of an electron. A mountain is a medium-sized object with a mass on the order of 10^{43} times that of our smallest mass, the mass of the electron. The ratio of the mass of the known universe to the mass of a mountain is about equal to the ratio of the mass of the mountain to that of an electron.

One cannot directly perceive or determine the mass of an object if it is much more than 10 times the mass of a man or if it is much less than 1 millionth of the mass of a man. Also shown in the figure is the estimate of the range of primitive sensory perception, hence the effective range of precision of our language and our primitive concepts. Our language is molded by an experience limited to that small sector; our way of thinking is constrained by our experience in that sector. We cannot expect that either the microscopic or the macroscopic character of the universe can be easily described with the language we use or easily understood in terms of the elementary concepts we meet in everyday life.

The Reality of Physics

Few would question the subjective character of art criticism or the view that great art, music, and literature are molded by the culture in which they are formed. Surely history is defined both by events and by those who record and describe the events. Even the history of Switzerland is described (quite) differently in Lucerne and in Zurich. Similarly, there are few absolutes accepted uncritically by all clinical psychologists. Sociology and anthropology, too, are affected deeply by the culture in which sociologists and anthropologists are imbedded.

In like fashion, the validity of physicists' descriptions of the physical universe has been questioned. Are those descriptions an accurate map of an underlying reality, or is this map a construct (perhaps one of many equally valid constructs) shaped more by the prejudices and limitations of physicists than by underlying design of nature (if such a design even exists)?

Clearly, the development of the description of the physical universe encompassed by the label physics is a process conducted by human beings and as such is subject to all the complexities and frailties of human behavior. And although physics is based on observations, and especially on the sharply controlled observations called experiments, these observations are themselves certainly guided and influenced by the intellectual climate inhabited by the observer. It is easy to see the expected and most difficult to see the unexpected. Hence, experiments are influenced strongly by theories, and the (conjectural) theories developed at any time and place are influenced strongly by an entire set of views held at that time and place. If the process is subjective, ill-defined, and prey to prejudice, can the results be trusted? Does the accepted body of physics describe nature, or is physics but a vast metaphor?

This is not the place to thoroughly plumb each side in such a debate, but it is important to note that virtually all working physicists believe firmly that there is a real universe and that their admittedly imperfect procedures converge to a description of that reality. Moreover, although they know that description to be incomplete, they are confident that it is not seriously in error. It may be useful to restate this in a fanciful but operational way. If, somehow, a flying saucer from a planet of the star Betelgeuse landed beside the Institute for Advanced Study in Princeton, New Jersey (where Einstein worked during the last half of his life), and if the equally arcane inhabitants of saucer and Institute were to compare their physics texts, no physicist would doubt that the two books would be in complete agreement. It is much less clear that the two groups would agree on the probable character of that which they did not know—indeed, they might disagree strongly on the best direction of research.

Why are physicists so certain they are right and even think that the very question regarding reality is silly? Another scenerio may suggest the reason for physicists' certainty and impatience with such a question. Imagine that you have been working for a long time on an enormous jigsaw puzzle of 10,000 pieces all laid out on a ballroom floor. After a month you have constructed a center containing a few thousand pieces, which is beginning to show the pyramids about Gizeh. Then someone who has never worked a puzzle questions the validity of the part you have reconstructed, suggesting that the scene really should show the London docks. No way!

If a description is real and correct, it must withstand the winds of time. And that of physics does. In my active lifetime as a physicist (dating from the first paper carrying my name published in 1947)—which I can claim covers half of the history of physics—I have never had to renounce anything I once knew was true. Which is not to say that a host of conjectures, some held dearly, have not vanished with the snows of yesteryear.

The Goals of Physics

Research in physics can usefully be divided into several categories, although the categories themselves do not have completely definite boundaries and blur into each other. First, there is fundamental and applied physics. The design of a device that will produce detailed information about the structures of the human body through a series of X-ray photographs taken from different aspects (the CAT scanner) can be considered applied physics—and very important applied physics, recognized by the award of Nobel Prizes. The study of those states of atoms which contribute to the line spectrum part of the X-rays is fundamental research of an intensive character. The characteristics of the X-rays are to be understood in terms of known fundamental properties of atoms and radiation. Here, details of nature are to be deduced from the application of well-established principles. Then, a study of the circular polarization of the X-radiation produced by an atomic transition might be expected to throw light on the relation of electromagnetism and the weak nuclear forces and contribute to the construction of a unified field theory linking all the fundamental forces of nature. This fundamental research, directed toward the analysis and construction of new basic structures of physics, we might call extensive research.

This extensive, fundamental research in physics is the subject of this book. What are the goals of such research? They can be stated in a myriad of ways. We will say that we must study the discrete and the continuous, and we must consider *variance* and *invariance,* or *change* and *conservation.* The consideration of these sets of antonyms leads us to the study of the character of elementary particles and fundamental fields or forces, and to the analysis of space and time, the structure of our universe, and the evolution and origin of that universe: indeed, to an understanding of the Grand Plan of the Master Architect.

2
Invariance and Conservation Laws

> Nature as a whole possesses a store of energy[1] that cannot be in any way either increased or diminished.... The quantity of energy in nature is eternal and unalterable.... I have named this general law "The Principle of the Conservation of Energy."
>
> Hermann von Helmholtz

The Special and the Particular in Observations: Boundary Conditions

Within the accuracy of our observations, and over time intervals small compared to the age of the universe (circa 15 billion years), the basic character of the universe appears to be independent of the time and position of observations. Observers who live at different times or in different places certainly see some aspects of the universe differently, but they will see some in the same way. The description of any particular event or series of events can be divided into two parts. One part depends upon local circumstances and thus upon the particular time, place, and history of the observation, while another part is more nearly universal and constitutes a fundamental rule of behavior or law of physics.

As an illustration of the separation of the particular and the universal, let us once again consider falling bodies and, for simplicity, restrict ourselves to the motion up and down of heavy bodies projected differently from different heights. We might determine the position of a heavy body (perhaps a ball) by taking a time exposure of the flight of the ball in conjunction with a stroboscopic light source that flashes once a second. The resulting picture might look like Figure 2.1. We might take a set of such photographs of different balls thrown in different ways from different heights and examine the data in a search for some universal element.

In one of the photographs we might search for regularity by determining

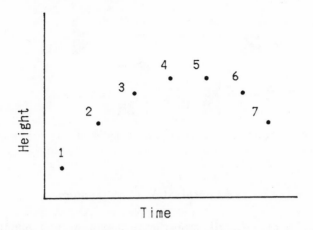

Figure 2.1 The results of a stroboscopic time-exposure photograph of a ball in flight. The light is flashed once a second.

the vertical position of the ball every second by making measurements on the photograph and recording the numbers in our notebook. Then we might inspect groups of three adjacent numbers—for example, the heights at 1, 2, and 3 seconds—labeling the numbers A, B, and C, or we could choose the set at 6, 7, and 8 seconds and label them A', B', and C'. The two sets of numbers will be quite different. Moreover, the difference between adjacent numbers of the two sets such as $B - A$ and $B' - A'$ will be different. But the difference between the differences of the first two and the last two of each set will be the same and equal to -10 meters (more accurately, -9.8 meters, or about -32 feet): $(C - B) - (B - A) = C - 2B + A = -10$ meters and $(C' - B') - (B' - A') = C' - 2B' + A' = -10$ meters.[2]

We could examine many sets of such stroboscopic pictures, taken at different places, showing different objects tossed from different heights, up or down, with different initial velocities or with different *initial* or *boundary conditions*. However, if we operate from the surface of the earth and consider only heights that are small compared with the radius of the earth, we find that the same simple recipe for the difference between velocities describes all the different observations: the change in velocity in 1 second, the acceleration, is about 10 meters per second squared (m/sec^2) or 32 ft/sec^2 directed downward. The equality of the accelerations or the differences between distance (defining a differential equation) and the value of 10 m/sec^2 constitute the universal attributes; the different boundary conditions represent the particular.

Of course, the universality of this illustrative example is limited. On the moon the acceleration would be only about ⅙ as great; on the sun, about 27 times greater. The broader generalization required to cover almost all gravitational interactions was found by Newton, who showed that the gravitational force depended in a definite manner on the masses of the relevant bodies, in this case the earth, moon, and sun.

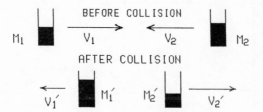

Figure 2.2   Volumes (masses) and velocities of quantities of water before and after a collision interaction.

Conservation Laws—The Conservation of Momentum

In our assembly of systematic properties of nature, we find quantities that have the property that their magnitudes are unchanged in the face of general change; that is, we find *conserved* quantities. Moreover, we find aspects of nature that remain constant or invariant as our view changes. We may refer to such invariances as *symmetries.* The unchanged, conserved quantities are always related to these unchanged, invariant symmetries. Moreover, these conserved quantities and the symmetries that define the conservations constitute specially useful bases for descriptions of the physical universe. They are important in themselves and important as examples of a class. It is then useful to discuss the simple, familiar conservations of energy, momentum, and angular momentum in an elementary—but nonetheless sophisticated—fashion.

Consider first the conservation of momentum. For clarity we will construct a set of thought experiments that are idealized versions of experiments that we could actually carry out. In the course of describing these experiments, we will define quantities that we will call *mass* and *momentum.*

Our set of experiments consists of the observations of collisions between quantities of water. Although we are interested at this time in pure samples of water, we bow to practicality and assume that the water is held in very light containers designed to move on a flat surface with almost no friction. (In a real experiment we would have to make corrections for the mass and friction of motion of the containers. In the spirit of the thought experiment, however, we neglect these complications.)

A collision between such volumes of water is shown in Figure 2.2. For simplicity we consider the special case of a collision between two volumes of water such that all motion lies in one direction.[3] Here a volume of water m_1 moving with a velocity v_1 collides with a second volume of water m_2 moving with a velocity v_2. During the collision, water may spill from one container to the other, so that the volumes held by each container may not be the same after the collision. Moreover, the collision may be elastic—perhaps the containers make contact through a spring—or completely inelastic—perhaps the containers stick together after the collision. Or the collision may have a character between these two extremes; we place no restrictions on it. Then, typically after the collision, we find a new volume of water in the first container m_1', moving with a new velocity v_1' and a new volume in the second container m_2' moving with a changed velocity v_2'.

We may make many observations of such collisions; in each case we will notice that there is a particularly striking relation between certain combinations of the measured quantities before and after the collisions. In all cases,

$$m_1v_1 + m_2v_2 = m_1'v_1' + m_2'v_2'$$

The sum of the "mass times velocity" quantities is unchanged by the collision or interaction of the volumes of water. We give the special name *momentum* to the product mv of the volume of water and velocity and summarize the results of the set of experiments by saying that *momentum is conserved.*

Of course we would hope that this conservation law holds, in some form, for materials other than water: what can be so special about water? For example, let us arrange a collision between a piece of iron and a volume of water. We will find that the original equations, developed to describe collisions of water, hold if and only if the volume of iron is counted as equal to about 7.6 volumes of water. Because volume is not a universally useful measure of iron (and other materials) as it takes part in collisions, we invent a name, *mass,* defined so that 1 liter of water is now called a mass of 1 kilogram and about 1/7.6 liters of iron have the same mass as a liter of water or a mass of 1 kilogram. Of course, it is conceivable that the equations describing collisions will not be satisfied for many interactions of iron, or some other material, using this new formulation. But in fact, the equations are satisfied, and we then extract from our set of observations a definition of a quantity, mass,[4] that allows a further definition of a conserved quantity, momentum.

Moreover, we need not limit our considerations to collisions such that two initial elements collide to form two final elements. The conservation law holds for interactions (or collisions) of any kind involving any number of initial entities going to any number of final entities. If we extend our observations (or experiments) to multiple collisions involving interactions of all kinds, we still find that the conservation law holds. The momentum of a closed system is an invariant or constant of nature.

The Conservation of Energy

Even very simple collision systems are not completely defined by the law of the conservation of momentum. This suggests to us that there must be further conservation laws or constraints in nature that might be revealed by the examination of some simple systems.

Figure 2.3 illustrates a familiar and very simple collision phenomenon. The objects could be a set of similar billiard balls aligned on a pool table or a set of similar coins set on a smooth, low-friction surface. Here we examine a system consisting of a number of identical hard balls, such as billiard balls,

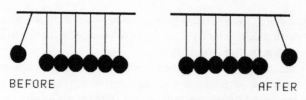

Figure 2.3   Elastic collisions of balls of equal mass.

each of mass m, suspended by strings so that they touch one another in a manner such that the centers of contact are in a straight line. If one of the balls at the end of the group is drawn back a distance s and let fall so that it collides with the set of balls with its velocity in line with the line of centers, one ball on the other end of the set will rebound to approximately the height and distance s' characteristic of the release of the first ball: $s = s'$. It is almost obvious, even without more direct measurements, that the velocity of the final ball just after the collision is equal to the velocity of the initial ball just before the collision. Because the balls are identical, the mass of the first ball times its velocity before the collision is equal to the mass of the second ball times its velocity after collision, which is what we expect from the law of conservation of momentum.

If we pull two balls back from the group, hold them together, and then let them fall and collide with the set, the two original balls will stop dead, and two balls from the opposite end of the set will rebound together almost as a continuation of the motion of the first two balls. Again the velocity of the first set of balls, v, just before the collision will be equal to the velocity of the second set, v', just after the collision. This is consistent with the law of conservation of momentum. If the mass of a ball is m, then

$$\text{before } (2m)v = (2m)v' \text{ after}$$

If we hold three balls back from the set and release them, they will stop upon colliding with the set, and three balls from the far end will rebound together with about the same velocity as the initial three. We will find the same kind of pattern with any number of balls.

Although these results are consistent with the law of conservation of momentum, they are not uniquely determined by that invariance. In the first case, where one ball was released and struck the row of balls and another ball rebounded from the far end, momentum would have been equally well conserved if two balls had rebounded from the far end, each with half the velocity of the first ball. Or three balls could have rebounded, each with one-third the velocity of the first ball. An infinite number of combinations can be found such that momentum is conserved. But the experimental result is unique and well determined. Another rule must operate to define the rather specific condition that is observed. An analysis of collisions of balls of different mass[5] shows that a rule to the effect that the quantity mv^2 is conserved will serve to select the processes that are actually observed. For convenience we will use $\frac{1}{2} mv^2$, to which we give the name *kinetic energy* as the conserved quantity.

Elastic and Inelastic Collisions

Most collision phenomena will not fit the simple conservation law just cited. Kinetic energy is seldom conserved and almost never exactly conserved. What good is a law that describes (imperfectly) only a very narrow class of phenomena? If we put a little putty between the balls we are using, as in Figure 2.3, we will find that the collisions take quite a different course. Indeed, no simple pattern is evident. If we make careful measurements, we will find that momentum is still conserved but that we have lost kinetic energy; the kinetic energy of all the balls after the collisions will be less than before the collisions.

Physicists call these interactions where kinetic energy is lost *inelastic* collisions and ascribe the deficit to poor accounting. That is, the elastic collisions we observe are believed to constitute a special set of interactions that exhibit, with special simplicity, a conservation law that *must* hold, when properly described, for all interactions. Physicists believe that energy is conserved, and the kinetic energy that is missing will be found in some other form if we learn to do our accounting better. Part of the kinetic energy lost from the macroscopic balls will be found in the form of increased kinetic energy of the microscopic particles that make up the balls and the putty between the balls. We can measure this energy in the form of heat and equate heat energy and mechanical energy. We may also improve our accounting by including changes in potential energy, discussed next.

Potential Energy

The concept of potential energy has been invented to allow us to generalize from the narrow concept of the conservation of kinetic energy in certain special, limited cases to the more important and general concept of the conservation of (all) energy. Figure 2.4 suggests the character of the concept of potential energy. A ball falls from a height onto a spring; the ball compresses the spring by falling on it; then the ball is projected by the spring into the air back to its original height. We consider the energy balance as follows: the ball has an initial gravitational potential energy with respect to the base height of the spring. In the course of the fall of the ball, this potential energy is converted to kinetic energy. When the ball strikes the spring, compressing the spring, the kinetic energy of the ball is converted to the potential energy of the compressed spring. This potential energy is released by the expansion of the spring and transformed into the kinetic energy of the rebounding ball. That kinetic energy is in turn changed back into the gravitational potential energy of the ball as it regains its original height. Neglecting small transfers of energy to the atmosphere (heating the atmosphere) as a consequence of the air's resistance to the flight of the ball, and neglecting some small inefficiencies of any real spring (which absorbs some energy in the heating of the spring), the total of the ball's kinetic energy, the ball's gravitational energy

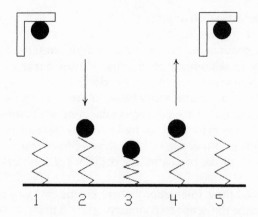

Figure 2.4   The history of a ball falling from a height onto a spring.

with respect to the base height of the spring, and the spring's potential energy is a constant. Energy is conserved.

It might seem that the introduction of potential energy is only a book-keeping trick and that the thesis of the conservation of energy would be a definition rather than a consequence of observations of nature. This is not the case; potential energy is real energy, always convertible to kinetic energy. The conservation laws work: no energy is ever left over after a cycle is completed, and the ball will never rebound to a height greater than its initial height. Moreover, when the ball rebounds to a lower height than its original one, the loss can be found quantitatively—in heat, for example. With more knowledge of forces such as gravity, we are able to understand the storage of potential energy in terms of the establishment of force fields, and we find, as we should, that the potential energy V has a mass M such that $V = Mc^2$, even as Einstein's relation $E = Mc^2$ holds for other energies.

In certain specific cases it is possible to show that the potential energy change is actually a change in the kinetic energy of microscopic particles. If the spring discussed in the foregoing example were a pneumatic cylinder, it would be easy to show that the change in the potential energy is just a short-hand way of describing a change in the kinetic energy of the gas molecules in the cylinder. If we extend our notion of particles and kinetic energy to encompass the energies of quantum particles contributing to force fields, we may be able to consider all changes in potential energy as changes in the kinetic energies of particles—but we will not pursue this rather subtle view further at this time.

The Conservation of Angular Momentum

Our day-to-day experience of nature suggests to us that there may be other constants of nature besides energy and momentum. A spinning wheel persists in its motion; not only does the rate of spin tend to remain constant, but the direction of the axis tends to remain fixed, as with a gyroscope. Although the conservation of angular momentum is most obvious when we consider circular motion, the concept is in fact more generally applicable. This invariance of nature may be introduced by analyzing a characteristic of motion in a straight line.

The trajectory of a particle moving in a straight line at a distance a from a point i is shown at the left in Figure 2.5. If the particle moves at a constant velocity v, it will traverse equal distances in equal intervals of time. Two such displacements are displayed in the figure, which can be considered, for example, distances traveled in 1 second: ds is equal to ds'. The shaded areas subtended by the displacements ds and ds' from the point i are then equal. Both areas have the form of triangles. The bases of these triangles, ds and ds', are equal from the condition that the particle is moving with constant velocity, and the displacements represent displacements traversed in equal times; the height of each triangle is the distance a, and the area of each triangle is therefore equal to $\frac{1}{2} a\, ds$. The path of the particle subtends equal areas with respect to any arbitrary point in equal intervals of time. If we multiply this area subtended per unit of time by the mass of the particle (and multiply by 2 as a convention), we have a quantity referred to as angular momentum, which we find is universally conserved.

The same particle moving with the same velocity but held by a string so as to move in a circle about the same point (displaced in the figure) is shown at the right in Figure 2.5. Because the area swept out about the point in a unit of time is the same as in the straight-line trajectory at the left, the angular momentum is the same. Indeed, if the string should break at the right time, the particle would continue in a straight line just as shown at the left. However, if the axis of the circle were directed otherwise—for instance, so that the plane of rotation was out of the plane of the paper—the angular momentum would surely be different; no break of the string could send the particle

Figure 2.5 The trajectory of a particle moving in a straight line with constant velocity is shown to the left. The trajectory sweeps out equal areas with respect to an arbitrary point i in equal times. Both ds and ds' are distances that the particle travels in 1 second. At the right, the particle moves in a circle with the same constant velocity and the same angular momentum with respect to the point i.

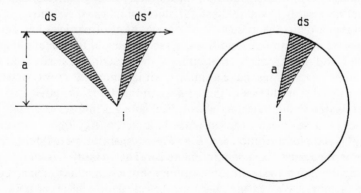

into the same trajectory as that shown at the left. Thus, angular momentum has a direction, or is a vector, and both the magnitude and the direction are seen to be conserved. By convention, the direction is taken as the direction of the axis of rotation such that the direction of the angular momentum in Figure 2.5 is into the paper.

For the simple situation shown in Figure 2.5, the magnitude of the angular momentum of the particle with respect to the point i can be expressed by the relation

$$j = mva$$

That is, j, the angular momentum, is equal to the mass m times the velocity v times a, the perpendicular distance between the path of the particle and the joint. Again, the angular momentum has a direction that is, in this case, into the paper.

Angular momentum is seen to be generally conserved—both in magnitude and direction. If we take a set of particles and follow them through a history of collisions, fissions, fusions, and any other kind of interaction, the total angular momentum about *any* point is constant; that is, the angular momentum of the particles about that arbitrary point is conserved. Johann Kepler noticed this conservation in his analysis of the orbits of planets about the sun; planets sweep out equal areas in equal times in their paths about the sun, a result that is codified as Kepler's Second Law.

The Principle of Least Action—The Lagrangian

One might suspect that the conservation laws constrain phenomena so that classical systems behave in a unique fashion, but this is not self-evident. Although the trajectory of the ball shown in Figure 2.1 (redrawn here as Figure 2.6) is such that the total energy, the sum of the potential and kinetic energies, is conserved and has the same value at each instant of time, is it clear that the conservation of the total energy—and the initial velocity— determine the curve defined by the snapshots? Perhaps another trajectory would also conserve energy. For so simple a phenomenon, it is not difficult to establish the unique character of the trajectory through direct calculations; it is not necessarily trivial to do so for more complicated systems.

Although the paths in space and time of a set of particles are uniquely determined by the initial conditions, an accounting of the forces among the particles, and the three conservation laws, this description seems a bit inelegant. Surely Nature must have a simpler, wittier, more clever way of describing Her world than this set of three rules. And, indeed, She has.

Let us divide the trajectory shown in Figure 2.6 into seven equal times. For each of these times we can calculate the kinetic energy T and the potential energy[6] V and plot the difference, $L = T - V$, against time, as shown by the points on the graph to the right in Figure 2.6. The quantity L is known as the *Lagrangian* (after the eighteenth-century French mathematician, Joseph Louis Lagrange). We can then perform the calculation for every point so as

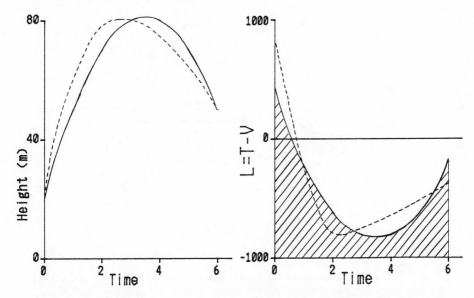

Figure 2.6 At the left, the solid line shows the height of the ball as a function of time—as shown in Figure 2.1. The dotted curve shows an arbitrarily chosen different trajectory. The solid line on the graph at the right shows the Lagrangian, $T - V$, of the ball as a function of time; the cross-hatched area under that curve is proportional to the action. The dotted line shows the value of $T - V$ for the deviant dotted trajectory at the left, and the area under that line, which is larger than the cross-hatched area, corresponds to the action for that path.

to get the smooth curve shown and to measure the area under the curve[7] from time $t = 0$ to time $t = 6$. That area, A_{06}, is called the *action* for the interval between t_0 and t_6. Any different path between the initial and final positions, such as that shown by the dotted line to the left in Figure 2.6, will lead to a larger action than that chosen by nature. The area under the dotted line on the graph to the right is greater than that under the solid line corresponding to the path chosen by nature, even as the action, defined by that area, is greater. The action will have the smallest value for the trajectory that Nature chooses. This is the *Principle of Least Action*.

Using the Principle of Least Action and knowing the form of the Lagrangian—and almost nothing else—we can deduce the correct motion of the ball. In particular, we need never have heard of the conservation of energy. We can proceed by trial and error and plot a curve such as that shown at the left in Figure 2.6, and then plot the corresponding curve of the Lagrangian versus time, as shown at the right. Then, by adjusting the trajectory curve so that the area under the Lagrangian curve (the action) is as small as we can get it, we should end up with a very good fit to the actual path. With enough mathematics (the Calculus of Variations), we can calculate first the trajectory and then the *equations of motion* of the ball if we know the form of the Lagrangian. (We will also find that the motion can be described by the simple relation $a = g$, and the acceleration is equal to a constant, $g = -9.8$ m/sec^2, the acceleration of gravity.) In mechanics, if we know the Lagrangian, we can in

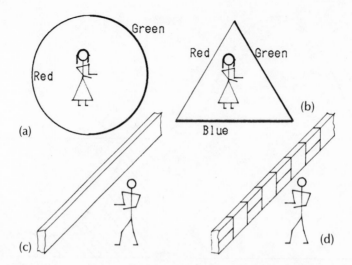

Figure 2.7 Various symmetries are shown with respect to rotation (*a*) and (*b*) and translation (*c*) and (*d*); (*a*) and (*c*) are continuous, whereas (*b*) and (*d*) are discrete.

principle calculate the equations of motion. In short, if we know the Lagrangian, we know everything—in principle. The conservation laws follow from the Lagrangian and are not separately invoked. (The "in principle" is stressed inasmuch as a very simple Lagrangian can be very difficult to solve.[8])

If Nature has defined the mechanics problem of the thrown ball in so elegant a fashion, might She have defined other problems similarly. So it seems now. Indeed, at the present time it appears that we can describe all the fundamental forces in terms of a Lagrangian. The search for Nature's One Equation, which rules all of the universe, has been largely a search for an adequate Lagrangian.

The Classical Conservation Laws and Space-Time Symmetries

As stated earlier, the existence of conserved quantities in nature is related to the existence of symmetries or invariances in nature. Before we explore the consequences of such symmetries or invariances and their explicit relations to conservation laws, it is useful to analyze the meaning of these concepts by considering the character of some elementary symmetries.

Figure 2.7*a* shows a circle with a woman observer standing at the center. Assume that the circle is painted half green and half red. If the observer is color blind, the enclosure is *invariant* with respect to rotation of the observer: the observer will see the enclosure in just the same way no matter how she is oriented. Of course, the symmetry is broken if she can detect the color in some way. For some purposes the circle is symmetric; for other purposes it is not. The character of the symmetry depends upon the tool used to measure or determine the orientation of the observer. Conservation laws based on a partial symmetry may be broken by processes that do not observe that symmetry.

The observer will see the circle in the same way for an infinite number of orientations. Moreover, the invariance is continuous; the view is the same as she slowly turns around. The woman in the center of the triangle in Figure 2.7b sees a threefold, discontinuous symmetry. If she whirls around with her eyes closed and then stops and opens her eyes and looks around, she will not know what direction she is facing up to a threefold uncertainty. The view is invariant with respect to a rotation of 120° by the observer (or by the triangle). Again, the triangle would exhibit a symmetry to anyone concerned with form but would not be symmetric to anyone who could observe that the sides were painted different colors.

The symmetry group need not be closed and finite. Figure 2.7c shows a man in front of a wall that extends forever without change in size, shape, or color. Of course, he cannot tell where he is by inspection of the wall. There is a complete and continuous invariance with respect to position along the wall. The observer is hardly better off as far as geographic clues are concerned when he is standing in front of the wall in Figure 2.7d. Here the wall is figured with a recurring pattern: the observer can tell where he is with respect to the pattern, but there is an invariance with respect to discrete intervals along the wall.

It seems possible to understand many of the conservation laws developed in our description of nature in terms of such symmetries or invariances of nature. The classical conservation laws—the conservation of momentum, energy, and angular momentum—appear to be related to invariances in the space-time description of nature. In particular, we are able to understand the law of conservation of energy in terms of the invariance of nature with time. If the results of *any* experiment are to be independent of the time when the experiment is performed, energy must be conserved. If time is relative, if no experiment can detect absolute time, if all the universal constants—the velocity of light, Planck's constant setting the scale of quantum mechanics, the charge of the electron, and so on—are to be independent of time, the conservation of energy will follow as a consequence.

In a similar fashion the conservation of momentum follows naturally from the invariance of nature with respect to place. If position is wholly relative, if no conceivable experiment can give results that depend upon the position of the laboratory, if the constants of nature do not vary with position, then conservation of momentum follows.

The conservation of angular momentum is a consequence of the rotational symmetry of the universe. Nature appears to be invariant with respect to angle or direction. There seems to be no natural absolute zero degree of angle, no intrinsically preferred plane or direction. The velocity of light is the same in all directions; the gravitational force from a sphere or the electrostatic field from a large sphere is the same in all directions.

All these statements can be true only in closed systems or isolated laboratories. The momentum of a ball thrown upward from the earth's surface is not itself conserved: the ball comes to a stop, reverses its motion, and falls to the ground. It is the total momentum of earth and ball that we believe is conserved. Our laboratory must either include the earth or be closed to the earth, excluding any of its effects.

The reasoning behind the connections between conservation laws and invariances follows from analyses of symmetries or invariances of the Lagrangian formulation of mechanics in 1918 by the mathematician Amalie Emmy Noether, who showed that every invariance (that is, symmetry) of the Lagrangian function must correspond to a conserved quantity. The invariances of nature under translation, time-displacement, and rotation, which are part of the entire set of symmetries of relativistic mechanics labeled Poincaré symmetries, should be noted here. While Emmy Noether's analysis is not easily accessible to us, we can construct some examples of asymmetric "universes" in which energy, momentum, or angular momentum are not conserved, which may serve as useful illustrations of the concepts involved. These universes will consist of rooms or laboratories holding an observer and some suitable apparatus, as suggested by Figure 2.8. Except for the influences that are manifestly introduced in the figure and the discussion, we assume that nothing external to the rooms affects the observers or their equipment. (For simplicity, in these examples we use electrostatic forces as the "universe-distorting" forces, and we need only note that like charges repel each other and unlike charges attract.)

In Figure 2.8a we are looking at a universe where electric charge varies with time. Assume, for definiteness, that it increases. Then the attractive force between the two unlike charges will increase with time, and the spring between them will be compressed, thereby adding to the potential energy of the spring—energy will not be conserved. We could arrange for the increasing force between the charges to drive an engine. Energy is not conserved in this universe, where the fundamental description of the universe—such as the magnitude or strength of electric charges—varies with time.

The observer in Figure 2.8b is watching the rotation of a dumbbell about its center, where the two weights hold equal and opposite electric charges. This universe is placed between the plates of a capacitor so that an electric field is directed across the universe. Of course, this is an anisotropic universe: a positively charged particle is attracted to the left, a negatively charged particle to the right. The dumbbell will accelerate when the positively charged part is at the top of the cycle and decelerate when that charge is at the bottom. Angular momentum will not be constant or preserved in that anisotropic universe.

In the universe of Figure 2.8c the potential energy of the dipole set of two charges will vary as the observer moves the system toward or away from the wall to the left. The electric field generated from the charge to the left of the laboratory wall (universe boundary) will attract the negative charge, which is closer, more strongly than it will repel the positive charge. If the observer lets go of the pair of charges, the system will accelerate to the left. Even as the

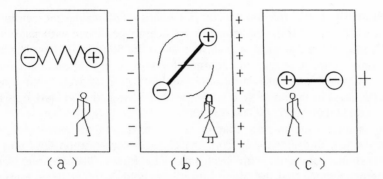

Figure 2.8 Universes that change (*a*) with time inasmuch as the strength of the electric charge changes with time, (*b*) with direction, and (*c*) with position. Throughout, the "+" signs indicate positive electric charges, and the minus signs "−" indicate negative charges.

velocity increases, the momentum increases, and momentum will not be conserved in this universe.

Although we have shown that in these very particular "universes"—which are not invariant with respect to time, direction, or position—energy, angular momentum, and momentum are conserved, we have not proved (and will not prove) the much more important converse—the conservation laws follow from the invariances. This proof can be demonstrated with special clarity in the framework of quantum mechanics.

One might object that if the whole system were included in our analysis—the external action that supplied extra charge in the universe of Figure 2.8*a*, or the capacitor plates of Figure 2.8*b*, or the external charge of Figure 2.8*c*—the conservation laws would still be valid. That is, energy, momentum, and angular momentum would be conserved in our universe but not in the observer's universes. This is true, but is it relevant? If we discovered that energy was not conserved in our universe, we could postulate an extra-universal source of energy and say that energy was conserved overall. Such a model might even be useful, but it would not supplant the observation that energy is not conserved in our universe.

This raises the question of the conservation of energy in our universe: does this universe change with time? If our universe is expanding from a beginning 15 billion years ago, a conclusion supported by astronomic evidence, do we not have an absolute time; is the universe not changing with time; and is energy really conserved in our universe? Perhaps not! On a cosmological scale we do not know that energy is conserved: our measurements of energy do not extend over such times. Some models of the origin and character of the universe contain implicitly a violation of the conservation of energy. The conservation laws pertain to an ideal universe for which our actual universe may stand only as a good local approximation. The symmetries are not necessarily absolutely valid, nor do the conservation laws necessarily hold absolutely.

Similarly, is the universe exactly isotropic so that angular momentum is exactly conserved? If the universe is more heavily populated with galaxies to the North, in the direction of Polaris, than to the South, will there be some nonconservation of angular momentum? Perhaps! At any rate, measurements of the conservation of angular momentum have been conducted that suggest that the mass of the universe is isotropically distributed to at least 1 part in 1,000,000,000,000,000,000,000,000,000,000.

It is interesting to note that in all the examples presented here, the observer will find deviations from the classical conservation laws only in observations connected with electromagnetic forces. The observer could properly conclude that the conservation laws hold for other forces (gravitational and nuclear), not for electromagnetic forces. In general, different forces do support different conservation laws even as the forces observe different symmetries.

Notes

1. Hermann Helmholtz used a word that presently denotes force, but he used it to label the concept we now call energy—hence the use of the latter word here.

2. The acceleration of the ball is shown in more detail in Table 2.1, a table of distances and their differences as a function of time. The kinetic and potential energies (measured in joules), T and V, and the Lagrangian, $L = T - V$, are shown for a ball with a mass of 1 kilogram.

The first column of Table 2.1 gives the time. The position or height of the ball is given in the second column. The third column shows the differences between the entries in the first column, which represent the change in s, Δs, divided by the elapsed time, Δt (which is 1 second in this case), or $\Delta s/\Delta t$. And the fourth column presents the difference between the entries in the second column, which are the differences in the velocities $\Delta(\Delta s/\Delta t)$, divided by the difference in time Δt, which is written compactly as $\Delta^2 s/\Delta t^2$. Note the interesting result that all the entries in the fourth column are the same, $\Delta^2 s/\Delta t^2 = -10$ m/sec²; the acceleration is constant. The kinetic energy T, potential energy V, and Lagrangian $L = T - V$ are listed in the last three columns for later reference. We have extracted a universal property from the set of observations, which is expressed formally by the *difference equation* $\Delta^2 s/\Delta t^2 = -10$ m/sec². In the calculus limit of infinitesimal increments of distance and time, this equation becomes the *differential equation $d^2 s/dt^2 = -10$ m/sec².*

3. We need not consider only collinear collisions. Indeed, in general the quantities **v** are vectors with components v_x, v_y, and, if the motion is not constrained to a surface, v_z. This vector equation is equivalent to three scalar equations holding for each component of the vector velocities. Then we can consider the motion in the three directions separately, and momentum in any direction is conserved.

$$m_1 v_{x1} + m_2 v_{x2} = m_1' v_{x1}' + m_2' v_{x2}'$$
$$m_1 v_{y1} + m_2 v_{y2} = m_1' v_{y1}' + m_2' v_{y2}'$$
$$m_1 v_{z1} + m_2 v_{z2} = m_1' v_{z1}' + m_2' v_{z2}'$$

4. Note that we define mass without considering weight. The equivalence of mass defined inertially (through the conservation of momentum) and mass defined by the force of gravity—the Equivalence Principle—is basic to Einstein's General Theory of Relativity.

Table 2.1 The position of the ball as a function of time together with the first and second differences of position with respect to time. The kinetic energy T, the potential energy V, and the Lagrangian $L = T - V$ are also shown.

t	s	$\Delta s/\Delta t$	$\Delta^2 s/\Delta t^2$	T	V	L
0	20		-10	612.5	200	412.5
		30				
1	50		-10	312.5	500	-187.5
		20				
2	70		-10	112.5	700	-587.5
		10				
3	80		-10	12.5	800	-787.5
		0				
4	80		-10	12.5	800	-787.5
		-10				
5	70		-10	112.5	700	-587.5
		-20				
6	50		-10	312.5	500	-187.5

5. For example, in a setup similar to that shown in Figure 2.3, we might use only three balls, glueing two balls together to have a mass of $2m$ and then allow the secured pair to collide with the one remaining ball. We could measure all velocities before and after the collision, recording the initial velocity of the two-ball set as V and the final velocity as V'; we might call the final velocity of the single ball u. After the collision we will find that

$$u = 4V \quad \text{and} \quad V' = \tfrac{2}{3}V$$

These results follow from the assumption that the conserved quantity is mv^2; that is, $\tfrac{1}{2}(2m)V^2 = \tfrac{1}{2}mu^2 + \tfrac{1}{2}(2m)V'^2$.

6. In this simple case the calculation of T, V, and L for each position of the ball is straightforward, and the values are presented in Table 2.1 for a ball with a mass of 1 kilogram. The kinetic energy will be $T = \tfrac{1}{2}mv^2$, and the potential energy will be $V = -mgh$, where v is the velocity, m is the mass of the ball, g is the acceleration of gravity (taken here as -10 m/sec^2 or -32 ft/sec^2, where the minus sign indicates the acceleration is downward), and h is the height of the ball. Because the total energy $E = T + V$ must be equal to $-mgh_m$, where h_m is the maximum height the ball reaches (when the ball is stopped and the kinetic energy is zero), $T = -mg(h_m - h)$ and $L = T - V = -mg(h_m - 2h)$.

7. The action A_{06} is described in the notation of integral calculus as

$$A_{06} = \int_{t_0}^{t_6} L \, dt$$

8. Consider, for example, the famous problem of three bodies moving under their gravitational attractions. The Lagrangian takes the simple form

$$L = \frac{1}{2}m_1 v_1^2 + \frac{1}{2}m_2 v_2^2 + \frac{1}{2}m_3 v_3^2 + \frac{Gm_1 m_2}{r_{12}} + \frac{Gm_1 m_3}{r_{13}} + \frac{Gm_2 m_3}{r_{23}}$$

In this relation G is the gravitational constant, m_1 and v_1 are the mass and velocity of mass 1, r_{12} is the distance between mass 1 and mass 2, and so on. Consider this problem: given the initial positions and velocities of the three particles, where will (any one) particle be at some specific later time? Although solutions are known for special cases, the general problem is quite intractable.

3

Covariance, Scalars, Vectors, and Tensors

> [We should] search for those equations which in *their general covariant* formulation are the *simplest possible.*
>
> Albert Einstein

The Dimensional Consistency of Equations

Distance, area, velocity, and acceleration have specific dimensions as well as quantitative values. Distance has the dimension of length; we speak of a distance of 1 foot, 2 meters, ½ mile, or 3.2 kilometers. Area has the dimension of length squared; we speak of an area of 17 square inches, 20 square meters, or 3,500,000 square miles. The area of a rectangle is calculated by multiplying the width (a length) by the height (a length) to deduce an area measured in units of length times length. Similarly, velocity has the dimension of length divided by time, and acceleration has the dimension of length divided by time squared.

Relations that are to be culturally invariant must connect quantities that have the same dimensions. For example, consider the equation that describes the path, in time, of a body moving in the vertical direction in the earth's gravity. We want to know the position of the body s at any given time t if the body is thrown from an initial position s_0 with an initial velocity (up or down, plus or minus) v_0. If the acceleration of gravity is g, we can express the relation of s and the other quantities by the equation

$$s = s_0 + v_0 t + \tfrac{1}{2} g t^2$$

Next, we append the dimensions of the quantities in parentheses, where L is the dimension of length (in inches, meters, etc.) and T is the dimension time (in seconds, centuries, etc.). Then the distances s and s_0 have the dimension of length (L); the velocity v_0 has the dimension of length over time (L/T); the acceleration has the dimension of length over time squared (L/T^2); and, of course, time is time (T). We then have

$$s(L) = s_0(L) + v(L/T) \cdot t(T) + \tfrac{1}{2} g(L/T^2) t^2(T^2)$$

If we multiply the dimensions together for each term, then each term has the total dimension of length (L). It follows that the equation is valid independent of the (consistent) set of units that are used. That is, the benighted American can safely use the equation although he or she measures everything in feet, whereas a Frenchman, the heir of the Enlightenment that led to the metric system, is happy with the equation although he uses meters. Moreover, if we should meet bug-eyed monsters emerging from flying saucers after their voyage from Betelgeuse, we will find that they use the same equation to describe falling bodies on their planet (albeit with a different value of g to describe the acceleration of gravity on that planet).

Sometimes relations (equations) that are not dimensionally correct are useful, but they will be bound to a specific set of units. For example, any heavy body will fall from rest a distance h in feet equal to 16 multiplied by the square of the time in seconds. We can codify this relation by the formula

$$h = 16t^2$$

But this equation will be useless (or wrong) for the Frenchman or the bug-eyed monster who does not use the foot as his unit of length. The results depend upon our cultural choice of the average length of a man's foot for our unit of length.

In a formal, logical sense the dimensional consistency of equations is important because it leads to a culturally dependent description of nature and this is important because we believe that the fundamental character of nature is independent of culture; we all believe that an American, a Frenchman, and our bug-eyed friend from Betelgeuse live in the same physical universe and that a description of that universe must exist that is the same for all.

Scalars, Vectors, and Tensors

As we have just stressed, classifications of physical quantities according to their dimensions must be independent of, or invariant with respect to, culturally determined choices of units of measurement if the relations of these quantities are to be culturally invariant. An equally important but much deeper concept is that the basic descriptions of the quantities used to describe natural phenomena reflect the symmetries of nature, so that observers in fundamentally equivalent situations describe the regularities of phenomena, the laws of physics, in equivalent ways. In order to accomplish that aim, it is important to classify quantities according to the precise way in which their description depends upon the viewpoint of the observer who measures the quantities. We use "viewpoint" first in a narrow sense, where we consider only the invariances of the universe with respect to the direction or orientation of the observer. Later, when we consider the theories of relativity, we will expand our use of viewpoint.

To consider such invariances, we introduce two observers, Oliver and Olivia—the names themselves define both similarities and differences—and place these observers such that they view the universe in different ways that reflect the symmetries we ascribe to nature. If Oliver and Olivia, who may differ in place, time, direction, velocity, or in other ways that we believe cannot affect the laws of nature, are to describe the natural order in equivalent ways, the relations they construct must have a property called *covariance*. We may approach the concept of covariance by considering a simple example of the character of the descriptions constructed by Oliver and Olivia using differently oriented coordinate systems. The special importance of covariance is really evident, however, as we reflect upon the structure of the Special and General theories of relativity. When we discuss the Special Theory of Relativity, we will consider covariant expressions relating space-time four-vectors that define the relations among measurements made by observers moving with different velocities. And the structure of the curved space-time that describes the effects of gravitation in the General Theory of Relativity is sensibly described only by covariant tensor equations.

To illustrate covariant classifications of various quantities, let us examine the observations made by Oliver and Olivia, who have summarized their measurements by constructing different maps of the northeastern part of the United States, as shown in Figure 3.1. In constructing his map, Oliver has defined north as the direction toward the North Pole; Olivia has used a magnetic compass bearing, made while she was in Boston, to define north for her map. Although we usually adopt the convention that north is the direction toward the geographic North Pole rather than the direction of the compass needle pointing to a nominal magnetic North Pole, neither convention is wrong—both are admissible. The two maps then differ in the orientation of their coordinate systems: magnetic north will lie about 15° west of polar north.

On each map the distance between New York and Boston will be 200 miles. The two observers measure the distance to be the same even though they use different coordinate systems. In general the distance measured between any two points is independent of, or invariant with respect to, the orientation of the coordinate system. Quantities that are described in a manner that is independent of the orientation of the coordinate system of the observer are called *scalars*. Only one number, in this case 200, is necessary to completely define a scalar.

According to Oliver, who uses the map where polar north defines the coordinate system, Boston is 200 miles northeast of New York or, more precisely, Boston is 200 miles from New York in a direction that is 45° east of north. Using her map based on magnetic north, Olivia also finds that the distance is 200 miles, but she determines that the direction is 60° east of north. Using a Cartesian coordinate system and his map, Oliver finds Boston to be (about) 140 miles north of New York and 140 miles east. Olivia, using her map, finds that Boston is 100 miles north and 173 miles east of New York. The two observers describe the position of Boston and New York differently in that the two numbers used to describe that difference in positions are defined dif-

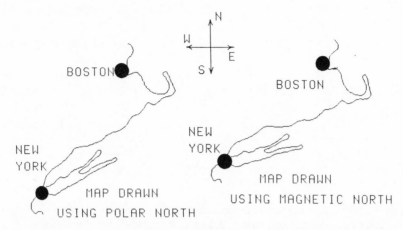

Figure 3.1 Maps showing the distance between New York and Boston using two different coordinate systems: Oliver uses polar north for his map, and Olivia uses magnetic north for hers.

ferently. Obviously, the two descriptions are rationally related to each other and to the angle (15°) that defines the difference between the coordinate systems used by the two observers.

Quantities that have this character are called *vectors*. Relative position is a vector. Two numbers are required to describe a vector in a two-dimensional space: in this case, the distance north and the distance east, or the total distance and the directional angle. Three numbers are required to describe a vector in a three-dimensional space: for example, north, east, and up. Four numbers are required for a four-dimensional space, and so on.

If we write the vector representing the distance of Boston from New York, measured by Oliver as **A** and by Olivia as **A′**, the two numbers that make up the vector (the distance north and the distance east) can be conveniently labeled A_N and A_E, or A_N' and A_E'. Usually, it is more convenient to label the axes of a two-dimensional coordinate system x and y; the components of a vector **B** can then be written as B_x and B_y.

Scalars and vectors are but the first two of an infinite hierarchy of quantities called, in general, *tensors*, which have increasingly complex descriptions. Tensors provide an economical way to describe curvatures and distortions of space. They are basic to theories of General Relativity.

Consider a simple tensor. Figure 3.2 shows at the left a small square of rubber sheeting with two arbitrarily selected points marked on the rubber. The relative position of one with respect to the other is a vector that we describe by two numbers, s_x and s_y, as shown in the figure. The same piece of rubber is shown at the right, now distorted by some stress. The positions of the two points are now different, and the new relative-position vector is described by two new numbers, s_x' and s_y'. The two sets of numbers, which are

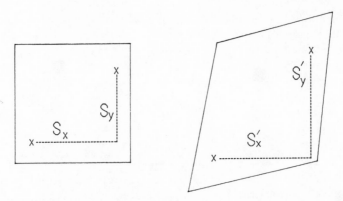

Figure 3.2 The relative position of two nearby points on a piece of rubber sheeting—at the left, before the rubber is stretched, and at the right, after it is stretched.

the results of two sets of measurements—one taken when the rubber was undistorted, one taken after the rubber was stretched—are related by a set of equations, as follows:

$$s'_x = u_{xx}s_x + u_{xy}s_y$$

and

$$s'_y = u_{yx}s_x + u_{yy}s_y$$

If the square of rubber is a small part of a larger sheet, we can expect to find that the four numbers u_{xx}, u_{xy}, u_{yx}, and u_{yy} are almost the same for any pair of points in the small square and thus completely represent the distortion of the rubber at the position of the small square. This distortion, which follows from a combination of stretching (or compression) of the rubber, shear, and rotation, is represented by a tensor, described by four numbers.

In a manner similar to that described for vectors, the description of a tensor depends on the orientation of the coordinate system. If Olivia observes the rubber sheet using a different coordinate system, she will describe the stress tensor that represents the stress in the rubber by a similar but different set of four numbers, u'_{xx}, u'_{xy}, u'_{yx}, and u'_{yy}.

In 3 dimensions a tensor of this character is described by 9 numbers, in 4 dimensions by 16 numbers, and so on. There are also tensors of higher rank[1] that are even more complicated, such as tensors of the third rank, which require 8 numbers for a description in 2 dimensions.

The Transformation Properties of Scalars, Vectors, and Tensors

A quantity described in a manner that is independent of the coordinate system is called a scalar. As the orientation of the coordinate system changes, the detailed description of a vector quantity changes; however, the changes are well defined, and they take place in a very particular way that is independent of any property of the quantity in question except that it is a vector. Hence, all vectors transform in exactly the same way. The changes in description are summarized in *transformation equations,* which are equations that relate the measurements of a quantity (scalar, vector, or tensor) by one observer to the measurements of a second observer who is using a different coordinate system.

Using, for convenience, x and y as labels for the coordinates of a rectangular (Cartesian) coordinate system, we write the results of Oliver's measurements (in miles) of the vector quantity **A** as the two numbers (components) A_x and A_y. Olivia's measurements of the same vector quantity, here written as **A′**, are labeled A'_x and A'_y. Taking x as north and y as east, and **A** as the vector distance from New York to Boston, from Oliver's notebook we have $A_x = 140$ and $A_y = 140$—and Olivia's notes read $A'_x = 100$ and $A'_y = 173$. Knowing the results of his own measurements and the angle θ defining the difference between the two systems, Oliver can determine the results of the measurements of Olivia (that is, the values of A'_x and A'_y) using the transformation equations and his own measurements A_x and A_y. In general, the transformation equations have a form such as

$$A'_x = \alpha_{xx}A_x + \alpha_{xy}A_y$$
$$A'_y = \alpha_{yx}A_x + \alpha_{yy}A_y$$

The form of the equations is most important; the new components depend on the old through the values of coefficients α. (For this simple transformation, $\alpha_{xx} = \cos\theta$, $\alpha_{xy} = \sin\theta$, $\alpha_{yx} = -\sin\theta$, and $\alpha_{yy} = \cos\theta$.) As we should expect, the only parameter that enters the relation is the angle θ, which Oliver knows to be 15°; the four coefficients of α are completely determined by the value of θ. Of course, if she knows that $\theta = 15°$, Olivia can deduce the results of Oliver's measurements from her own using the same set of transformation equations.

Now we have a more complete definition of a vector:[2] in a two-dimensional Cartesian coordinate system, a vector is a quantity described by a set of two numbers that transform into each other under a rotation of the coordinate system. More generally, a vector in an N-dimensional space is described by a set of N numbers that transform into each other according to transformation equations that depend upon $N - 1$ parameters. A vector in three dimensions is therefore described by a set of three numbers for which the transformation equations contain two parameters that can be considered to correspond to two angles or rotations in the three-dimensional space.

The components of tensors also transform into each other under a coordinate change.

Covariance

It appears that aside from purely local effects, observers moving at different velocities, observers at different times, observers at different places, and observers who look in different directions or use coordinate systems that are differently oriented, view the universe in the same way. Any general description of the universe must therefore be independent of the velocity and position of the observer and must not depend on the time of observations or the directions chosen to define the coordinate system. Any general relations or equations must then be independent of the choice of coordinate system.

By requiring that all general equations be dimensionally consistent, we are able to write equations that are valid or invariant regardless of the choice of units. Our equations for accelerated motion were shown to be dimensionally consistent and therefore equally valid for the metric system, the English system, or any other system of consistent units. Similarly, a general relation or equation describing nature should not depend on an orientation chosen by the observer and should then be independent of the choice of orientation of a coordinate system. The separate quantities or terms of the equation must transform in the same way under an operation changing the coordinate system. In short, the equations must hold equally for Oliver and Olivia, who use differently oriented coordinate systems, though the components of vectors and tensors are measured differently by the two observers.

When all the terms of an equation transform in the same way, the equation is *covariant*. Equations that are not covariant can be useful and correct if used in an appropriately limited manner, as they are generally valid in only one particular coordinate system. For example, we might state that a ball in free fall near the surface of the earth will fall straight down with an acceleration of 10 m/sec^2. If we adopt a standard coordinate system such that the direction "up" is labeled y, we can state the relation more elegantly by writing

$$a_y = g$$

where g is a constant equal to -10 m/sec^2. The quantity a_y is a component of a vector. Assume that the observations are made by an observer in Peoria, Illinois, and that the coordinate system is set up so that the y-direction is up in Peoria. But then the description of falling bodies expressed by the equation is hardly adequate for observers in Montevideo or Osaka. The y-direction that is up in Peoria does not represent up in Osaka or Montevideo. However, if we represent the relation describing the acceleration of the falling body as

$$\mathbf{a} = \frac{g}{r}\mathbf{r}$$

where $\mathbf{a}$ is the vector acceleration, r is the magnitude of the radius of the earth, and $\mathbf{r}$ is the vector distance between the center of the earth and the observer, we have a relation that is correct and meaningful to any observer on the surface of the earth, be he or she a resident of Osaka, Montevideo, or Peoria. Thus we say, *covariantly,* the ball falls with an acceleration of 10 m/sec^2 toward the center of the earth. This equation is a covariant relation between the vectors $\mathbf{a}$ and $\mathbf{r}$. The previous equation, which singles out one component of $\mathbf{a}$, the acceleration vector, is not covariant and is therefore valid in only one coordinate system.[3]

This example of a covariant description of the acceleration of a falling body under the influence of the earth's gravity serves to illustrate the relation between covariant descriptions of nature and the existence of natural symmetries. If the universe were intrinsically anisotropic, there would be important descriptions of nature that would depend upon the values of a specific component of a vector, and covariant descriptions of such phenomena would not be useful. If we restrict ourselves to a "universe" that consists only of a small region on the surface of the earth, such as a laboratory in Peoria, that universe is definitely anisotropic; down is quite different from north or east. Falling bodies will not be described correctly by an observer who uses a map that is rotated so that down is replaced by east. But in the entire universe as we know it, isotropy rules, and descriptions based on differently oriented maps are equivalent if each map is complete and contains both the earth and the falling ball. Universal symmetry is contained or expressed by the covariance of general descriptions of nature. Each term in an equation must have the same transformation properties.

Notes

1. Indeed, the word tensor can be used generally for all the categories we have mentioned. Here, a scalar is a tensor of zero rank (represented by only one number, whatever the dimensionality of the space). A vector is a tensor of rank one, with a number of components equal to the dimension of the space; the tensors of rank two, such as the stretch tensor we looked at, have a number of components equal to the square of the dimension of the space. In general, tensors of rank n have d^n components, where d is the number of dimensions of the space.

2. An arithmetic of scalars, vectors, and tensors can be defined. Quantities can be added covariantly only if they are of the same character; scalars can be added to scalars, vectors to vectors, and tensors to tensors. Scalars add simply, $a + b = c$; the addition of two vectors (or two tensors), $\mathbf{C} = \mathbf{A} + \mathbf{B}$, proceeds by adding each component separately: $C_x = A_x + B_x$. A vector can be multiplied by a scalar to produce a vector simply by multiplying each term in the vector by the scalar: $c \cdot \mathbf{A} = \mathbf{B}$, $B_x = c \cdot A_x$. In general, vectors and vectors, vectors and tensors, and tensors and tensors can be "multiplied" in various ways such that the product is a tensor of a rank that is not smaller than the difference in the ranks of the multiplicands and not larger than their sum.

3. The full equations for the motion of a body at an initial position s_0 with an initial velocity v_0 where the acceleration is $\mathbf{a}$ can be written covariantly as

$$\mathbf{s} = \mathbf{s}_0 + \mathbf{v}_0 + \tfrac{1}{2}\mathbf{a}t^2$$

where each term transforms as a vector.

4

The Discrete in Nature—
The Atoms of Demokritos

The Atoms of Democritus
And Newton's Particles of light
Are sands upon the Red Sea shore
Where Israel's tents do shine so bright.

William Blake

Prescientific Concepts

It is possible to divide or cut a macroscopic quantity of homogeneous material into two parts such that each part retains the basic character of the original. The division of a piece of gold into two parts leaves each part gold. But how far can such divisions be carried? Either the divisions can go on forever, generating smaller and smaller portions of gold, or there is a limit such that no further divisions can be made—or at least no further divisions leaving the parts as gold. Matter is either continuous or particulate.

The question of the particulate or continuous nature of matter was considered early in the intellectual history of man, certainly by the Greeks, but not settled until this century. Indeed, very able scientists such as Ernst Mach, Wilhelm Ostwald, and Pierre Curie considered seriously the continuum model of matter as late as the first decade of the twentieth century. Mach never accepted the reality of the atom.

As far as we know, the Greeks first entertained the possibility that all matter was made up of a small number of different kinds of fundamental pieces. More precisely, the Greeks considered the view that a particular kind of matter could not be divided indefinitely; there was a smallest, indivisible piece called an "atom" (from the Greek word ατομοσ, literally non-cuttable or indivisible). In the forms most congenial to us, these ideas appear to have been originated by Leukippos and his disciple Demokritos in the fifth century B.C. We cannot be certain about the precise character of these men's ideas because we know their views only from secondary sources such as the long Latin poem of Lucretius, *De Rerum Natura (On the Nature of Things).*

It would seem that the atomistic school of Demokritos and his followers considered matter as made up of very small (beyond perception) atoms that fit or lock together to form solids. Atoms also existed in a void, moving about and sometimes colliding very much as in the modern picture of a gas. There were many kinds of atoms differing in shape and size. Various kinds of gross matter followed from assemblies of differently shaped atoms arranged in different patterns. Indeed, it is possible to select a set of ideas from the Greeks that might be interpreted as similar to our modern conceptions of atoms. However, such a selection and interpretation of Greek concepts is probably unjustified. The Greek ideas of the atom are idealistic and nonscientific: they are too vague to be put to any well-defined operational test.

However, because every educated European of the seventeenth, eighteenth, and nineteenth centuries had read the Latin of Lucretius, atomic ideas were pervasive and may well have influenced the construction of the better defined atomic concepts that developed scientifically in that period. Indeed, it can be argued that Leukippos, Demokritos, and their followers established an intellectual tradition or climate that made it easier for later scientists to define and accept atomic concepts per se, as well as the view that important aspects of reality lie beyond access by our senses and therefore those senses must be extended through the fruits of rational inquiry.

Certainly, the influence of atomic ideas was important and widespread. Most of the founders of the Royal Society in the seventeenth century subscribed to a "corpuscular philosophy" that was recognizably an extension and refinement of the atomism of Demokritos. Galileo's views were rather similar, and Newton even extended the corpuscular concepts (incorrectly) to light. Yet there was certainly no firm evidence for atomism at this time, and the corpuscular description of matter was properly considered a working hypothesis attractive to many but not all natural philosophers. Leibniz, Denis Diderot, and Jean Lamarck are eminent examples of those who preferred to look in other directions, and Immanuel Kant seems to have concluded that matter is infinitely divisible.

The Chemical Bases of Atoms

The first compelling scientific evidence for an atomic character of matter derives from early work establishing the chemical properties of matter. By the early years of the nineteenth century, quantitative analyses of chemical compounds had strongly suggested that true chemical compounds contained fixed ratios of elementary quantities. Moreover, when the same two elementary quantities combined to form two different compounds, the quantities combined in different ratios, and these ratios bore simple integral relations to one another. An example are some reactions from later developments in chemistry, such as chemical reactions between hydrogen (H) and oxygen (O):

$$2H + O = H_2O \qquad 1 \text{ gm H} + 8 \text{ gm O} = 9 \text{ gm } H_2O$$
$$\textit{ratio } H:O = 1:8$$

Two atoms of hydrogen plus 1 atom of oxygen combine to form 1 molecule of water. In terms of mass, 1 gram of hydrogen plus 8 grams of oxygen produces 9 grams of water: the ratio of hydrogen to oxygen is 8 to 1 by weight. Hydrogen plus oxygen also produces hydrogen peroxide:

$$2H + 2O = H_2O_2 \qquad 1 \text{ gm H} + 16 \text{ gm O} = 17 \text{ gm H}_2O_2$$
$$ratio \text{ H:O} = 1:16$$

The ratio for the two compounds of the hydrogen-to-oxygen ratio of constituents, $(1/8)/(1/16) = 2/1$, is a fraction such that the numerator and denominator are small integers.

Similarly, consider the products of the reaction of sulfur and oxygen to produce sulfur dioxide,

$$S + 2O = SO \qquad 32 \text{ gm S} + 32 \text{ gm O} = 64 \text{ gm SO}_2$$
$$ratio \text{ S:O} = 1:1$$

and sulfur trioxide,

$$S + 3O = SO_3 \qquad 32 \text{ gm S} + 48 \text{ gm O} = 80 \text{ gm SO}_3$$
$$ratio \text{ S:O} = 2:3$$

The ratio of the ratios of constituents is $(1/1)/(2/3) = 3/2$, again a fraction composed of small integers.

The occurrence of such small integers is easily understood if matter is made up of discrete units of identical elements, and most difficult to understand otherwise. The recurrence of small integers in such ratios strongly suggests that the elements oxygen, sulfur, and hydrogen are made up of discrete units. Moreover, if we look at a large number of compounds containing sulfur, hydrogen, and oxygen, such as sulfuric acid (H_2SO_4), we find that a hypothesis that these elements come in discrete units with relative masses (H = 1, O = 16, and S = 32), and form compounds that consist of simple, unchangeable combinations of such units, fits all the observations we have made.

By 1803, sufficient information on chemical reactions had accumulated so that John Dalton could use these data to support his hypothesis that the chemical elements consisted of assemblies of atoms, and the compounds formed from chemical reactions of these elements consisted of specific combinations of these atoms, which we now call molecules. Of course, both atoms and molecules were recognized as being very small; small beyond direct perception. Moreover, the calculations of Dalton gave some indication of the relative masses of the different atoms and molecules.

Then, in 1808, Gay-Lussac observed that when two gases combined chemically to form another gas, the ratios of the volumes of the original and final gases could also be expressed as ratios of small integers. For example, two volumes of hydrogen plus one volume of oxygen combine to form one volume of water vapor. These observations led Amedeo Avogadro, in 1811, to suggest that under conditions of equal pressure and temperature, equal volumes of any gas held equal numbers of atoms or molecules. Because his peers lacked an appreciation of the concepts of the kinetic theory of gases, Avogad-

ro's conclusions were not generally accepted until a better understanding of
the character of gases was developed about 40 years later.

It was evident in the early years of the nineteenth century that the ratios
of the atomic masses of different elements derived from chemical experiments, together with Dalton's atomic hypothesis, could often be expressed by
integers within the accuracy of the measurements. Carbon, oxygen, nitrogen,
and hydrogen seemed to have masses in the ratio of 12:16:14:1. Observation
of these simplicities lead William Prout, in 1815, to make a proposal that
suggested that the atoms of Dalton had a subatomic structure (of what we
would now call elementary particles). Prout suggested that the masses of *all*
elements were whole numbers or, more properly, integral multiples of some
basic mass that he sensibly identified as the mass of the hydrogen atom. The
"indivisible" atoms must then be made up of some simpler, more fundamental building block. But Prout's conjectures, largely validated more than a century later, were rejected when accurate measurements of some of the heavier
elements led to values of the relative atomic masses (the atomic weights) that
were far from integral. In particular, the very accurate measurements that led
to an atomic weight of chlorine of 35.45 contradicted Prout's theory.

Today, we know that elemental chlorine is made up of about 75 parts Cl_{35}
to 25 parts of Cl_{37}, and these different "isotopes" of chlorine do have almost
integral masses on Prout's scale of 35 and 37. Prout was largely correct. The
different elements have masses that are almost integral multiples of the mass
of the hydrogen atom, and Prout was essentially correct in considering that
all elements are made up of hydrogen—just as we now know that the nuclei
of atoms, which hold most of the mass, are made up of neutrons and protons
(the neutron is now deemed to be a neutral proton), and the proton is the
hydrogen atom's nucleus.[1]

The Kinetic Theory of Gases—Boyle's Law

Although Dalton's analysis of the character of chemical reactions strongly
supported the hypothesis that matter consisted of assemblies of atoms, the
hypothesis actually revealed more of the character of chemical reactions than
of atoms. Soon, however, the atomic view was given important further support, in that the properties of gases were shown to be plausibly explained, in
great detail, as a consequence of the kinetic motions of particles (or atoms)
that might be supposed to make up the gas.

Perhaps the first important observation contributing to the kinetic theory
of gases was made by Robert Boyle in about 1660. Boyle demonstrated that
the volume of a gas was inversely proportional to the pressure exerted on the
gas. (Edmé Mariotte, in France, also proved this relation about the same
time.) Writing the pressure as P and the volume as V, we can write Boyle's
law as

$$PV = K \qquad \text{or} \qquad P = \frac{K}{V}$$

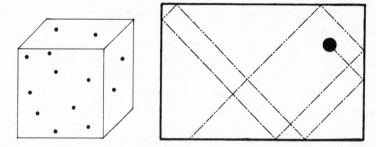

Figure 4.1 At the left is a box containing gas particles; at the right, a billiard table with the trajectory of a single ball marked on the table.

where K is a constant depending on the quantity of gas, the kind of gas, and the temperature. According to Boyle's law, if we hold the temperature constant and reduce the volume of the gas by a factor of two, the pressure will increase by a factor of two. Although Boyle considered, correctly, that this result must follow from the motion of atoms making up the gas, a quantitative understanding of Boyle's law in terms of the kinetic properties of the gas particles was first developed by Daniel Bernoulli much later in the 1730s.

It is possible to follow Bernoulli's reasoning rather straightforwardly. Consider a quantity of gas in a rectangular box such as that shown in Figure 4.1. In general we ascribe the pressure or force of the gas on the walls of the box to the result of collisions of gas molecules on the walls. To illustrate the character of the effect, assume a two-dimensional model where one billiard ball rolls without friction on a billiard table (with no pockets!). In the course of its motion, the ball will strike the far end rail and rebound elastically. This impact provides an increment of force to the rail that will be proportional to the mass of the ball times the velocity normal to the end rail. If the rail were detached from the table and mounted on springs, the springs would be compressed as a consequence of the force exerted on the rail by the rebounding balls.

The ball, rebounding with its velocity along the table reversed, then travels down to the near end rail, rebounding from that rail, and then back to strike the far rail again. With no friction it will repeat the entire pattern over and over again. Although the ball will carom off the side rails in the course of these passages, the elastic collisions with the sides will not affect the motion up and down the table. The average force on each end rail will be proportional to the frequency of hits, the velocity of the ball at each hit, and the mass of the ball. Because the frequency of the hits is proportional to the velocity, the force will be proportional to the square of the velocity. And because the kinetic energy of the ball is proportional to the mass times the square of the velocity, the mean force on the rails is proportional to the energy of the balls. (Although we are discussing only the end rails specifically, it is clear that we can make the same argument for the sides.)

Now if we have the same number of balls, with the same mean energy, on a table half as long, the frequency of collisions of any ball with the end rails will double even as the balls travel half as far between collision with the ends, and the force will then double. If the width of the table is reduced by a factor of 2, the force on the end rails will be unchanged, but the force per unit length will be doubled as the same number of collisions occur on a shorter rail. Indeed, in general, we see that the force per unit length on the end rail—and on any rail—will be inversely proportional to the area of the table. Moreover, if we had two balls on the table, with the same average energy, the forces would be doubled; indeed, the forces must be proportional to the number of balls.

We can extend the two-dimensional argument of balls moving on a pool table to consider atoms moving in a three-dimensional container. The results are similar;[2] the force per unit area or pressure will be proportional to the average energy of the balls (or atoms); the pressure will vary proportionally with the number of atoms (or the amount of gas); and the pressure will vary inversely with the volume of the container.

We have now explained Boyle's result from the assumption that the gas molecules move freely (as in the void of Demokritos), and the observed pressure of the gas on a barrier is the consequence of collisions of the molecules with that barrier.

In light of the two-dimensional argument, we can describe the behavior of gases constrained in a container in a very simple way. If the length of the box in Figure 4.1 is reduced by a factor of 2, the volume will be reduced by a factor of 2. Because the molecules will have to travel only half as far between collisions, there will be twice as many collisions per second with the face bc, and the pressure will double. The force on the four sides perpendicular to bc will not change, but their area will be reduced by a factor of 2; hence, the pressure on those sides will also increase by a factor of 2. The pressure is then doubled if the volume is reduced by one-half. Or, in general, the pressure is inversely proportional to the volume if the quantity of gas (the number of particles, n) is held constant and if the temperature (proportional to the kinetic energy of the particles) is held constant.[3]

If the average kinetic energy of a molecule is increased by a factor of 4, the average velocity must increase by a factor of 2. With twice the velocity, the momentum change upon collision with a wall will double, and the frequency of collisions will also double, thereby increasing the force and then the pressure by a factor of 4. With the volume constant, the pressure increase is the same as the increase observed experimentally by doubling the absolute temperature. Thus we are led to associate absolute temperature (measured from absolute zero, $-273°$ Celsius) with the average kinetic energy of the particles making up the gas.

It is therefore evident that if we warm a container of gas, the pressure increases. Or, if we keep the pressure constant, the volume will increase. This relationship defines a quantitative temperature. If we hold the pressure constant, the volume will be proportional to the temperature and a measure of the absolute temperature. On this scale, if we equate the volume of the gas in

equilibrium with melting ice to 273° and equate the volume in equilibrium
with boiling water to 373°, we have thus defined the absolute (Kelvin) temperature scale with the size of the degree equal to that of the Celsius (or cen-
tigrade) scale. On any scale, absolute zero is defined as the temperature such
that the mean kinetic energy of the particles is zero; the particles are at rest
at absolute zero. (Such a statement is modified by quantum mechanical considerations, but that is not important at this time.)

In the above discussions we have simplified matters in certain respects
that do not, however, affect the conclusions. In particular, we have neglected
the effects of collisions of the gas molecules with each other. Essentially, as a
consequence of the conservation of momentum, the neglect of the effect of
the collisions is justified. Also, the individual collisions with the walls are not
in fact elastic as we postulated. However, the collisions can be shown to be
elastic on the average if the gas and the walls are at the same temperature, so
that no energy is exchanged on the average. There are effects resulting from
the forces between atoms or molecules, and from the finite size of the molecules, which can be important if the density of the gases is high, and Boyle's
law holds, as we have described it, in the limit of low density of the gases.

Although the results of Dalton and Boyle could be seen to follow from an
atomic hypothesis, the absolute size or mass of an atom could not be deduced
from either the chemical data or the gas laws. Without a measure of the mass
of individual atoms or, equally, the number of atoms in a given amount of
matter, the entire atomic theory could be seen (and was by some) as a useful
calculational procedure with no definite footing in reality. However, the
sophisticated calculations of Clausius and James Maxwell, conducted
between 1857 and 1866, defining the *transport properties* of gases led to a
measure of the size of the atom (or molecule) and then to a much more solid
and compelling basis for the corpuscular view of matter.

If our billiard table contains a large number of white balls, and red balls
with the same average energy as the white balls are injected at one end, it will
be some time before the red balls are distributed evenly about the table, as
the collisions with the white balls will impede their *diffusion* throughout the
table. Clearly, if all the balls are very small so that ball-ball collisions are rare,
the red balls will be quickly distributed about the table. However, if the balls
are large, so that the mean distance a ball travels between collisions is small—
much smaller than the length of the table—the red balls will be impeded by
the collisions, and some time will elapse before red balls will be found at the
far end of the table. Hence, the rate of diffusion of the red balls from one end
of the table to the other depends on the probability of collision, and that probability in turn depends on the size of the balls. If we transfer the poolroom
argument to gases in a container and analyze diffusion times in gases, we will
acquire information concerning the sizes of the atoms or molecules that make
up the gas according to the kinetic theory.

Similar arguments apply to the viscosity of gases and heat conduction in
gases. If the near rail of the table were set to vibrating, this "hot" rail would
transmit energy to the balls that struck it (heating the assembly of balls near

that rail). After a large number of collisions, the balls at the far end would be found to have a somewhat higher mean energy. "Heat" would be transferred from one rail to the other. Again, if there were no collisions, the extra energy would be transferred easily as the "hot" balls sped quickly to the far end. The existence of the collisions reduces the rate of transfer of energy from one end to the other. Going again from poolroom to physics laboratory, the thermal conduction of gases depends on the collision frequency of the atoms or molecules that make up the gas, and inversely, measurements of the conduction of heat can be used to define the size of the molecules. The radii of the molecules (or atoms) are found to be on the order of 10^{-10} meters, or about 1/10,000 times the smallest radius particle discernible in a microscope.

Then, from measurements of the diffusion of gases and the thermal conductivity of gases, the collision cross sections of molecules in the gases can be determined. By comparing this cross-sectional area of the molecules with their volume in the liquid state,[4] the value of the number of atoms in a gram-atomic-weight of matter, which is called Avogadro's number, is determined. For example, for hydrogen with an atomic weight of 1 unit (by definition), the number of atoms in one gram will be about 10^{24}. This program, which resulted in an evaluation of N, was largely completed in the last half of the nineteenth century.

At this time other estimates of the size of molecules, and thence the value of Avogadro's number N, were made by determining the minimum thickness of oil and soap-bubble films and assuming that these minima were 1 molecule thick. These measurements also suggested that N was on the order of 10^{23} or 10^{24}.

Although the detailed agreement between the chemical arguments of Dalton and the atomic concepts contained in the development, by Clausius and Maxwell, of the kinetic theory of gases brought most scientists to an acceptance of the reality of atoms and molecules, some eminent physicists were still unconvinced of the reality of atoms. Philosophic considerations played a part in their reserve. In particular, Mach, who was a pioneer in emphasizing a logical positivism that excluded nonoperational concepts from science, could never be convinced that any theoretical construct of the atom could ever be anything but a symbol standing for sets of procedures; it could not be "real" in the sense of quantities that can be perceived directly by the senses. Ostwald, more pragmatic, was concerned that although the atomic picture was attractive and a useful working hypothesis, a better understanding of matter might still show that the hypothesis was seriously in error.

A Direct Observation of Molecules—The Brownian Movement

Although the success of the atomic theory of matter in chemistry and in the kinetic theory of gases satisfactorily established the concept of atoms for most scientists, these concepts were collective and relied on observations of enormously large numbers of particles—numbers on the order of 10^{23}. It seemed that there was no evidence for the interaction of single atoms. However, in 1827 Robert Brown published an account of his observation of the motion of pollen grains in a quiescent liquid. Although the material could be set aside in a quiet place for a long time (in fact, years), the pollen grains continued to move irregularly as observed in a microscope. In 1905 Einstein published a mathematical exposition of the Brownian effect showing that the motion of the very small pollen grains was due to the bombardment of the grains by the individual atoms or molecules of the liquid.

We can gain some insight into the character of the Brownian movement by considering the behavior of a small cube of light material placed in a box filled with gas at atmospheric pressure. For simplicity, assume that the cube has dimensions of 1 centimeter (cm) on a side and that we can neglect gravity so the cube is floating in the box. A cube placed in this way will not move perceptibly even though it is being struck very often by individual molecules. Indeed, it will be struck about 10^{24} times per second on each face by individual molecules. Now we understand why the cube does not jump around under the bombardment. The force on each face is the result of a very large number of very weak collisions. The number of collisions is so large that in any short time, almost exactly as many gas molecules hit one side as the other, and there is no net force.

However, if the pressure were applied not by collisions with 10^{24} molecules per second, each with a mass of about 10^{-22} gm, but by 100 random collisions per second with molecules that weighed 1 gm, we might expect the cube to be buffeted around a great deal even though the average force would be the same on each side. The discrete character of the interactions, hence the discrete character of the gas, would be directly observable.

One can see qualitatively that the amount of random motion incited by the irregular buffeting must be related to n, the number of molecules per unit volume; a measure of the motion might then be used to determine the value of n. The pollen grains seen by Brown were very small, about 10^{-4} cm in diameter, and the number of collisions is sufficiently small that the fluctuations in the number hitting one side or the other produces motion that can be easily detected. Einstein calculated that the random motion (known as a random walk) led to an average displacement of a pollen grain from its original position that increased as the square root of the time. For a set of typical pollen grains, he showed that the mean displacement from an original position would be about 10μ in 1 second, 20μ in 4 seconds, 30μ in 9 seconds, and so on, for the values of N estimated from the consideration of transport phenomena discussed earlier. According to Einstein, the degree of motion varies as $1/N$, where N is again Avogadro's number.

Following these calculations, a series of very precise measurements by Jean Perrin established the value of Avogadro's number as $N = 6 \cdot 10^{23}$. This confirmation of Einstein's theory by the measurements of Perrin dispelled all doubts about the reality of atoms. Twenty-four hundred years after Demokritos suggested that matter may be made up of discrete parts, his hypothesis was completely confirmed.[5] In the next 75 years, three more layers of divisibility—nuclei, nucleons, and quarks—were uncovered, and subquarks are now under discussion.

Notes

1. There are still effects on the order of 1% that follow from mass differences associated with the different binding energies (mass and binding energy are equated through the Einstein mass-energy relation, $E = mc^2$) of different nuclei, but this would not have been perceived in measurements made in Prout's time.

2. A more nearly quantitative analysis is not difficult. We first follow one particle of gas, atom or molecule, where the velocity of the particle in the x-direction is v_x. If the x-dimension of the box is a and the collisions with the ends can be taken as elastic (so the sign of the velocity component v_x changes but the value stays the same), the particle will make

$$\frac{v_x}{2a} \text{ collisions per second}$$

Because, in this description, the particle rebounds elastically, the change in momentum of the particle upon hitting the wall, dp_x, which is the momentum transferred to the wall, is simply the difference between the momentum in the x-direction before the collision, $p_x = mv_x$, and the momentum after the collision, $p'_x = -mv_x$:

$$dp_x = p_x - p'_x = 2mv_x$$

The momentum transfer per second to the wall will be that individual transfer multiplied by the number of collisions the particle makes with the wall each second:

$$\frac{dp_x}{dt} = 2mv_x \cdot \frac{v_x}{2a} = \frac{mv_x^2}{a}$$

If there are n particles in the box, the total momentum transfer will then be

$$\frac{dp_x}{dt} \text{ (total)} = \frac{nm\bar{v}_x^2}{a} = F_{bc}$$

where the bar over the quantity indicates the average taken over the n particles and where F_{bc} (the momentum transfer per unit time) is the force on the end of the box with area bc. From the Pythagorean theorem, the square of the total velocity is equal to the sum of the squares of the components,

$$v^2 = v_x^2 + v_y^2 + v_z^2$$

and because there is no preferred direction,

$$\bar{v}_x^2 = \frac{\bar{v}^2}{3}$$

and then

$$F_{bc} = \frac{nm\bar{v}^2}{3a}$$

Pressure is force per unit area, so the pressure on the side bc is the force on that area divided by the area bc:

$$P = \frac{F_{bc}}{bc} = \frac{nm\bar{v}^2}{3abc} = \frac{2}{3} \cdot \frac{1}{V} \cdot \left(n \cdot \frac{1}{2} m\bar{v}^2 \right)$$

where $V = abc$ is the volume of the box. In summary, considering that the total kinetic energy of the gas E is merely the sum of the kinetic energies of the individual particles, $nmv^2/2$,

$$P = \frac{\frac{2}{3} E}{V} \quad \text{or} \quad PV = \frac{2}{3} E$$

3. We are now in a position to define *temperature* in terms of the average kinetic energy of the gas particles. It is convenient here to consider the properties of a "mole" of gas, that is, an amount in grams equal numerically to the molecular weight of the gas. A mole of hydrogen gas, H_2, is then 2 grams; a mole of CO_2 would be $12 + 2 \cdot 16 = 44$ grams. We then replace the number n of gas particles in our equations by $N = 6 \cdot 10^{23}$, Avogadro's number, the number of molecules in a mole of gas. Then we write

$$PV = N \cdot \tfrac{2}{3} E = PV = NkT \quad \text{where } kT = \tfrac{2}{3} E$$

Here we define T as the absolute temperature, and k is Boltzmann's constant, relating this temperature to the average kinetic energy of the gas molecules.

4. For example, 1 gram-atomic-weight of water, H_2O, will be 18 grams, which will occupy 18 cubic centimeters (cm^3). Spread in a layer 1 molecule thick, this water will cover an area of about Na^2, where N is Avogadro's number and a is the effective diameter of the water molecule. Taking $a \approx 10^{-8}$ centimeters and using $N = 6 \cdot 10^{23}$, the area will be about $6 \cdot 10^7$ centimeters squared (cm^2), or something like 2 acres. Working backward, better with oil than water, N can be estimated reasonably from a measurement of the maximum area covered by a small amount of oil.

5. To my mind the most exciting, readily accessible evidence for the atomicity of matter is revealed by looking at the radium-zinc sulfide mixture painted on the dials of luminous watches (made before 1970). One can see individual nuclear decays if one holds the dial up to the dark-adapted eye. The passage of individual alpha particles from the decay of the radioactive nuclei of radium and its daughter products produces enough light in striking the zinc sulfide phosphor to be clearly visible. Shown this phenomenon in Ernest Rutherford's laboratory in Cambridge, even Mach was almost convinced that matter is divided into atomic sizes.

5

The Continuum in Nature—
Faraday's Fields

> I do not perceive in any part of space, whether
> (to use the common phrase) vacant or filled
> with matter, anything but forces and the lines
> in which they are exerted.
>
> Michael Faraday

The Character of a Field

There are two complementary faces to our description of the fundamental microscopic character of physical reality. Like Demokritos, we can consider that reality is constructed of sets of discrete source particles (such as electrons, protons, or quarks), with the modern addition that the forces between these particles take place through the exchange of field particles (such as photons, mesons, or gluons) created and absorbed by the source particles. Conversely, we can consider reality to be made up of a set of interacting fields (or better, one *unified* field) in Faraday's sense, where the field is the ultimate reality and the particles are quantum condensates of the field. In such a description, source fields of electrons, protons, and quarks interact with force fields of photons, mesons, and gluons. Of course, there is but one reality, and the different aspects—particle or field—are different pictures of that reality. Indeed, the two views, particular as particles, and continuum as fields, are largely fused in modern theory.

Although the concept of a particle is primitive, hence congenial—we know of pebbles and cherrystones from earliest experience—fields are much more intellectual constructions and much less accessible intuitively. However, the concept of field is so basic to any consideration of fundamental forces that we must consider the properties of fields if we are to gain insight into the character of these forces. For more than half a century, physicists have believed that one force field (with accompanying source particle or matter fields) was enough for God and the four fields we now observe—gravity, electromagnetism, and the weak and strong nuclear interactions—are but four different aspects of God's one field. Einstein spent the last 30 years of his

49

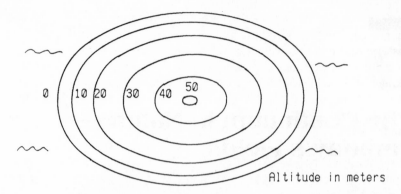

Figure 5.1 A map is shown of a hypothetical island in the sea, where the topography of the island is described by contours that define the values of the scalar altitude field.

life searching for, but not finding, the one universal unified field that was to describe all the forces of nature. At present, important fundamental conjectures concerning the character of the universe center on Grand Unified Theories called, inelegantly, GUTS. Because this one, all-encompassing field is the Holy Grail of physics, we can justify a close look at what is meant by field.

A field is defined by the value of a quantity at each point in space. That quantity may transform as a scalar and require only one number at each point for its evaluation, or as a vector or tensor requiring several numbers for a complete definition. A map of the United States showing the surface temperature at each point describes a two-dimensional scalar field. The pressure at each point in a fluid defines a three-dimensional scalar field; that pressure is completely described by one number (such as pounds per square inch). A moving fluid will have a velocity at each point in three dimensions described by three numbers that give the direction and velocity of the flow at that point. Even more complex fields can be important; the description of the distortions of a sheet of rubber must define the rotation, stretching, and shear of the rubber at each point. Hence, at each point four numbers, transforming as a tensor, define the two-dimensional tensor field required to describe the distortion of the sheet. In a block of rubber, a three-dimensional field of nine numbers at each point, which transform as a three-dimensional tensor, would be required to describe the distortion.

The framework of our understanding of the forces of nature is constructed from these three varieties of fields. It now seems possible—even likely—that the universe was born in a fluctuation of a scalar field. Three of the four fundamental forces—electromagnetism, the weak nuclear interaction responsible for beta decay (and the heat of the sun), and the strong nuclear interaction holding nuclei together—are described naturally by vector fields. Four-dimensional tensor fields are required to consider the distortion of space and time, induced by the presence of mass, which leads to gravity in Einstein's General Theory of Relativity.

Because vector fields describe all fundamental forces except gravity (and even gravity in Newton's approximation which is valid for most considerations), we must consider vector fields in some detail if we are to gain insight into the character of the fundamental forces of nature. We might begin by looking at a specific example of a vector field that is familiar to us. Assume that we need a description of a hypothetical island in the sea which will be especially useful to hikers. A topographical map showing the altitude at each latitude and longitude represents the scalar field that defines the altitude over an area. Such a map of the island is shown in Figure 5.1. (The smoothness and symmetry of the island, hardly to be found in nature, are introduced for simplicity.) Although we only label certain specific contour lines on the map as a matter of convenience, we are describing a situation such that an altitude is ascribed to *every* point. At a location at which the longitude is x and the latitude y, the altitude can be labeled $H(x, y)$. We will usually refer to this scalar field simply as H, where H represents the infinite set of numbers that define the altitude at each point on the island.

In many cases the same physical reality can be described by fields with different transformation properties. It is important to consider such different, though nearly equivalent, descriptions, inasmuch as one may display deep simplicities or lead to important generalities obscured by another. The topography of the island contained in the values of the scalar field H, shown in the contour map of Figure 5.1, can also be described by a map of the gradient of the altitude **G**, a vector field, as shown in Figure 5.2. The gradient of the altitude, **G**, is a vector in the direction of maximum slope of the land (opposite to the direction that a ball placed at the point would roll) and with a magnitude equal to the slope. For example, $\mathbf{G}(x, y)$, the vector gradient at a location given by latitude x and longitude y, might be described as 22 meters per kil-

Figure 5.2 A map of the island is shown where the topography is described by the vector field **G** defining the direction and magnitude of the gradient of the altitude at each point. The lengths of the arrows show the direction and value of the gradient at representative points.

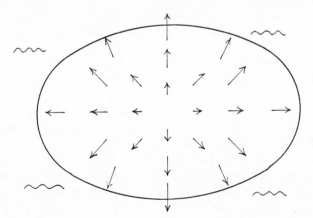

ometer directed northeast, which is a 2.2% slope in a direction such that a ball placed at the point (x, y) would roll southwest. It is conventional to use the symbol **G** to represent the entire field of number pairs for each point. The vector field **G** is the gradient of the scalar field H: in the notation of vector calculus,[1] $\mathbf{G} = \text{grad } H = \Delta H$.

Obviously, a map of the gradient field can be constructed completely from a knowledge of the scalar field. The direction of the gradient at a point is the direction of the shortest distance between nearby contours of the kind shown in Figure 5.2, and the magnitude is inversely proportional to the distance between those contours. As any hiker knows, the slope is steepest when the contours are close together.

The scalar map can also be reconstructed from the vector map except for the overall height. From the vector map we would not know if the island were in the Dead Sea, at a base altitude of -400 meters, or in Lake Titicaca, where the water level is at $+4000$ meters. This invariance of the character of the hiker's path with respect to the base altitude of the island is a very simple example of a *gauge invariance*. Such gauge invariances appear to play a large part in defining the character of the fundamental force fields.

The field **G** is a special vector field, a conservative field. If we were to roll a ball around any closed path on the island, we would find the force (resulting from gravity) on the ball along the path was sometimes against the ball (as the ball was forced uphill) and sometimes with the ball (as the ball was moving downhill). The sum of all the forces in the direction of the path multiplied by the distances over which the forces act is the work done on or by the ball, and the total work done, around the closed path, must be zero. If there were no friction and the ball were confined to a track, given a sufficient push the ball would traverse the course defined by the track and come back to the starting point with the same velocity as was given at the release; the kinetic energy of the ball would be conserved. Similarly, a cyclist traveling a closed-course path on the island will gain precisely as much energy from gravity going downhill as he or she would lose going uphill. It is only such conservative vector fields that can be defined as the gradient of a scalar potential field, which contains the same information.

Although gravity, as a conservative field, will not help a cyclist on a round trip, nonconservative fields are not so constrained. There might be a local cyclonic disturbance or whirlwind such that the wind would be at the cyclist's back through the whole trip. The diagram of Figure 5.3 shows a wind velocity pattern **v** such that a cyclist would be helped around any clockwise path. For path (a) the wind is always at her back; for path (b) the wind is strongly at her back during the "3:00" leg and weakly against her for the "9:00" leg (as positions on a clock face) with a net gain. If the cyclist were to travel in the opposite direction, she would lose energy to the wind.

It is useful to define precisely the property of the wind velocity field **v** that allows a supporting (or hindering) wind throughout a closed-loop path. If, for each small segment of path, we multiply the component of wind velocity v_s along the path by the length of the path segment ds and sum the values

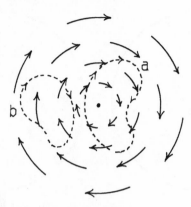

Figure 5.3 Cyclonic wind velocities and the paths of two cyclists who are assisted overall by the wind pattern, which exhibits curl.

of $v_s \cdot ds$ around the path—and then divide this sum by the area enclosed by the path—we will have calculated the average value of curl **v**, a measure of the wind circulation, over that area.

The label "curl" denotes circulation. If we make such a computation over a path encompassing a very small area about a point (x, y), we may consider that we have determined the value of curl **v** at that point. The whirls that develop when a cup of coffee is stirred with a spoon illustrate a velocity field that exhibits curl. If one placed very small cork rings on the surface of the coffee, the rings would drift with the overall current but would also rotate about their axes as a consequence of the circulation of the fluid. That rate of rotation would be proportional to the curl of the liquid velocity at the position of the ring. (Numerically, the curl of the velocity would be just 4π times the number of rotations of the ring per second.) The character of such drift and rotation is suggested by the left-hand sketch of Figure 5.4.

In the case of the ball rolled about a closed path on the island, the sum of all the forces multiplied by the distances over which they acted is zero. This summation is proportional to an average of the curl of the field **G** = grad H over the area enclosed by the path; hence, the average value[2] of curl **G** = curl (grad H) = 0. This is true for any path, however small; thus, at any point, curl **G** = 0. The result we have described follows generally for any vector field that can be expressed as the gradient of a scalar field; the curl of the gradient of a scalar field is always zero. (The next time a cyclist or jogger repeats the hoary joke to the effect that she is looking for a round-trip path that is downhill all the way, you can reply, "But that is impossible—the curl of a gradient is zero!")

There can only be a net gain or loss of wind aid in going around a path in the wind; the rings in the coffee rotate either clockwise or counterclockwise; there is only a twofold ambiguity in the sign of the curl in a two-dimensional field. However, the rotation of weighted cork spheres floating at equilibrium

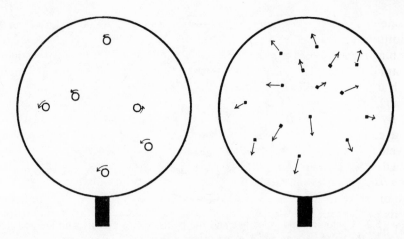

Figure 5.4 The sketch at the left shows small cork rings floating on the surface of a cup of coffee that has been stirred to mix the cream. The rings drift with a velocity **v** and rotate on their axes with a frequency that is just 4π times curl **v**, where **v** is the field defining the velocity of the surface current at the position of the ring. The sketch at the right shows the drift of small bits of cork on the surface of a liquid in a region where the velocity field **v** diverges. The foot of the arrows show the position of the pieces of cork at a time t, and the head shows the position at a later time, $t + dt$.

in the coffee defines both the magnitude of the curl of the liquid velocity and a vector direction—along the axis of rotation. In a three-dimensional system, the curl has a direction (defined to be along the axis of rotation). Hence, the circulation of the liquid at each point defines a vector, and the curl of a vector field is itself a vector field.

The Electrostatic Field—Faraday's Lines of Force

Electromagnetism, the interactions of moving and stationary electric charges, was the first fundamental force to be well understood,[3]—largely in terms of the field concepts of Faraday expressed by, and enhanced by, Maxwell's mathematics. Because there are deep similarities among all the fundamental forces, a careful consideration of electromagnetic fields is important in illuminating not only the field description of that particular force but also the description of all forces. Specific differences notwithstanding, electromagnetism is quite similar to the strong nuclear forces responsible for the binding of nuclei and the weak nuclear interactions responsible for beta decay (and the heat of the sun) and somewhat similar to gravitation.

It is useful to begin a broad consideration of electromagnetism by analyzing the electrostatic forces between stationary charges. A concept of electricity has existed at least from the time of the Greeks. (The word "electricity" is derived from the Greek word ηλεκτρον [electron], meaning "amber.") Amber is one of the best non-conductors of electricity or insulators available to a primitive technology. Such an insulator can store electric charge generated by rubbing material such as fur. The charged amber can then be used to demonstrate some of the simple, familiar phenomena of electrostatic induction, such as the attraction of small pieces of dried leaf. It was these properties that brought amber, and thus electricity, to the attention of the Greeks.

By the beginning of the nineteenth century, many people had contributed to a considerable knowledge of electricity, including important concepts developed by the extraordinary Benjamin Franklin, much of which was published in his *Experiments and Observations on Electricity Made at Philadelphia in America*. In particular, Franklin, considered by his contemporaries as one of the great scientists of his day, first noted that there are two different kinds of electricity. (Our electrical terminology—plus and minus, positive and negative electricity—stems from Franklin.) Franklin's work was interrupted by the demands of public service in about 1770, never to be renewed. Sometime after that, in 1777 in France, Charles Coulomb demonstrated that the forces between two stationary electric charges—electrostatic forces—were proportional to the product of the two charges and varied inversely with the square of the distance separating the charges, very much as with Newton's gravity. This result, Coulomb's law, for the force F between two electric charges Q_1 and Q_2 separated by a distance R is written as

$$F = \frac{Q_1 Q_2}{R^2}$$

where F, Q, and R are measured in consistent units.[4] As given in the equation, the force is repulsive, pushing the charges apart, and is directed along the line between the two charges. If the charges have different signs, the force will be attractive.

Coulomb's law describes the force between electric charges but provides no insight into what mechanisms might actuate that force across the distance between the charges. From this description, electrostatic force is very much a force-at-a-distance, a concept that creates a certain philosophic unease over an inferential incompleteness.

Although virtually all electrostatics is contained implicitly in Coulomb's law, the description of electrostatic interactions through the concept of the electrostatic field leads to concepts and results that are hardly evident in a purely particular description. Michael Faraday invented the modern concept of the field to explain and correlate the results of the experiments he performed between 1820 and 1845 concerning the properties of electricity and magnetism. This remarkable man, with only an elementary education and little mathematics, developed the concept of the field, which constitutes the basis of Maxwell's electromagnetic theory, Einstein's general theory of relativity, and the structure of quantum chromodynamics, which describe the forces between the quarks which are bound up to make elementary particles. Actually, Faraday's ignorance of mathematics probably contributed to his development of the idea of the field; his lack of formal mathematical training forced him to devise a strongly ph sical concept. Although the theory of fields lends itself to powerful mathematical techniques, the highly physical nature of Faraday's fields is such that far-reaching and subtle conclusions can be drawn with a minimum of mathematical apparatus.

Coulomb's law can be used to define the electrostatic field and to introduce Faraday's concept of "lines of force." Consider an assembly of charges and a very small "test charge" q that we can place at any point, and assume that we are able to measure the force on the test charge resulting from the charge assembly. Then, at any point, we *define* the direction of the field as the direction of the force **F** on the test charge, and the strength of the field as the force per unit charge on the test charge. Writing the electrostatic field as **E**, where boldface type indicates that the field is a vector quantity,

$$\mathbf{E} = \frac{\mathbf{F}}{q}$$

We may now follow Faraday and define lines of force as hypothetical lines that lie in the direction of the field such that the density of lines is equal to the field strength. A note of caution is in order: no meaning can be attached to a discrete line of force. Only the direction and density of lines carry meaning; the lines constitute a "handle" for us to grasp reality, not reality itself.

Even as the field is defined in terms of the force on a test charge, a charge Q in a field **E** is acted on by a force **F** that lies in the direction of the field lines and is proportional to the magnitude (and sign) of the charge and the density of lines.[5]

$$\mathbf{F} = \mathbf{E} \cdot Q$$

Both the concept of the field and the idea of line-of-force or field line are seen by considering the field about a single, isolated charge, $+Q$, and $-Q$, as shown in the left and central sketches of Figure 5.5. Note that for a negative charge $-Q$ the direction of the field lines is reversed from that derived from positive charges. In this picture, a positive charge Q emits $4\pi Q$ lines, and a negative charge $-Q$ absorbs $4\pi Q$ lines. The choice of positive and negative, emission and absorption, is a convention and carries no physical significance. The precise number of lines is also a convention dependent upon the choice of units.

If the lines of force do not fade away and begin or end only on charges, the density of the lines from an isolated charge will fall off inversely with the square of the distance. Hence, the force on a test charge will also vary with the inverse square of the distance as expressed by Coulomb's law. Conversely, Coulomb's law requires that the lines are conserved, beginning or ending only on charges.

A simple example will illustrate the relation between the line-of-force picture and Coulomb's law. Consider the lines through a second sphere at twice the radius of the first. At twice the distance, the density of lines is reduced by a factor of 4 even as the force on a test charge is reduced by that factor from Coulomb's law. But the area of the second sphere, with double the radius, is

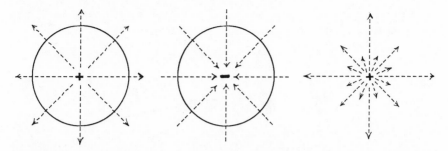

Figure 5.5 At the left, electric field lines are emitted by a charge at the center of a sphere; in the center, field lines are absorbed by a negative charge; at the right, lines are emitted by the charge of a short-range force.

greater than the area of the smaller sphere by a factor of 4, so that the product of density times area remains the same. Then an isolated charge Q can be considered to emit a definite number of lines of force (with our choice of units, $4\pi Q$ lines[6]) that go out radially from the charge indefinitely. No matter how large the sphere, $4\pi Q$ lines of force will pass through the surface of that sphere. We say that the electrostatic force is a long-range force or that the range is indefinite.

Gravity is also a long-range force, while the standard nuclear force holding neutrons and protons together in nuclei has a range of only about 10^{-13} centimeters, and the weak interaction force responsible for beta decay of nuclei (where electrons and neutrinos are given off in the radioactive decay) have a range of less than 10^{-15} centimeters. For such short-range forces, the field strength falls off with distance much faster than $1/R^2$, and at a distance on the order of the range, the field lines can be considered to gradually fade away, as suggested by the right sketch in Figure 5.5.

Further Properties of the Electrostatic Field— Linearity, Divergence, and Curl

To a very, very good approximation, electrostatic fields add linearly as superimposed; at any point in space, the electric field derived from two different sets of charges is just equal to the sum of the fields from each set, as might be measured in the absence of the other. The fields from the two sources do not affect each other. Figure 5.6 suggests the character of this addition for electric fields.

To the extent that electric fields are additive (or add linearly), light beams, which are oscillating electric and magnetic fields, do not scatter from one another, nor will light be affected in passing through static electric fields. But there is some breakdown of this linearity for very large fields. The enormous electric fields at the surfaces of atomic nuclei have been seen to scatter light somewhat (Delbrück scattering), and phenomena very similar to light-light

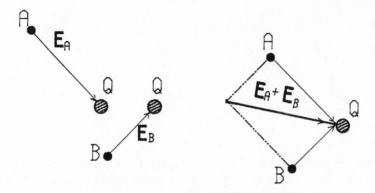

Figure 5.6 An example of the addition of electric field vectors. In the absence of charge B, the force on Q is $Q \cdot \mathbf{E}_A$; in the absence of B, the force will be $Q \cdot \mathbf{E}_A$. In the presence of both, the force will be $Q \cdot (\mathbf{E}_A + \mathbf{E}_B)$ where the fields add vectorially as shown.

scattering has been observed in the course of collisions between very high-energy electrons and positrons.

Figure 5.7 shows enclosed regions that contain a charge and a region that does not but is adjacent to a charge. For each region, lines enter and lines leave, but the total number of lines of force that leave a region minus those which enter is seen to be $4\pi Q$, the number of lines emitted by the enclosed charge Q. If Q is positive, more lines leave than enter, and the lines diverge from the region. If the charge is negative, more lines enter the region than leave, and the lines converge—or the divergence of the lines is negative. If the charge is zero, as many lines leave the region as enter, and the divergence is zero.

If the average density of charge in a small volume dV is ρ (charge per unit volume), the number of lines that leave the volume will be $4\pi\rho \, dV$. In the language of vector calculus,[7] the divergence of the vector field $\mathbf{E}$ at that point is written

$$\text{div } \mathbf{E} = 4\pi\rho$$

This equation, which states that the divergence of the electric field at any point in space is proportional to the charge density at that point, is known as Gauss's law.

Just as the rotation of small cork rings floating on the surface of a cup of coffee serve to define the curl of the vector field describing the velocity of the surface of the liquid, the motion of a set of small corks floating on the surface of a liquid can demonstrate the divergence of the vector velocity field. If the corks near some point move away from that point and away from each other—presumably as a consequence of a flow of fluid upward to the surface from some inlet—we know that the divergence of the velocity field at that point is positive. The right side of Figure 5.4 suggests the character of a positive divergence of the field representing the surface velocity of a liquid. If the corks begin moving toward a point, crowding together—presumably as a consequence of fluid from the surface moving downward to leave the system through some drain—the divergence of the fluid velocity at the point of convergence is negative. A positive divergence occurs in a region where there is a source of the vector field; a negative divergence occurs in a region where there is a sink of field.

The electric field produced by stationary charges has a further special property: the curl of the field **E** is zero. This property of the field, a complete lack of circulation, which is derived from more detailed analyses, is suggested by the performance of a thought experiment on the two-dimensional representations of the fields shown in Figures 5.5 and 5.7. Consider these to be fields of a surface fluid velocity with small cork rings afloat on the surface. The lines of electric field will then be flow lines charting currents in the fluid. Form the symmetry of the field, we can see that the rings will not rotate—there is no more reason for them to rotate clockwise than counterclockwise—although they will be carried along in the direction of the lines of force by the flow of the hypothetical liquid. There will be no circulation of the field; the curl of the field will be zero everywhere. Although the symmetry will be less evident in fields derived from the effects of several charges as a consequence of the superposition principle, the sum of fields will not induce rotation denied to any one; cork rings will still not rotate, and the curl will still be zero.

This lack of circulation of the electric field derived from stationary charges has the important consequence that the field can be expressed as the gradient of a scalar field, the electrostatic potential field ϕ; we can write **E** = grad ϕ. We recognize the parallel between the field **E** and the gravitational field **G** experienced by hikers on the island in Figure 5.1 and the further parallel between the scalar electrostatic potential field ϕ and the scalar altitude field H.

Figure 5.7 Lines of force passing through regions. At the left, a region holds a positive charge; in the center, a region holds a negative charge; at the right, lines of force pass through a region holding no charge. Although the figures are two-dimensional, the implications hold for three-dimensional enclosures.

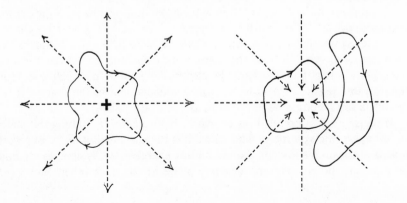

In summary, the nature of electrostatics can be defined by the properties of an electric field **E**. The coupling of the field to nature through the electric charge is defined by the equation relating force **F** on a charge q to the value of the field: $\mathbf{F} = q\mathbf{E}$. The vector field **E** is itself defined, at any point, by the divergence and curl of the field at that point, hence by the relations

$$\text{grad } \mathbf{E} = 4\pi\rho \quad \text{and} \quad \text{curl } \mathbf{E} = 0$$

where ρ is the charge density. Moreover, because the curl of the electrostatic field is zero, that field can be described as the gradient of a scalar field, the electrostatic potential ϕ:

$$\mathbf{E} = \text{grad } \phi$$

This description of electrostatics in terms of the properties of an electric field **E** is little more than Coulomb's law rewritten—and rewritten in a seemingly more obscure fashion. But the field description admits generalizations, obscured in the simple force law, which encompass all nature.

The Energy Stored in the Electrostatic Field

We have discussed the electric field **E** as if it were essentially a clever notational construction useful in calculations. Is there anything to the concept of the field beyond that? Is the field real? Considerations of the meaning of reality are more the province of philosophers than of physicists; however, the field can arguably be considered as real as the charges. We have inferred that the field is a secondary manifestation generated by the charges as primary elements; we can also consider the field to be the primary element where the charges are singular regions in the field. The field goes to infinity at the position of a point charge.

Moreover, the electric field carries energy and, hence, has mass. To show this we analyze the energy stored in a parallel plate capacitor. The capacitor is constructed of two parallel conducting plates separated by a distance d as shown in Figure 5.8a; the left-hand plate holds a positive charge, the right-hand plate an equal negative charge. Both charges are evenly spread over the plates with a density σ.

Figure 5.8b shows a cross section of the capacitor together with the electric field lines emitted by the positive charge and absorbed by the negative charge. From the symmetry of the system, the electric field lines must be normal to the plates as shown. Far from the edges of the capacitor, no direction toward one edge is better than another. The electric field strength, which is proportional to the density of field lines, is then independent of the spacing if the spacing is small compared to the dimensions of the plates.

If the spacing is doubled (as shown in Figure 5.8c) by pulling the plates further apart, the volume of field which fills the space will double. But work must be done on the system, thereby adding energy to the system, to separate the plates as the positive and negative plates attract each other. Moreover,

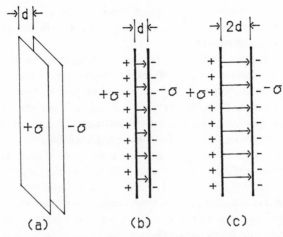

Figure 5.8 (*a*) A parallel plate capacitor. (*b*) A section of the capacitor showing the electric field lines. (*c*) The capacitor with the spacing between plates increased.

this energy must go into the field. Where else? Nothing is changed but the spacing—and then the amount of field. The charges on each plate "see" the same electric field after the change in position as before; no measurements by a microscopic observer on the plates can detect any effect of the change in positions of the plates. Thus, the field must store the energy[8] used in increasing the separation of the plates. Quantitatively, there are $E^2/8\pi$ units of energy per unit volume.

Although the result was found from the consideration of a particular situation chosen because of its simplicity, the conclusions are general. Any electric field contains, holds, or simply has stored energy proportional to the square of the field strength. Moreover, from Einstein's mass-energy relationship, $E = mc^2$, the energy stored in the field has mass; hence, the field has mass. The mass of a cubic centimeter of field of strength E is simply

$$m = \frac{E^2}{8\pi c^2} \text{ grams}$$

where c is the velocity of light.

For a common parallel plate capacitor charged with practical charge densities, the mass of the stored field is much less than we can detect. However, the mass of the electrostatic energy stored in nuclei is typically on the order of 1% of the total mass of the nucleus and is easily discernible.

The Magnetostatic Field

It is useful to extend the consideration of static electromagnetic phenomena from the analysis of the effects of stationary charges to situations where there are stationary currents—that is, where charges move but in a closed system, so that the distribution of charge and the electric field are not changed. Aside from fundamental considerations, gross matter abounds with natural currents, and magnetism, a consequence of these currents, is part of our primitively accessible environment.

It appears that the Greeks were familiar with magnetism from the properties of an iron compound, Fe_3O_4, found in Magnesia near the Aegean coast. The ore came to be known as magnetite, and small pieces of the mineral were called magnets; they had the property we now call magnetism.

In a universe like ours, constructed of electrically charged elements, magnetism and the magnetic field can be considered a relativistic consequence of the electric field. If the speed of light were infinite, or if all charges moved very slowly, there would be no magnetic field and no magnetism. But in the universe we live in, where the speed of light is finite and electrical charges do move, magnetic fields accompany electric fields. The other vector fields associated with weak and strong nuclear forces have similar magnetic counterparts that derive from relativistic effects.

Although magnetism is now viewed as a consequence of a change in the electric field or the motion of electric charges, the relations between magnetism and electricity were not understood until the nineteenth-century work of Faraday and Maxwell. Indeed, the unification of electricity and magnetism in Maxwell's equations describing an electromagnetic field should be considered the first step in the program to understand all forces as aspects of one unified field. We consider here magnetism, then, because of our interest in that unification; because of our interest in an example of a field that displays both interesting similarities and interesting differences with respect to the electrostatic field; and as an example of similar counterparts to the weak (beta-decay) and strong nuclear forces.

Even as the concept of the electric field was generated to provide a powerful conceptual extension of Coulomb's law describing the forces between charges, the magnetic field can be thought of as such an extension of the description of the forces between two currents of moving charge.[9] Although it will be evident that the description of the magnetic field we invent to describe the effects of currents will be quite similar to the description of the electric field, there are also sharp differences. The circulation of the electric field derived from stationary charges, expressed as curl **E**, is zero; the magnetic field can be defined as a field with a circulation generated by electric currents according to the relation

$$\text{curl } \mathbf{B} = \frac{4\pi \mathbf{j}}{c}$$

where **j** is the current density, and the factors 4π and the speed of light c follow from the choice of units for the field **B**. Figure 5.9 suggests the character of the magnetic field generated by a current uniformly distributed over the cross-sectional area of a conductor flowing in a direction out of the paper.

At any point within the conductor, there is a circulation of the field proportional to the current density. (If this field described the velocity of the surface of the cup of coffee discussed earlier, cork rings floating on the surface

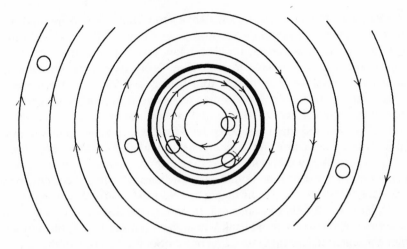

Figure 5.9 The magnetic field **B** in a cross section cutting a cable carrying a uniform electrical current. If the field represents the velocity of the surface of a fluid, small rings floating on that part of the surface representing the conductor will rotate about their axes with a frequency equal to the electrical current density. Rings floating outside that area will drift in a circle about the conductor but will not rotate on their axes. The diagram also describes the vector potential **A** generated by a long solenoid, where the solenoid current travels clockwise in the region of the heavy circle. Again, there is a circulation inside of the circle (solenoid) and no circulation outside; hence, there is a magnetic field inside the solenoid and no field outside.

in the region corresponding to the conductor area would all rotate about their centers with a frequency proportional to the current density at the place they occupy, as well as circulating about the center of the conductor.) Outside the conductor there is no current, hence no circulation (here, the cork rings would move around the center of the conductor but would not rotate on their axes).

If there are magnetic charges—generally called poles—those poles would provide a source of magnetic field just as an electric charge provides a source of electric field. However, in spite of careful searches, no magnetic poles have been found, and we are justified in describing the field in the absence of poles as having no sources or sinks; that is

$$\text{div } \mathbf{B} = 0$$

The lack of circulation of the electrostatic **E** field (curl **E** = 0) allows us to describe electrostatic phenomena in terms of a scalar field, the potential ϕ, so defined that **E** = grad ϕ. We have noted that the circulation or curl of a vector field is also a vector field. Such a field will have no divergence. There are no sources or sinks of the curl of a vector field.[10] Hence, the lack of divergence of the magnetic field **B** allows us to express that field as the curl (or circulation) of another vector field, which we commonly label **A** and call the *vector potential:*

$$\mathbf{B} = \text{curl } \mathbf{A}$$

Why do we bother to do this? We have gone from the description of the forces on moving charges to the construction of a field **B** that does not obviously carry more information, and then to another step of abstraction in the design of the field **A**. Similarly, from Coulomb's law describing forces among stationary charges, we have moved to a field **E** and then to an even more abstract field ϕ. All to what end? Moreover, ϕ and **A** appear to be especially artificial because for certain charge and current configurations, the vector potential **A** field and the scalar potential ϕ field will have nonzero values in regions where the electric and magnetic fields, and all forces on charged particles, are zero. As an especially interesting example, consider the fields of a solenoid magnet, a long tube wound with wire through which a current flows. Figure 5.9, which illustrated the magnetic field near a current, serves equally well to show the **A** field of a solenoid. Interpreted in this manner, the figure illustrates the vector potential field **A** in a plane normal to the solenoid axis where the solenoid current is traveling in a clockwise direction in the heavily marked circle. We recall that **B** = curl **A**; the magnetic field is the circulation of the vector potential field. There is circulation of the **A** field inside of the solenoid; hence, there is a magnetic field inside the solenoid. Outside the solenoid the **A** field is still large, but there is no circulation of the **A** field and thus no magnetic field.

If there is no magnetic field in the region outside the solenoid, has the vector potential field **A** any meaning in that region save as a calculational trick? Are there any physical effects of the A-field in a region where there is no magnetic field? Yes, there are such effects; the existence of the A-field affects the phase of the quantum-mechanical de Broglie wave of charged particles leading to observable consequences even in regions where there is no magnetic field (the Bohm-Aharanov effect, which is discussed more fully in Chapter 17). Hence, the A-field is the logical vehicle to describe quantum effects.

Subtle is the Lord,[11] and we suspect that His grand design is couched in terms of ϕ and **A** rather than in a language of fields and forces, which might seem closer to us.

Notes

1. For the two-dimensional scalar field H, the gradient **G** has the form

$$\mathbf{G} = \operatorname{grad} H = \frac{\partial H}{\partial x}\mathbf{i} + \frac{\partial H}{\partial y}\mathbf{j} = \nabla H$$

where **i** and **j** are unit vectors in the x and y directions.

2. For a path about an infinitesimal area da, the work done divided by the mass of the ball and the acceleration of gravity would be equal to $da \cdot \operatorname{curl}(\operatorname{grad} H)$. Considering the field **E** rather than grad H,

$$\operatorname{curl}\mathbf{E} = \left(\frac{\partial E_z}{\partial y} - \frac{\partial E_y}{\partial z}\right)\mathbf{i} + \left(\frac{\partial E_x}{\partial z} - \frac{\partial E_z}{\partial x}\right)\mathbf{j} + \left(\frac{\partial E_y}{\partial x} - \frac{\partial E_x}{\partial y}\right)\mathbf{k} = \nabla \times \mathbf{E}$$

where **i**, **j**, and **k** are unit vectors in the x, y, and z directions.

3. Newton's description of gravity is much earlier and quite accurate if the fields are not too strong. Einstein's General Theory of Relativity, required for the description of strong gravitational fields, was developed 50 years after Maxwell.

4. When units are not specified, we use the c.g.s. system most commonly used by scientists. Electrical quantities are measured in e.s.u. (electrostatic units).

5. As an illustration, consider the force on a charge Q_2 from a field $\mathbf{E}_1$ generated by a charge Q_1:

$$E_1 = \frac{Q_1}{R^2} \qquad F = Q_2 \cdot E_1 \qquad F = \frac{Q_1 Q_2}{R^2}$$

We have recreated Coulomb's law as we should. (We write the vector quantities without the boldface when we consider only the magnitude and not the direction.)

6. We can write the density of lines of force, $d = E$, as

$$d = \frac{Q}{R^2}$$

The total number of lines N at that distance R will then be equal to the density times the area of the sphere of radius R, $A = 4\pi R^2$:

$$N = d \cdot A = \frac{Q}{R^2} \cdot 4\pi R^2 = 4\pi Q$$

Hence, the number of lines of force at a distance from a charge is equal to $4\pi Q$ and is independent of the distance from the charge.

7. For a three-dimensional field describing the velocity of a fluid, the divergence in a small region is proportional to the flow of liquid into the system from that region divided by the volume of the region. For example, assume that a volume of 1 cubic centimeter (cc) contains the orifices of a set of fine tubes that are discharging 10 cc/sec of fluid into the system. Then the average value of div v over that region will be 10 per second (10 cc/sec flow divided by 1 cc volume). In the notation of vector calculus

$$\text{div } \mathbf{E} = \frac{\partial E_x}{\partial x} + \frac{\partial E_y}{\partial y} + \frac{\partial E_z}{\partial z} = \nabla \cdot \mathbf{E}$$

8. The contribution to the field strength between the plates from each plate is $2\pi\sigma$, as each unit area, holding a charge of $+\sigma$, emits or absorbs $4\pi\sigma$ field lines ($2\pi\sigma$ in each direction). From the superposition principle, the field strength between the plates is the sum of the fields resulting from the individual plates, or $4\pi\sigma$. Elsewhere, the fields are zero from cancellation.

The field from the charge on the positive plate, $E_+ = 2\pi\sigma$, generates a force on the charge, $-\sigma$, residing on each unit area of the negative plate, which is equal to

$$f = \sigma \cdot E_+ = 2\pi\sigma^2$$

where f is the force per unit area on the plate. The total force F between the plates is just the force per unit area multiplied by the plate area, or

$$F = fA = 2\pi\sigma^2 A$$

To pull the plates apart, perhaps to a spacing $2d$, requires work, an input of energy W that is the product of the force multiplied by the distance the plates are pulled apart, $W = Fd$. That is,

$$W = F \cdot d = 2\pi\sigma^2 Ad = 2\pi\sigma^2 V, \text{ since } Ad = V$$

where V is the added volume between the plates and then the added volume of electric field. From the expression for the total energy stored W and the expression for the field strength E, with a little algebra we can derive the energy stored per unit volume in terms of the field strength:

$$\frac{W}{V} = 2\pi\sigma^2 = 2\pi\left(\frac{E}{4\pi}\right)^2 = \frac{E^2}{8\pi}$$

9. A flow of I units of electric charge per second is described as a current **I**, a vector of magnitude I in the direction of the flow. The fundamental description of the magnetic force between two currents is complicated by the impossibility of creating isolated elements of current and by the dependence of that force on three directions, the relative direction of each current with respect to each other and with the direction of the distance between the two elements. For current elements $d\mathbf{i}_1$ and $d\mathbf{i}_2$ separated by a distance r which are part of current loops (as they must be), one can describe the force between two elements by the relation

$$d\mathbf{F} = -\frac{di_1 di_2}{c^2 r^2}\cos\theta_{12}$$

where θ_{12} is the angle between the elements, and the force is attractive and along the direction from element 1 to 2.

10. This conservation of the curl or circulation of a vector field led to a "vortex" theory of indestructable, conserved atoms popular in the late nineteenth century. We know now that such a model is quite inadequate in detail, but the spirit of such a conjecture is similar to that of modern views on the character of elementary particles.

11. This is a paraphrase of a comment by Einstein and the title of Abraham Pais's definitive scientific biography of Einstein (Oxford University Press, 1984).

6

The Nature
of Space and Time—
The Special Theory
of Relativity

> It appears, from all that precedes, reasonably
> certain that if there be any relative motion
> between the earth and the luminiferous ether,
> it must be quite small.
>
> Albert A. Michelson and Edward W. Morley

The Relative Character of Time and Position

We have emphasized the importance of invariance properties of the universe: descriptions of general properties of nature appear to be independent of some aspects of an observer's situation. The physical laws, which are statements that describe relations among the results of observations of the universe, appear to be independent of time, position, and direction.

We must conclude, then, that there is no preferred time, position, or direction for observations. Moreover, only time differences and differences in position and orientation have meaning; there is no absolute time and position, and no fundamentally primary direction. In an enclosed laboratory, with no access to the external world, the question whether the laboratory clock is correct can have no meaning. Similarly, nothing in nature depends on the position of the laboratory or the alignment of the apparatus in the laboratory, and no measurement conducted by that apparatus can give results that depend on the position and alignment.

Although there seem to be no absolute origins of position or time, physicists are still interested in establishing relations among measurement in different systems. If the clock in the laboratory where Oliver works is one hour ahead of the clock in Olivia's laboratory, we know we have no way of determining whether Oliver or Olivia are reading the correct time except by defining "correct" in an arbitrary fashion (such as deciding that a clock in Greenwich, England, is correct). But we can determine relative time, and if Oliver's clock is an hour ahead of Olivia's clock, we can relate the time t of an observation by Oliver of an incident (perhaps an eclipse of the moon) from his laboratory to the time t' at which Olivia records her observation of the same incident by the equation

$$t = t' + 1 \text{ hour}$$

The observational character of the quantities used in this equation must be stressed; t and t' are recorded times, perhaps written down in notebooks by Oliver and Olivia.

Similarly, we can determine the relative position of the two laboratories—but again, there is no absolute position except by convention. Ordinarily we use a latitude-longitude scale based on the position of the poles and an arbitrary point in Greenwich, and, for a third dimension, mean height above sea level. But these are hardly fundamental coordinates of the universe. We establish coordinate systems in each laboratory, however, and find relations between the two systems so as to relate observations made by Oliver and Olivia in their different laboratories. If Oliver's laboratory is ten kilometers north of Olivia's, observations by Oliver of an event at position d can be correlated with observations by Olivia using a latitude equation

$$d = d' - 10 \text{ kilometers}$$

with a similar relation for the east-west position. Again, the equation relates definite observations, perhaps surveys made by Oliver and by Olivia using their particular coordinates, where the results of the survey of the event's position are written down in their respective notebooks.

In this discussion we need not—and should not—concern ourselves with any metaphysical considerations of the nature of space or time. Rather we are considering the relations among observations made in space and time.

Invariance with Respect to Velocity— The Galilean Transformation

Are the laws of physics invariant with respect to velocity? Each of us has been seated in a train waiting to depart from the station and wondered whether the train on the next track or our own train was moving. Only the relative velocity of our train and the train on the other track is easily noticed. Furthermore, when our train is definitely in motion, there is no gross difference between our relation with our near environment and the relation with that environment when the train is stationary. Nothing near us is obviously changed by any motion of the train; if we were juggling oranges, the pattern of our movements and the flight of the oranges around us would not be different if the train were stationary or moving smoothly. If the roadbed were ideally smooth and the windows were opaque, there would be no obvious way in which we could tell if we were moving or not. This suggests that a general principle may exist to the effect that we cannot, under any circumstance, determine whether the train is moving. If this is true, the motion of the train can have no absolute significance.

Indeed, if we establish a laboratory on the train, we will find that within the accuracy of our present techniques, there are no experimental results that depend upon the velocity of the train. The laws of physics, our description of the universe, then seem to be invariant with respect to velocity. Relative velocity, of course, still has a meaning, even as relative time and relative position are meaningful. However, just as it appears that absolute time and absolute position have no significance, it seems that absolute velocity has no meaning. Such moving systems that are not accelerating are called *inertial systems,* and physicists postulate that the description of nature is the same in any inertial system.

We can determine the relative velocity of moving systems or moving laboratories by various conventional means. There is no intrinsic absolute velocity, although it is convenient for most terrestrial applications to consider that the earth's surface is stationary. But this is a convention of the same kind as the conventions of Greenwich time or the Greenwich zero of longitude.

To relate the measurements of events in one laboratory with the measurements in another laboratory when the laboratories are moving, we need relations between the coordinates and times used in the two laboratories. Figure 6.1 shows such a pair of laboratories. Olivia's laboratory S' is on a train moving with a velocity v in the x-direction with respect to Oliver's laboratory S, located on the platform beside the railroad tracks. Olivia is equipped with a clock that reads time t' and a meter stick of length L', and Oliver has a meter stick of length L and a clock that reads time t. Previously, when the train stopped beside the platform, Olivia and Oliver compared their instruments and found no differences. Now we make the *classical* (nonrelativistic) assumption that a second in laboratory S is the same as a second in laboratory S', and a meter in S is equal to a meter in S'. This is a very good approximation if v is much less than the speed of light c.

We can now consider relations between the measurements of Oliver and Olivia of the distance between two events and the time elapsed between the events. The first event will be a spark that occurs when an electrode on Olivia's train at the position marked x_0' passes an electrode, marked x_0, on the platform where Oliver works. The second event is a similar spark that occurs when an electrode at x_1' passes the electrode at x_1. (This rather odd choice of events was made to emphasize the equivalent standing of the two coordinate systems; however, events that might nominally be considered a natural part of one system or the other could be chosen with no change in principle.)

Each observer will record the time that elapses between the events with his or her clock and will measure the distance between the events with his or her meter stick. If an instantaneous signal is available, the observer, in either system, need simply read his clock when the signals come in. Perhaps, either

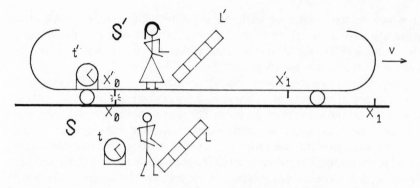

Figure 6.1 Oliver and Olivia are in different coordinate systems, S and S', measuring the distance and elapsed time between two events with their clocks and their meter sticks L and L'.

for practical or fundamental reasons, no instantaneous signal is available. Then a signal moving at a finite speed must be used. This could be the light of the spark itself, which moves at a speed of about $3 \cdot 10^8$ meters per second, or a message sent by a pet turtle that waddles at a pace of 0.1 meters per second. In either case the time recorded by the clock must be corrected for the time it takes the signal to travel to the observer. Of course, calibration measurements must be made of the velocity of light or of the turtle. But however the information is obtained, by light beam or turtle, the results of the time difference between the two events are definite numbers that the observers write down in their notebooks.

If t_1 is the time Oliver reads when he sees the second spark and t_0 is the time of the first spark, the time difference he records will be

$$dt = t_1 - t_0$$

Similarly, Olivia will record the time difference dt', where

$$dt' = t'_1 - t'_0$$

If instantaneous signals exist, Oliver and Olivia could use the signals to continually compare their two clocks; if velocity is relative, neither system can be privileged, and the clocks must agree on the time interval between the events:

$$dt = dt'$$

The possibility of such instantaneous signals and then the equality of time intervals measured in different systems is an implicit precept of classical physics.

Oliver and Olivia will not record the same distance, however. Each will measure, at his leisure, the distance between the two electrodes in his system by counting the number of times he must lay down the meter stick to cover the distance between the two electrodes in his system. An inspection of Figure

6.1 shows that the distance measured by Oliver between the two electrodes on the platform

$$dx = x_1 - x_0$$

will be greater than the distance

$$dx' = x_1' - x_0'$$

that Olivia measures between the electrodes mounted on the train.

Oliver may take an egocentric position and say that his measurements of the distance between events is correct and Olivia's measurement on the train is in error because she did not take into account the "fact" that the train was moving during the time that elapsed between the two events. Thus, he might argue, the electrode x_1' moved during this time a distance $v \cdot dt = v \cdot dt'$. He could then conclude that the proper relation between the measurements recorded in the two notebooks is such that

$$dx = dx' + v \cdot dt'$$

Independent of the question of which observer is "correct," the equation does express correctly the relations between the measurements made and recorded in their respective notebooks by the two observers.

However, if absolute velocity has no meaning, it is equally permissible for Olivia, on the train, to prefer her system of reference. She might claim that her measurements are correct and Oliver's results on the platform are wrong because he did not take into account the "fact" that the platform was moving backward with a velocity v. From Olivia's viewpoint the platform and electrode x_1 moved backward a distance $v \cdot dt' = v \cdot dt$ during the time dt' that elapsed between the two sparks. According to Olivia, Oliver on the platform should have subtracted that distance from his measurements to get the "correct" result, and the records in the notebooks would then be related as

$$dx' = dx - v \cdot dt$$

Again, independent of any judgment as to which observer is "correct," the equations do correctly express the relations between the two sets of measurements. Indeed, if the observers are wise, each will agree that the other is proceeding competently and correctly in making the measurements.

The three equations relate the measurements, made and recorded by observers in differently moving coordinate systems, of distances and elapsed times between events. These equations, which relate the measurements of Oliver and Olivia residing on coordinate systems that are moving with respect to each other, are similar to the equations introduced in Chapter 3, which relate the measurements of x and y distances by Oliver and Olivia using coordinate systems oriented at different angles with respect to each other. Such equations are called transformation equations, and this set, connecting measurements of time and space differences, comprises the equations of the *Galilean transformation*. These are the equations of Galilean or classical relativity, where the term relativity refers to the relative character of reference frames moving at different velocities. When we include, properly, the finite velocity of light, these relations will be superseded by the equations of the Lorentz transformation of Einstein's Special Theory of Relativity.

Signals Carried by a Moving Medium

During the last decades of the nineteenth century, it became evident that some aspects of nature could not be understood using intuitive or classical meanings of space and time. In particular, evidence that the speed of light was the same for all observers, independent of the character of their motion, seemed to demand a review of the classical concepts of space and time. These problems were largely resolved by the formulation of a new description of nature by Einstein in 1905: the Special Theory of Relativity.

To understand the consequences for measurements of space and time of constraints on the properties of signals (such as light signals), we must consider further the relations among measurements in systems that are moving relative to one another; to do so, we need the Galilean transformation relations. In particular, we might examine measurements when instantaneous signals are not available. It is convenient to discuss particularly simple cases such that

$$dx = dx' \quad \text{and} \quad dt = dt' = 0$$

That is, the two signals are simultaneous as noted by observers who have available to them instantaneous signals. The relative configuration of the two systems, the railroad car and the platform, is shown in Figure 6.2.

Olivia, on the train, uses her meter stick to measure the distance between the two electrodes as 100 meters, and she stands exactly between the two. She uses a coordinate system such that the first electrode, labeled i, is at zero, and the second, labeled ii, is at 100 meters:

$$x'_i = 0 \quad \text{and} \quad x'_{ii} = 100 \text{ meters}$$

Of course, we are interested primarily in measurements in which the signals are carried by light. However, to emphasize both the generality of our results and problems concerning the media in which the signal travels in a manner that is close to our intuitive understanding, we will require our observers to use another familiar carrier of signals, the pigeon. Olivia, on the train, will receive signals from the two electrodes by two calibrated pigeons (calibrated pigeons!) that are released at the instant the spark occurs and that fly at a speed of 10 meters per second. The velocity of the railroad car with respect to the platform is $v = 5$ meters per second. The car is enclosed and carries its atmosphere along with it. Olivia finds that the two birds reach her simultaneously. Because each bird flies at a speed of 10 meters per second and has traveled 50 meters, she knows the events took place simultaneously just 5 seconds before the birds reached her.

On the platform Oliver sees a different, though simply related, picture. He

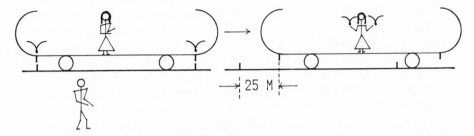

Figure 6.2 Oliver and Olivia are in different inertial systems watching the flight of pigeons bringing messages of the occurrence of sparks. At the left, the pigeons are released; at the right, Olivia receives the pigeons.

sees (here, "see" represents the reception of instantaneous signals) the two birds leave promptly at the time of the sparks, and he sees the birds reach Olivia simultaneously, 5 seconds later. However, according to Oliver, Olivia will have moved with respect to the coordinate system of the platform during the 5 seconds. At the time of the sparks, Oliver determines the positions on his platform system of the first spark, the observer Olivia, and the second spark as

$$x_i = 0 \qquad \text{Olivia} = 50 \text{ meters} \qquad x_{ii} = 100 \text{ meters}$$

From Oliver's view, at the time the pigeons reach Olivia she will have moved a distance of 25 meters down the track—a distance equal to the velocity of the train, 5 meters per second, multiplied by the elapsed time, 5 seconds. By that time bird i, which started from the electrode at x_i', will have traveled 75 meters, and bird ii, starting from the electrode x_{ii}', will have traveled only 25 meters in the platform system. According to Oliver's measurements, the velocity of bird i is 15 meters per second, and that of bird ii is 5 meters per second.

In general we can express those velocities as

$$u = u' + v$$

where u is the velocity of the pigeon as determined by Oliver and recorded in his notebook, and u' is the velocity of the pigeon Olivia measures and records in her notebook. Velocities of the signals in one system are added to the relative velocity of the two systems to determine the velocity of the signal in the second system. The velocities add.

We might as easily discuss the inverse of the previous situation. Perhaps Oliver could hire two pigeons to bring him signals of the sparks where the pigeons would fly over the platform. Then, from the view of Olivia, observing the scene from the railroad car,

$$u' = u - v$$

The minus sign is used because the platform is going backward from Olivia's view in the train. The description is symmetric; neither Oliver or Olivia, platform or train, is privileged.

In preparing for an examination of the velocity of light, we should consider a variation of the conditions analyzed above. It is possible that the railroad car is an open flatcar and does not carry air with it. (Assume a calm day with no breeze.) In this case Olivia will note that a pigeon flies 50 meters in the positive x-direction (forward) in 10 seconds, so its average velocity is only 5 meters per second. But the pigeon flying in the negative x-direction (backward) covers the 50 meters in only 3.33 seconds for a velocity of 15 meters per second. Whether Olivia concludes that there is a wind of 5 meters per second blowing over her (stationary, as far as she is concerned) flatcar or that pigeons perversely like to fly faster in one direction than the other, is of no practical relevance in her analysis of the observations.

Birds i and ii do not reach Olivia at the same time in this case. Bird i, which carries the message from spark i, reaches Olivia 6.67 seconds after bird ii, which carries the message of spark ii. Knowing the different velocities of the birds and that each bird flies 50 meters, Olivia concludes that the two sparks are simultaneous.

Oliver views the efforts of his colleague from a different viewpoint. He says that the two birds travel at the same speed of 10 meters per second but that the movement of the car changes Olivia's position. In the platform system Olivia is at the point $x = 50$ meters at the moment the sparks are emitted. Because Olivia is moving at the rate of 5 meters per second in the positive x-direction, the bird from spark i must fly farther to overtake her, standing on the flatcar; the bird takes 10 seconds for that trip. But as Oliver sees it, Olivia is traveling towards the spark ii so the bird that brings that message has a shorter distance to travel and reaches Olivia after only 3.33 seconds.

How to Measure the Ether Wind

If light is used to propagate signals and if the relations of the Galilean transformation are correct—that is, if they correctly describe the relations between observers in different inertial systems—we must expect to have to consider the detailed character of the propagation of light. It is necessary to ask two questions: (a) what is the medium that transmits light, and (b) does the earth pass freely through the medium such as an open flatcar through the atmosphere, or does the earth carry the medium with it like an enclosed railroad car? Suppose that we call this hypothetical medium *ether;* of course, the assignment of a name does not increase our knowledge of the medium. The classical concept of space and time, as coded in the relations of the Galilean transformation equations, allows no fundamental alternatives to those listed in question (b).

The problem of determining which of the alternatives is valid (and we shall find, paradoxically, that neither is valid) differs from the analyses of the observers on the railroad car and platform because we have no platform; we are resident only on the train—our earth. Then, to illuminate the problem of measuring the ether drift and answering question (b), we consider the similar problem where Olivia, on the railroad car, must measure the velocity of the wind past her car. How can she measure the velocity of the car with respect to the air that carries the pigeon or even find out if such a wind exists? The simplest thing she can do is to measure the velocity of the pigeon in the direction of the suspected motion of the car and in the opposite direction. If the car is moving with velocity v through the air and the pigeons are known to fly with velocity c with respect to the air, the pigeon flying in the direction of the car's motion (and against the "wind" passing over the flatcar) will have a velocity

$$c_+ = c - v$$

The pigeon flying opposite to the direction of motion of the car (downwind) will have a velocity

$$c_- = c + v$$

The velocity of the car is half the difference between these velocities

$$v = \tfrac{1}{2}(c_- - c_+)$$

This is the same result we would obtain if the car were stationary and a wind were blowing from the $+x$-direction with a velocity v.

Before Olivia accepts this method as a basic procedure for measuring the relative velocity of the flatcar and the air, she must decide how she is going to measure the velocity of the pigeon. In practice she might establish a specific flight distance, perhaps 100 meters, station herself with a stopwatch at the finish line, and observe and time the start and finish of the flight. This procedure supposes the use of a signal that is instantaneous (or at least very much faster than the pigeon) and independent of the air flow. In practice she would see the pigeon start the race. She would use light as her signal.

If no such instantaneous signal is available or if she cannot be sure that a signal would be unaffected by the air flow, this procedure fails. How can she tell when the pigeon starts? When should she start her stopwatch? Certainly it would be ridiculous for her to use another pigeon to bring her the message that the racing pigeon had started. She must invent a more subtle scheme.

A method of measuring a pigeon's velocity that does not require instantaneous signals is to measure the time it takes the bird to fly back and forth over a course—from the observer to a goal at some distance L and back again. Using this method, Olivia will not have to know when the bird reaches the far goal, only that it gets there. That is easy to check in various ways. If there is no wind from the motion of the railroad car, the time elapsed between the departure of the bird and the return will be just the total round-trip distance divided by the velocity:

$$t = \frac{2L}{c}$$

If the car is moving, the round trip will take a little longer. If the velocity v of the car is much smaller than the velocity of the pigeon c, the fractional

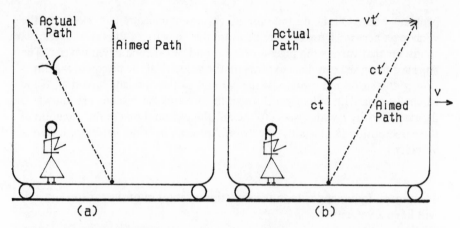

Figure 6.3 The paths of pigeons as seen from the flatcar. In (*a*) the pigeon is blown off course from the view of the observer Olivia on the car. In (*b*) the pigeon aims upstream but is blown on course.

increase[1] in time will be v^2/c^2; if $v = \frac{1}{10}$ of c, the time required for the trip will be 1% greater.

Hence, if Olivia knows how long the pigeon should take for the round trip when the train is halted by the platform or no wind is blowing, she will be able to detect the motion of the train (or the presence of a wind across her flatcar) by noting that the pigeon takes longer to make the trip. If she knows the value of c, the speed the pigeon will fly in the air, she can determine the wind velocity v by seeing how much more slowly the pigeon makes the round trip if the train is moving or the wind is blowing. However, if Olivia has spent her life on the flatcar, she may never have had the opportunity to observe the pigeon flying in still air and measure its velocity; she will not know the velocity c.

She can solve this problem of determining c, the air speed of the pigeon, by sending the pigeon back and forth along a course transverse to the motion of the flatcar—or motion of the wind, as suggested by Figure 6.3. An analysis of the pigeon's behavior on this course will show that the pigeon is still affected by the wind. Its times will still be increased by the existence of a wind, but not so much as for the other course;[2] the proportional increase in the trip time will be $\frac{1}{2}v^2/c^2$ if the train travels much slower than the pigeon or if v is much less than c. Then, if the train speed is one-tenth the pigeon speed, the extra time required for a round trip on the moving train will be 0.5%; the difference will be just one-half as great for the transverse round trip as for the round trip parallel to the railroad tracks. By analyzing the different slowing-down factors we can deduce both the velocity of the wind, v, and the air speed of the pigeon, c.

Both of the pigeon's round trips—the trip around the course parallel to the direction of motion of the railroad car, and the trip transverse to the direction of motion—take longer when the car is moving or the wind is blowing, but the parallel trip takes longer than the transverse trip. If Olivia starts one pigeon on the transverse trip and one on the parallel trip simultaneously (choosing pigeons that are identical twins—an interesting problem for biologists) using pigeons that have identical airspeeds, the pigeon on the transverse course will return before the pigeon flying the parallel course *if the car is in motion through the air*. The proportional time difference, $\Delta t/t$, will depend on the velocity of the wind (or, from the view of Oliver on the platform, where the air is still, on the velocity of the railroad car):

$$\frac{\Delta t}{t} = \frac{\Delta t'}{t} - \frac{\Delta t''}{t} = \frac{1}{2}\frac{v^2}{c^2}$$

At first thought it might seem that Olivia, on the railroad car, has not improved her situation very much, because she still needs to know c in order to find v. But the qualitative fact that one pigeon took longer than the other tells her that the train is moving, and quantitatively she may now measure the ratio v^2/c^2 instead of the difference between c^2 and $c^2 - v^2$, which she would have to use if she timed only the parallel round trip. If v is very much smaller than c, then even if c is known only approximately, the comparison of the two flights will give the value of v rather accurately, while the measurement of the time required to take just the one trip will be useless.

Of course, these measurements allow Olivia to determine if the flatcar is moving through the air and to measure the absolute velocity of the flatcar with respect to that air. The velocity of the flatcar is not truly relative: the earth and its atmosphere is a privileged reference system. Is there such a *universal* privileged reference system? Has nature constructed a platform from which the velocity of the earth can be defined? We can examine this question by measuring the velocity of light on the earth rather than of pigeons on a train.

The Michelson-Morley Experiment

The procedures used in the last section to determine the velocity of a flatcar through still air by measuring the flight times of pigeons can be used to detect the motion of the earth through an ether—to detect the ether wind expected by most physicists of a century ago. The railroad car is now the earth; the signal is carried not by a pigeon but by light; the wind we will try to detect is not air but the supposed carrier of light, the ether; and the observers are ourselves in the persons of Albert Michelson and Edward Morley at Case Institute and Western Reserve University in the Cleveland of 1886–1887. Because the velocity of the earth around the sun is only about 10^{-4} times the velocity of light, $v^2/c^2 = 10^{-8}$, and a very sensitive experiment must be designed. Of course, because the sun is moving with respect to the galaxy and the galaxy itself is moving, the effective velocity with respect to the mass of the universe is now known to be larger; but that velocity was not known in 1886.

Figure 6.4 is a simplified diagram of the apparatus of Michelson and Morley. A beam of light from a source designed to provide light that is nearly

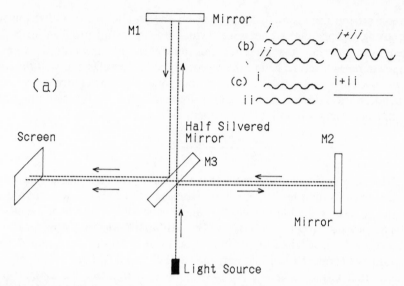

Figure 6.4 (*a*) The paths of the light beams in the Michelson-Morley experiment. (*b*) Individual amplitudes and sum of the amplitudes of two waves in phase. (*c*) Amplitudes and sum for two waves out of phase.

monochromatic is incident upon a half-silvered mirror M_3 that splits the beam; about half of the light is transmitted, and about half is reflected. The transmitted beam passes to mirror M_1 and is reflected back to M_3, where it is reflected again to a screen viewed by an observer. The reflected beam from M_3 passes to a mirror M_2 that reflects it back to and through M_3 so that it is also incident on the screen. If the total distance the two beams travel is the same and if the velocity of light is the same for the two beams, the light waves will be in phase at the position of the observer, as suggested by Figure 6.4*b*, and the observer will see a bright spot on the screen. If the waves are out of phase by one-half wavelength, as shown in Figure 6.4*c*, the two beams will tend to cancel, and there will be at best only a very dim spot on the screen.

By carefully moving one of the mirrors a very short distance and observing the shifts in brightness, Michelson and Morley convinced themselves that they could detect a shift of $1/100$ of a wavelength. (Their exact procedures were somewhat more complicated but the same in principle as described here.) The experiment was performed by setting the two mirrors M_1 and M_2 in a specific position—say, such that the beams were in phase—and then observing any change in the brightness of the spot that might result from a change in the relative phase of the interfering beams as the apparatus was rotated. Assume that the apparatus was set so that the times required for the two beams to traverse their respective courses were the same. This would be done by adjusting the mirrors until the brightness of the spot was a maximum. Then, when the apparatus was rotated 90°, the rays would take a different time to traverse the apparatus if there were an ether wind.

The light beams are analogous to pigeons, and the ether corresponds to the air through which the flatcar passes. The two orthogonal light paths to mirrors M_1 and M_2 correspond to the two orthogonal paths the pigeons take flying on the train. The number of waves per second in each light beam is equal to the frequency of the light, f. If the traversal time of two beams differs by an amount dt, the number of wavelengths in the two rays will differ by an amount $d\lambda$, where

$$\frac{d\lambda}{\lambda} = f\,dt$$

The wavelength shift would result in a difference in phase of the two beams at the screen, hence a difference in the intensity of the spot seen by the observer.

The Michelson-Morley apparatus was designed to detect a velocity of $\frac{1}{30}$ the velocity of the earth about the sun. No wavelength shift was noted at all. The experiment was repeated at different times of the year, and with other variations, but no effect[3] was ever observed. There was no ether wind.

Although it seems improbable, it still might be possible that the earth swept the ether along with it, just as the closed railroad car swept the air along with it. If this were the case, the Michelson-Morley experiment would give a null result. Just as pigeons in a closed railroad car would fly at the same velocity with respect to the car in the longitudinal and transverse directions, light would travel at the same velocity, independent of direction, if the ether were pulled along with the earth. However, an astronomical observation made long before, in 1827, by James Bradley, then Astronomer Royal of Great Britain, disposed of such a view. Bradley noticed that he had to orient his telescope at a slightly different angle, depending on the time of year, in order to align the telescope on the various fixed stars. The positions of the stars appeared to vary slightly, in a regular fashion, during the year. The character of this *aberration* is suggested by Figure 6.5. The angle at which the telescope is oriented to view any fixed star changes by an angle equal to 2θ over a period of six months. Indeed, the apparent position of the stars moves about in a circle during the year such that the radius of the circle subtends an angle θ.

Upon careful analysis of his measurements, Bradley found that a particularly simple explanation fit all his observations of aberrations. The telescope moved slightly as a result of the earth's motion as the light passed down the tube. This is illustrated schematically in Figure 6.6, where the earth and telescope are moving to the left with velocity v. The light enters the telescope and takes a time equal to $dt = L/c$ to travel to the bottom of the tube, where L is the length of the tube. During this time, the telescope moves a distance $v \cdot dt$. If the light is to strike the center of the telescope mirror, the telescope must be tilted as shown in the figure. The angle of tilt θ of the telescope with respect to the true direction of the star is such that

$$\theta = \frac{v}{c}$$

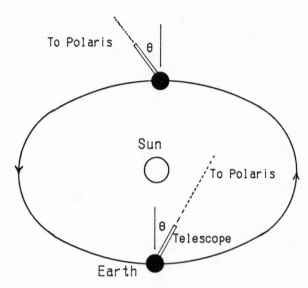

Figure 6.5 The orientation of a telescope viewing Polaris at different times of the year.

where θ is measured in radians (one radian = 57.3 degrees; v is the velocity of the earth about the sun; and c is the velocity of light).

If the ether were dragged along with the earth, the phenomena of aberration would not be observed. The light would be dragged along with the telescope no matter what the velocity of the earth, and the angular orientation of the telescope would not have to be changed during the year.

The observation of aberration shows that the ether (if there is an ether) is not carried along with the earth like the air in the enclosed railroad car. Yet the Michelson-Morley experiment showed that the ether *is* carried along with the earth. There is a paradox!

For completeness, we should also examine the "projectile" theory of light. Can we resolve the paradox just identified by using such a model of light? If light had properties like those of a classical particle, and the wave nature was somehow not important, we might expect that the relevant velocity of light would be defined with respect to its source—for instance, the muzzle velocity of a bullet from a gun. The velocity with respect to the ground of a bullet fired forward with a muzzle velocity of 2800 feet per second from a World War II British Spitfire plane traveling at 600 feet per second (about 400 miles per hour) is 2800 + 600 = 3400 feet per second. Such an addition-of-velocity model is consistent with both the Michelson-Morley experiment and the aberration results. There would be no Michelson-Morley effect if the experiment had been designed to use machine-gun bullets. And there would be an aberration result if light traveled like bullets.

However, this rather unattractive particle model of light is excluded by direct measurements on the velocity of light emitted by particles that are themselves moving with very high velocities—velocities near the speed of light. The emitted light travels with the characteristic speed of light independent of the velocity of the emitter. The projectile model is thus inconsistent with the results of observations and cannot be used to resolve the paradox posed by the Michelson-Morley experiment and the observation of aberration.

The existence of the paradoxical Michelson-Morley result contradicts our classical beliefs concerning the relations of time and space itself. The experiment shows that the velocity of light is not dependent on the motion of the earth—our metaphorical pigeon representing light flies at a constant velocity c with respect to Olivia on the railroad car. The railroad car must then be closed, sweeping the air—or ether—along with it. However, the aberration results show that the ether, or air in the car, is not swept along. The car is an open flatcar, and the pigeon flies at a constant velocity c with respect to Oliver, the observer on the platform. But according to Galilean relativity, the velocities of the pigeon, or light, in the two inertial reference frames are related such that

$$c = c + v$$

which is contradicted by observation.

Our logic is impeccable; our result, that *the velocity of light is the same for any observer in an inertial reference system,* is truly *relative,* because an observer cannot detect his or her absolute motion by measuring the speed of light. Yet that conclusion is in contradiction with the results derived from our logical constructions; therefore, some basic assumption must be wrong.

We will find that our implicit assumptions concerning the character of space and time must be revised, and a new description of nature must be constructed—a description that must include the absolute character of the

Figure 6.6 The path of light passing through a moving telescope.

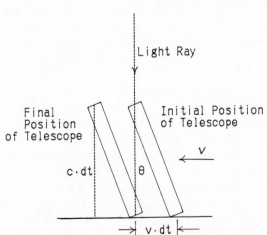

speed of light and the relative character of inertial coordinate systems. This description is the Special Theory of Relativity.

The Lorentz-Fitzgerald Relations

Observations concerning the character of light have led, inescapably, to the conclusion that the speed of light is observed to have the same value by an appropriate observer in any inertial coordinate system, where by inertial we mean any coordinate system such that we can neglect acceleration or gravity. How can we understand this (experimental) observation of nature and retain the description of nature—of space and time—that has worked so well in correlating other observations? This was an essential problem of physics at the end of the nineteenth century.

Following Einstein, we will proceed to analyze the measurements different observers make of space-time events and examine the relations forced on us by the results of those measurements. We can begin by again enlisting the aid of Oliver and Olivia and suggesting that they conduct a variant of their experiment with pigeons on a slow train. We direct them to a much faster train and ask then to use light to transmit messages rather than pigeons. As before, Oliver will set up his laboratory on the platform, and Olivia will make measurements from the train. This time, they will each measure the speed of light, rather than that of pigeons, along a set of two paths on the train and a set of two paths on the platform. On train and platform, courses will be set up parallel and perpendicular to the direction of the tracks.

Before the passage of the train past the platform in the last experiment, the train stopped at the platform, and Oliver and Olivia laid out courses on the train and platform together and satisfied themselves that the two sets of courses and the other equipment on the train and platform were identical. In particular, the measuring equipment was calibrated, the two meter sticks were found to be identical, and the two clocks kept the same time. Each course was prepared with a length of 300,000 kilometers (more precisely, 299,992 kilometers) so that light would take just one second to travel one way on the course. Moreover, the train was known to travel at a speed of 30,000 kilometers per second. (Obviously, so wide and fast a train required a special, poetic, license.)

Now, when the train is running past the platform, both Oliver and Olivia measure the time light takes to go back and forth over the transverse course of length L'_y laid out on the railroad car. On the platform Oliver measures the time required for the round trip and writes that result, $t_y = 2.01$ seconds, in his notebook. He then calculates the time he expects if the speed of light is c (300,000 kilometers per second) and finds the calculated result[4] is in agreement with his measurements. Although the transverse dimension of the course is 300,000 kilometers, the light must travel farther—as shown by the sketch in his notebook (Figure 6.7a)—because of the motion of the train; hence, the total time must be greater than 2 seconds. Indeed, Oliver uses that

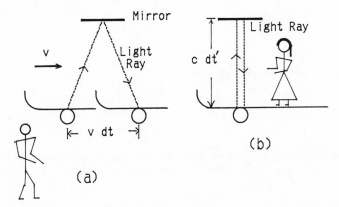

Figure 6.7 (a) Path of light traveling on the transverse course laid out on the train as seen by Oliver from his position on the platform and recorded in his notebook. (b) The path seen by Olivia from her place on the train and sketched in her notebook.

calculation to verify that the light he sees is traveling with the standard velocity, c.

However, upon observing Olivia, he is upset to see that she writes in her notebook a time t'_y of 2.00 seconds, which is smaller than his value by 0.5%. Moreover, that time which she measured is just what she expected from her calculation[5] and the sketch in her notebook of the path of the light (Figure 6.7b). She uses the calculation to verify her assumption that the speed of light is indeed equal to c, the standard value.

Why do the results of the measurements differ? Oliver can only conclude that the clock Olivia is using must be running slow; it must lose 0.01 second every 2 seconds. Because it was cross-checked with the platform clock when the train was at rest in the station, perhaps the motion affected it. At any rate, the relation

$$t_y = \frac{t'_y}{\sqrt{1 - v^2/c^2}}$$

expresses the difference[6] in the elapsed times Oliver and Olivia recorded in their notebooks.

Now Oliver, standing on the platform, watches Olivia, on the railroad car, measure the speed of light along the course that lies parallel to the motion of the train past the platform. Oliver also measures, with his stopwatch, the time for the light to go back and forth along the course as t_x. He finds that the time required is 2.01 seconds, which does not seem quite right to him, as he has calculated the time that the light should take to be 2.02 seconds. According to his calculations the motion of the train leads to a longer total distance traveled by the light. (The character of the increase is shown by Figure 6.8, taken from Oliver's notebook, where he uses $v = c/2$ to show the effect more clearly.) He calculates the time required for the first leg of the trip, where light

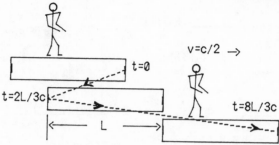

Figure 6.8 A diagram showing the distance Oliver sees the light travel back and forth, reflected by the mirror on the train. For clarity the ratio of train velocity to light velocity, v/c, is taken as one-half.

and mirror are moving toward each other, as the length of the course divided by the sum of the velocities, and he finds a time of 1.0/1.1 seconds. (He knows that $v = 0.1c$.) For the second leg of the course, where the goal is moving away from the oncoming light, the effective velocity is the difference between the train and light velocity, and Oliver finds that the time should be 1.0/0.9 seconds (just as $v = 0.1c$). With his pocket calculator, Oliver adds the two ratios and finds the total time to be 2.02 seconds.[7] Then why did his measurements show that only 2.01 seconds were required for the trip? Oliver can only conclude that the motion of the train must have caused the course length to shrink just enough to account for the 0.01-second difference. Indeed, the shrinkage[8] must then be 0.5%.

Now Oliver notes that Olivia has recorded a shorter time,[9] $t'_x = 2.00$ seconds—as he expected, because he "knows" that her clock is slow. But he is surprised to see that a remeasurement of the distance that she made gives the original value rather than the shrunken value he finds. Hence, the time she records agrees with her calculation that the required time should simply be the total distance, $2L'_x$, divided by the speed of light, c.

Musing over the difference in the length determinations, Oliver says to himself, "I expected that she would find a shorter time for the light trip back and forth along the x-course, as I noted that her clock is running 0.5% slow from the comparison of our results timing the y-path circuit. But I can only conclude that the motion of the train must have changed distances as well as times and that the parallel course on the train, the x-path, must have shrunk by about 0.5% since we laid out the paths when the train was stopped beside the platform. I know that Olivia just checked the length by laying down her meter stick and counting meters, but her meter stick must have also shrunk. The motion of the train must have caused both stick and path to shrink by 0.5%—and also caused her clock to run slowly by 0.5%."

Oliver, who knows only classical physics and Galilean relativity, continues: "I do find it strange, however, that the errors Olivia has made as a consequence of the failures of her measuring instruments act in such a way that she still gets the same, correct value for the speed of light that I do."

In summary, as viewed by the observer on the platform, clocks on the train moving parallel to the platform are seen to be running slow, and lengths on the train parallel to the direction of the motion of the train are seen to be foreshortened.

Now we can repeat our scenerio from the viewpoint of Olivia who watches Oliver make his measurements of the speed of light by determining the time the light takes to travel back and forth over courses laid out on the platform parallel to the railroad tracks and transverse to the tracks. For the first entry in her notebook, Olivia writes, "Although I am stationary, the platform is moving by me. My measurement of that motion gives me a result for the platform velocity of $-v$ meters per second. That is, the platform is moving backward." The rest of the entries in her notebook are quite similar to the entries in Oliver's notebook made when Olivia was measuring the velocity of light over the two courses on the train. And the results are similar, insofar as Olivia concludes, "It is clear to me, from my observations, that the clocks on the platform are running slow by 0.5% (perhaps from the motion of the platform) and that all lengths parallel to the direction of motion of the platform are foreshortened by 0.5%. However, I find it most curious that in spite of the errors in his measuring instruments, Oliver on the platform manages to get the standard value for the speed of light because the errors seem to conspire in such a way as to generate that result."

Each sees the other's clock running slowly, and each sees the other's meter stick shortened when laid out in the direction of relative motion. Moreover, both find that the speed of light has its standard value whether the light moves transverse or parallel to the direction of relative motion. The results, for both observers of both sets of measurements, are then in accord with the principle of the relativity of motion and with the results of the Michelson-Morley experiment.

The hypotheses of the foreshortening of lengths in the direction of motion was first postulated by Hendrick Lorentz, Professor at the University of Utrecht, and by George Fitzgerald at Trinity College, Dublin, as explanations of the results of the Michelson-Morley results. Their view was that the "ether wind" caused this contraction of the length. Lorentz then pointed out that if the ether wind also caused the clocks in the "moving" system to run slowly, the observer in the moving system could not detect either the wind or his motion and would get the same results in his measurements as a "stationary" experimenter.

The Relative Character of Observers in Different Inertial Frames

Explanations of the Michelson-Morley results that demand such special distortions of the measuring instruments are unsatisfactory because they are designed to fit one experimental result and thus have an artificial character.

The ether wind is undetectable because it affects the measuring instruments in precisely the way needed to fool the instruments. Logic like this is satisfactory as a beginning hypothesis, but carried further it is very close to the reasoning of a witch-doctor rather than a scientist. If the ether cannot be detected, it is simpler to assume that it does not exist. What meaning can the concept of the ether have if the ether is in principle undetectable? This was the view taken by Einstein.

In much the same spirit, if the ether is undetectable in principle, we must disregard the idea that the clock is incorrect or that the meter stick is incorrect because of this external agency, the ether. The clock measures time; the meter stick measures distance. Indeed, as we use the concepts to describe relations among events in this universe, time is *defined* as that which is measured by a clock, and spatial distance is *defined* as that which is measured by a meter stick. If the clock on the railroad car runs slowly from the view of the man on the platform, it must be time itself that is seen to pass slowly. If the meter stick is seen to be shortened, it must be space itself that is seen to be compressed. Time and space can have meaning only in observation and measurement.

In 1904 Henri Poincaré stated that a new dynamics must be developed that would not differentiate between observers in different reference frames (relativity) and would lead naturally to the conclusion that the speed of light was the same for all observers. In 1905 Einstein published his Special Theory of Relativity, which accomplished that aim. The basic equations of Einstein's theory were the equations of Lorentz and Fitzgerald. But Einstein gave them a new meaning.

If only relative velocity is meaningful, there can be no intrinsic difference between the view of the railroad car by Oliver, the observer on the platform, and the view of the platform by Olivia, the observer on the railroad car. Moreover, the speed of light must be the same as measured by each observer. The particular distortion of space and time suggested by Lorentz is uniquely determined by these requirements. The train system, moving in the x-direction as seen by a stationary observer, Oliver, will appear to be shortened in the x-direction, and time will appear to pass more slowly on the train. However, from the view of Olivia, the observer on the railroad car, the platform system will be appear to be shortened, and time will appear to pass more slowly on the platform.

There must be no contradiction between these views. In order to illuminate these conclusions, we now turn to a particular sequence of events using the scene and equipment shown in Figure 6.2; however, light now replaces the pigeons as the carrier of signals. Oliver and Olivia must set their respective electrodes 100 meters apart. They measure this distance using their meter sticks when the train is halted alongside the platform, and they recheck their measurements when the train (or platform?) is in motion. They also calibrate their clocks during the time the train is stopped. For the measurements of the two events, the spark emitted during the passage of the first set of electrodes (event i) and the spark emitted by the coincidence of the second set of electrodes (event *ii*), each observer stands exactly between the two electrodes and receives signals from the sparks that allow him or her to decide which spark appeared first. Furthermore, each observer measures the velocity of light by determining the time it takes for a ray to go a certain distance in the transverse direction and return.

The results are in the form of a dialogue between Oliver and Olivia.

Oliver: I measured the speed of light by timing the light on the transverse (to-and-fro) course, and I got the correct handbook value for the velocity. I noted that the electrodes on the railroad car moving past my platform with velocity v were not so far apart as mine on the platform. Therefore spark (*i*) occurred before spark (*ii*).

Olivia: I also measured the speed of light by timing the transverse course, and I got the correct handbook value. The electrodes on the platform, which was moving by my stationary train with a velocity of $-v$ (backward), were not as far apart as my electrodes on the train. Therefore spark (*ii*) occurred before spark (*i*).

Oliver: As for the measurements of Olivia, I know that she measured the spacing of her electrodes correctly (indeed, I helped her when the train was parked alongside the platform), but I see now that the distance has contracted. Her check of the distance while the train was moving was invalid because her meter stick has been shortened. The reason she saw event (*ii*) before event (*i*) is that she was moving toward (*ii*), so naturally the light didn't take as long to get to her from (*ii*) because it didn't have as far to go. As to her measurement of the velocity of light, actually the light took longer to go back and forth on the transverse course because the train was moving, but her clock was wrong. It was running just slow enough that she got the correct value, c.

Olivia: It was Oliver who measured his electrode distance incorrectly. His electrodes were closer together than mine when the platform passed by my train (probably as a consequence of the motion), and his check of the distance was invalid because his meter stick became shorter. Because he was moving backward with a velocity $-v$, the light from event (*i*) reached him before the light from spark (*ii*) even though, as I have reported, the spark (*ii*) came first. Because he was moving, the time the light took to run his transverse course was longer than it should have been. He got the right value, c, for the speed of light only because his clock was slow.

Notice that for each observer, both their measurements and their views of the other's measurements are symmetric. As a third observer we might ask, "Who really was right? Which spark occurred first? Which electrode-spacing measurement was really correct? Which electrode spacing was the greater?" A physicist's answer is that both observers were correct, but that neither the length of the meter stick nor the timing of clocks is an invariant independent of the condition of the observer.

The Lorentz Transformation

When Oliver and Olivia were working as surveyors in Chapter 3, using maps with different orientations, they found it important to develop a set of equations that could be used to relate their measurements. With such equations Oliver could use Olivia's measurements to check his own, and vice versa. These relations are called transformation equations. When Oliver and Olivia shifted their efforts from land surveying about New York and Boston to pigeon racing as viewed from platforms and trains, they constructed the transformation equations of Galilean relativity to relate their measurements of pigeon flights. Now that they have given up pigeons (and slow trains) and have become interested in measuring light velocities on fast trains, they find it again important to construct equations that relate their measurements. These are the equations of the *Lorentz transformation*.

To construct these equations in a convenient manner on the fast train, Oliver and Olivia redo the experiment they conducted on the slow train using pigeons, as illustrated in Figure 6.1, where the events are sparks produced when electrodes on the train pass electrodes on the platform. Indeed, their fundamental reasoning follows the same course. They consider first the relations between their measurements of the distance between the two events (the two sparks). The quantities x and x' represent, for example, the number of times Oliver and Olivia have laid down their meter sticks in marking off the distance between the two electrodes on platform and train. We first listen to Oliver, who says, "The distance x I measured is equal to the distance between the electrodes on the car (measured by Olivia as x' meters) plus the distance the train moved during the elapsed time t; that distance is just the velocity of the train multiplied by the time between sparks, or vt. Although Olivia says the distance is x' meters—that is, she put down her meter stick x' times along the distance—her meter stick was too short by 0.5% or, better, by a factor of

$$\sqrt{1 - v^2/c^2}$$

and the correct distance is not so great as x' meters. (If her meter stick were longer, the correct length, she would have needed to lay down her meter stick fewer times along the distance.)" According to Oliver the "correct" distance between the electrodes on the train is

$$x'\sqrt{1 - v^2/c^2} \text{ meters}$$

and the distance between the electrodes on the platform is then

$$x = x'\sqrt{1 - v^2/c^2} + vt \text{ meters}$$

and, with a little algebra,

$$x' = \frac{x - vt}{\sqrt{1 - v^2/c^2}}$$

This equation relates the values of the measurements x and x' made by the observers in their different inertial systems.

If we carry through the same kind of analysis from the view of Olivia, we arrive at the relation

$$x = \frac{x' + vt'}{\sqrt{1 - v^2/c^2}}$$

which has the same form as the previous relation except that the sign of v is reversed because the platform is traveling backward from the view of the train.

We have implied that the measurement of the elapsed time between the two events by the different observers will not necessarily result in the same recorded value; the time t recorded by Oliver will not necessarily be equal to the time t' that Olivia finds. This difference is established by the two equations relating distance. With a little algebraic manipulation[10] we have

$$t = \frac{t' + x'v/c^2}{\sqrt{1 - v^2/c^2}} \quad \text{and} \quad t' = \frac{t - xv/c^2}{\sqrt{1 - v^2/c^2}}$$

These four equations, the equations of the Lorentz transformation, relate the numerical results of the operations of measuring the distance and time between two events in one inertial system with similar results made in a second system that is moving with velocity v with respect to the first. Do the equations demand that time and distance, such as the time between two events or the distance separating the two events, actually be different as seen from the two different systems, or merely that they be measured differently? As physicists we can accept no meaning of time and space independent of measurements and the operations involved in measurement.

We might notice that for values of v that are very much smaller than c ($v^2/c^2 \approx 0$), the equations of Special Relativity become, or *correspond* to, the equations of Galilean relativity—as they must. Galilean relativity is not so much wrong as it is an approximation valid for small velocities.

Our discussion has been limited to displacements in the direction of motion. It is rather easy to show that distance measured perpendicular to the direction of relative motion of the two observers must be the same for both. In the sketch of train and tunnel shown in Figure 6.9, distance measured upward must be the same for Oliver standing by the tunnel and Olivia in her position as engineer of the train. This follows from the symmetry of the two observers. Consider the contradiction that occurs when we assume that the moving object expands in the transverse direction. Then Oliver, beside the tunnel, concludes that when the train passes through the tunnel at high speed, it will strike the tunnel because it will have expanded in size. Conversely, Olivia, knowing velocity is relative, says, "I am standing still and the tunnel is approaching at high speed. Since the tunnel must then expand, I will have no difficulty in passing through." Yet either the train will hit the tunnel or it

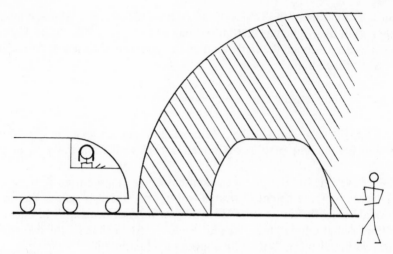

Figure 6.9   Observers in different inertial systems comparing the height of a tunnel.

won't; thus, Oliver and Olivia must see the same result. That is possible only if transverse dimensions are seen—in other words, measured—to be the same by both observers. We can then add to the Lorentz transformation equations concerning x and t the relations

$$y = y' \quad \text{and} \quad z = z'$$

where y and z are distances transverse to the direction of relative motion.

Time as a Fourth Dimension—the Geometry of Space-Time

In their past work as surveyors described in Chapter 3, Oliver and Olivia wrote down simple geometric transformation equations to relate their measurements. For any vector—for example, the vector distance from New York to Boston—they relate Olivia's measurements of the vector distance $\mathbf{A}'$ with components A_x' (magnetic north) and A_y' (magnetic east), with Oliver's measurement written in his notebook as $\mathbf{A}$ with components A_x (polar north) and A_y (polar east), using the transformation equations

$$A_x' = A_x \cos \theta + A_y \sin \theta$$
$$A_y' = A_x(-\sin \theta) + A_y \cos \theta$$

where θ was the angle of 15° between the polar coordinate system Oliver used and the magnetic coordinate system Olivia used—between polar north and magnetic north. Although they differed in their determinations of the components of the vector they measured, the east-west and north-south distance between the two cities, these components were related rationally through the transformation equations. However, they did agree on the distance between

New York and Boston. That distance, a scalar invariant, was the same for both coordinate systems. From the Pythagorean theorem, the distance D was found to be

$$D = \sqrt{A_{x'}^2 + A_{y'}^2} = \sqrt{A_x^2 + A_y^2}$$

In general, the distance between any two towns on their maps was the same, although the two observers did not agree as to how far north or east one town was from the other. The *sum* of the squares of the difference north and the difference east was always the same for both maps; the square root of the sum, the distance between the towns, was invariant with respect to the direction used for north.

Intrigued by the existence of such relations developed in the course of their surveying work, Oliver and Olivia became interested in the possibility that similar invariants might obtain for the measurements they made of the space and time intervals between events when Olivia was on the train and Oliver on the platform. Here they found that the *difference* between the squares of the space differences between events each measured and the square of the time difference between events multiplied by the speed of light, taken from their notebooks, was the same. Writing the space differences as x and x' and the time differences as t and t', all taken from their notebooks, Oliver and Olivia found that

$$c^2s^2 = c^2t^2 - x^2 = c^2t'^2 - x'^2$$

where the quantity s, called the proper time interval, is the same for both observers. The proper time between two events is a Lorentz invariant,[11] the same for any observer whatever his or her velocity.

For convenience we have neglected carrying along the third dimension in our discussion of geometric transformations and both transverse dimensions in our consideration of Lorentz transformations. Using all three spatial dimensions and the one time dimension, the invariant proper-time interval between two events, s, is often written as

$$ds = \frac{1}{c} \sqrt{c^2t^2 - x^2 - y^2 - z^2}$$

where t is the time interval between the events and x, y, and z are the components of the spatial difference in any inertial system.

Although ct plays a formal role as a fourth dimension in the relativistic description of space-time, nearly equivalent to that of the Cartesian spatial coordinates x, y, and z, the radical differences between time and space that are part of our experience are reflected in the formalism through the difference in sign between the squares of the spatial differences and the square of the time difference. Although there are striking similarities between the time dimension and spatial dimensions, there are also very deep differences.

In particular, although different observers see the space-time interval between two events as composed of different intervals of space and time taken separately, just as observers using different coordinate systems assign different values to the x and y components of a vector, space and time are not completely interchangeable as are x and y. If we rotate a coordinate system by 90°, we interchange x and y: $x \rightarrow y$ and $y \rightarrow -x$. By rotating the coordinate system 180°, we change x to $-x$ and y to $-y$. If we could perform the same kind of transformation on the (x, ct) coordinate system, we would be interchanging past and future. However, for better or worse, we cannot do this: time travel is no more allowed relativistically than it is in classical physics. Even as we can divide time into past and future classically, we can divide space-time into a part that is absolutely in the past and a part that is absolutely in the future. But we do have a third region of space-time—a part that is neither unchangeably past or unchangeably future but may be seen as past or future by observers in different inertial frames.

As a pragmatic definition of future, we will use the concept of causality by postulating that "if event A can effect or initiate event B, B is in the future with respect to A and A is in the past with respect to B." It is important to note that the equations of the Lorentz transformation together with the postulate that no signals can propagate faster than the speed of light—and therefore the basic concepts of special relativity—hold to this definition of past and future. If event A is situated in space-time such that for some particular inertial frame of reference a signal from A could effect event B in any manner, that condition or possibility will hold for observers in any possible inertial system.

This property can be demonstrated if we assume that in some inertial system S, two events take place. Event A occurs at a point $x = 0$ and a time $t = 0$ (we may arrange this by starting our clock when the event takes place and measuring distance from the point of the event); the other event, B, takes place at a point x and a time t. If A is to effect B, a signal from A must be able to reach the point x before the time t. If the signal travels at the speed of light c, this requires that

$$t > \frac{x}{c} \quad \text{hence } ct > x \quad \text{and} \quad s^2 = c^2t^2 - x^2 > 0$$

From the invariance of the proper time s, we see that if the distance between the two events x is less than ct, which is required if the first event can initiate or effect the second in one reference frame, that inequality must hold for any reference frame. Past and future are invariant with respect to the observer if the past can be causally related to the future. This situation will obtain if s^2 is positive. If this is the case, we say that the space-time interval is *timelike*.

In contrast, if the distance between events, x, is larger than the value of ct, no light signal can pass from event A to the position of event B before B occurs. Because we postulate that no signal can propagate faster than the speed of light (otherwise our arguments used to develop special relativity collapse), event A cannot cause event B. With $x > ct$, s^2 is negative, and we then say that the space-time interval s is *spacelike*. Events separated by spacelike proper-time intervals can be seen in different time orders by different observers, but this kind of reversal of past and future does not violate causality. The previous discussion of the priority of sparks observed by Oliver and Olivia serves as an example of a noncausal reversal of past and future.

The irreversibility of causal past and future is violated if signals can be transmitted with a velocity greater than that of light. Curiously, there is no logical objection to the existence of faster-than-light particles except for the violation of causality, if these particles—called *tachyons*—are constrained never to go slower than the speed of light. The violation of causality that follows from signals that travel faster than light can be illustrated by a simple example for which we postulate instantaneous signals.

Assume that when Olivia, traveling in her fast train at a velocity of 0.866c, passes Oliver on the platform at noon, they synchronize their watches. An hour later, at 1:00 P.M. according to the platform clock, Oliver sends an affectionate message to Olivia by his instantaneous telegraph. As a consequence of time dilatation, only one-half hour has passed on the train, and Olivia logs in the message at 12:30 by her clock. Deciding that she objects to his message, she telegraphs him by the instantaneous telegraph at 12:40 by her clock to ask him not to send the message. Of course, from her view the platform has been traveling at high velocity with respect to the train, and time is slowed on the platform. Hence, as a consequence of the time dilatation, Oliver receives Olivia's message at 12:20 by his clock, *before he sent her his message*. Being a good fellow, of course he does not send the message.

Even more exciting, if either Oliver or Olivia can obtain accommodations on a tachyon train, they could travel backward in time. (Models of tachyons have been constructed that are believed to avoid causality violations. There is no evidence for either kind of tachyons.)

The Addition of Velocities and $E = mc^2$

Olivia owns a greyhound that is running on the train in the x-direction with a velocity that Olivia measures as u'. When the dog's time trial occurs, the train happens to be traveling past the platform in the x-direction with a velocity v. What is the velocity u of the dog as seen (measured) by Oliver on the platform? What is the relativistic extension of the Galilean rule for the addition of velocities, where $u = u' + v$? Oliver and Olivia can work out the correct relationship using the Lorentz transformation equations.

In making the measurement of her dog's speed, Olivia first measures the length of the course she has laid out to be $x' = 300,000$ kilometers and then, with her stopwatch, measures the time, $t' = 10$ seconds, which elapses while the dog runs the course. She finds the dog's velocity simply by dividing the distance by the time: $u' = x'/t' = 30,000$ kilometers per second or one-tenth the speed of light (a champion dog!). Of course, being a careful person, she writes all these numbers down in her notebook.

Oliver proceeds similarly by making his own determination of the distance the dog travels past the platform and finds that the distance between the position of the dog (relative to the platform) at the start of the trial and the position of the dog as he reaches the end of the course has a value x. Then, watching the trial, he measures the elapsed time with his stopwatch as t. Of course, he then calculates the velocity by dividing the distance by the time to get $u = x/t$—and he, too, writes down all the numbers in his notebook. But he finds a different—slower—velocity than he expected. With the train traveling at one-tenth the speed of light, he finds the distance x to be a little longer than twice 300,000 kilometers (603,000 kilometers) and the elapsed time to be a little longer than 10 seconds (10.15 seconds); the velocity of the dog with respect to the platform is a little less than two-tenths of the velocity of light (19.80% rather than 20%), the sum of the train speed and Olivia's measure of the dog's speed. The velocity of the train and dog do not add. However, upon reflection, Oliver compares his measurements with Olivia's and sees that they are related, just as he might have expected from the Lorentz transformation equations. The two observers then derive a general formula:[12]

$$u = \frac{u' + v}{1 + u'v/c^2}$$

It is especially interesting to use this relativistic formula to consider the velocity of light in the two coordinate systems. Assume that Olivia was timing light instead of a dog; then $u' = c$. From the addition formula we find that the value of u determined by Oliver on the platform is

$$u = \frac{u' + v}{1 + u'v/c^2} = \frac{c + v}{1 + v/c} = c\,\frac{1 + v/c}{1 + v/c} = c$$

And, as we expect, the speed of light is the same in the two systems. This result was used as a postulate basic to the reasoning that led to the development of the Lorentz transformation relations and the formula for the addition of velocities. The result, $u = c$, is a check on the chain of reasoning.

The addition-of-velocity formula, along with the rest of the Special Theory of Relativity, is implicitly checked continuously in the course of measurements made in experimental studies of elementary particle reactions. For example, a particle called the π^0 decays, emitting two photons or gamma rays that travel at the speed of light (because they are high-frequency light) in the system (the train system) of the π^0. Regularly, in experiments, π^0 particles are produced at very high energies so that they travel at speeds very near the velocity of light. However, from the view of the experimenter in the laboratory (or platform), the velocity of the light emitted by the π^0 forward along its direction of flight is not $2c$ but, of course, just c.

The Energy-Momentum Four-Vector: $E = mc^2$

The breakdown of the classical addition-of-velocity formula suggests that classical descriptions of momenta and energy as conserved quantities will probably be invalid relativistically. Simple examples demonstrate that this is indeed the case.

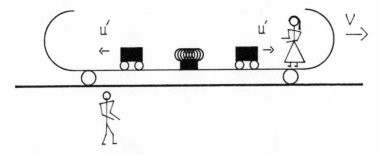

Figure 6.10 Observers in different inertial systems considering the motions of two carts pushed apart by a spring.

On the train shown in Figure 6.10, two similar carts, each with a mass m, initially at rest on the train, are pushed apart by the pressure of a light spring so that one moves forward and one back. Olivia, on the train, measures their velocities in the x-direction as u' and $-u'$, where $u' = c/2$. Before the spring is actuated, the total momentum of the carts in the train system is zero; the carts are not moving. Using classical physics and the classical expression for momentum as the mass multiplied by the velocity, Olivia finds that after the action, the momentum is

$$p' = mu' + m(-u') = 0$$

and momentum is conserved.

From the view of Oliver on the platform, also using classical physics, the momentum of the two masses before the spring pushes them apart is

$$p = 2mv = mc$$

where v is the velocity of the train—which has now speeded up so that $v = c/2$. After the spring sends the two carts moving apart, Oliver finds that one cart (moving backward from Olivia) is stationary in his platform system (and has zero momentum), whereas the other cart is moving forward with a velocity of c and, with a mass of m, has a classical momentum of mc. The sum of the two final momenta is again mc, which is less than before the spring operated, so that classical momentum is not conserved. Indeed, classical momentum is not, in general, conserved relativistically.[13]

A somewhat more complicated example shows that classical kinetic energy is not conserved relativistically. Again, we may analyze a simple interaction that takes place on the train from the view of Oliver, standing on the platform. In this case, assume that two objects with mass m (perhaps the same carts), are moving toward each other on the train in the transverse y-direction with velocities of $-u'_y$ and u'_y as suggested by Figure 6.11, where both velocities are of a magnitude $c/2$ and the velocity of the train is the same, $v = c/2$. The carts collide elastically, bouncing off at right angles to their original directions of motion, so that they are moving with the same speeds in the x-direction—that is, with velocities u'_x and $-u'_x$ where $u'_x = u'_y$

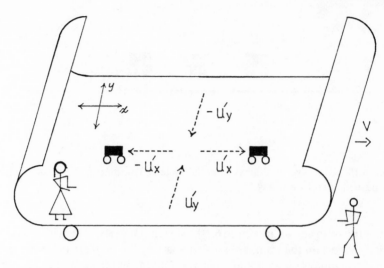

Figure 6.11 Elastic collision of two carts on a railroad car.

$= u'$. In the train system before and after the collision, Olivia calculates the classical kinetic energies of the two bodies as simply

$$2 \cdot (\tfrac{1}{2}mu'^2) = mu'^2$$

In this system, the (classical) kinetic energy is conserved; the carts are moving with the same speed, albeit in different directions, and have the same kinetic energy after the collision as before.

Oliver, on the platform, measures the initial kinetic energy of the two carts as $\tfrac{7}{16}mc^2$. After the collision he finds the kinetic energy to be $\tfrac{9}{25}mc^2$, and the classical kinetic energy is not conserved.[14]

With the breakdown of the classical conservation laws at high velocities (an experimental result predicted correctly by the application of the Special Theory of Relativity to velocity), can relativistic conservation laws be found to replace them? Because the classical laws do work for velocities very much smaller than light, such relativistic laws must reduce to the classical laws in the low-velocity limit. With such a clue and assumptions concerning the logic of nature, one can develop relativistic analogues to the classical laws of the conservation of momentum and energy. However, their validity must be established experimentally because they, like the original classical laws, must be regarded as descriptions of observations.

In short, we find experimentally that new quantities, which we label momentum and total energy, are conserved where the components of the momentum are defined as

$$p_x = \frac{mv_x}{\sqrt{1 - v^2/c^2}} \qquad p_y = \frac{mv_y}{\sqrt{1 - v^2/c^2}} \qquad p_z = \frac{mv_z}{\sqrt{1 - v^2/c^2}}$$

where v_x, v_y, and v_z are the velocities in the x, y, and z directions and

$$v^2 = v_x^2 + v_y^2 + v_z^2$$

In the limit of v very much smaller than c (so that v^2/c^2 can be set to zero in the above equations), the relativistic definition of momentum reduces to the classical definition:

$$p_x = mv_x \qquad p_y = mv_y \qquad p_z = mv_z$$

Similarly, we construct a relativistic energy that is conserved experimentally:

$$E = \frac{mc^2}{\sqrt{1 - v^2/c^2}}$$

In the limit of very small velocities, $v \rightarrow 0$, this expression reduces to $E = mc^2$ rather than the classical expression for kinetic energy, $E = \frac{1}{2}mv^2$. However, if we manipulate the expression above algebraically to the form

$$E = mc^2 + mc^2 \left(\frac{1}{\sqrt{1 - v^2/c^2}} - 1 \right)$$

we separate the quantity E into two parts, where we call the first term, mc^2, the *rest mass* energy, and the second term the kinetic energy. In the limit of v small compared to c, the second term expressing the kinetic energy T reduces[15] to $T = \frac{1}{2}mv^2$, the classical kinetic energy.

To describe nature covariantly, Oliver and Olivia found it necessary to describe space and time measurements in terms of four component vectors (four-vectors) that were designed so that Oliver could use appropriate transformation equations to compare his measurements of the space and time between events with Olivia's measurements even though Olivia might use a coordinate system that we rotated with respect to his and even though she might be moving with respect to him. The classical space components x, y, and z plus the time variable ct were shown to make up a four-vector in space-time such that the components of the vector transformed into each other as observers in different inertial systems described the results of their observations.

Just as Oliver and Olivia measured different values of distance and time in their different coordinate systems, they will find different values for the components of momentum and for the total energy of a system. If they are going to compare their measurements, they need, again, a description of energy and momentum such that transformation equations relate the results of their measurements (just as their measurements of space and time are so related). Using the space-time four-vector **s** as a model, Oliver and Olivia find that they can define an energy-momentum four-vector **p** with the three momentum components corresponding to the three space components and the relativistic energy (divided by c) corresponding to the time component

$$p_0, p_x, p_y, p_z \qquad \text{corresponding to} \qquad ct, x, y, z$$

where

$$p_0 = E/c = \frac{mc}{\sqrt{1 - v^2/c^2}}$$

Just as *all* three dimensional vectors transform in the same manner to relate Oliver's map and Olivia's, *all* Lorentz four-vectors transform in the same manner to relate Oliver's observations on the platform to Olivia's on the train. Hence, the four components of the four-momentum **p** transform according to the equations of the Lorentz transformation, as do the components of the space-time four-vector **s**.

We may notice that the relativistic expression for a component of momentum has the same form as the classical expression if we replace the classical mass by a relativistic variant, as follows:

$$p_x \to mv \qquad \text{as} \quad \frac{m}{\sqrt{1 - v^2/c^2}} \to m$$

Everywhere that mass appears in classical physics, the quantity

$$\frac{m}{\sqrt{1 - v^2/c^2}}$$

occurs in relativistic mechanics. We can well lump the whole term as mass and, noticing that this expression becomes larger as v, in the denominator, becomes larger, say that *mass increases with velocity*.

Does mass really increase with velocity, or is such a statement simply a convenient way of expressing the results of the Special Theory of Relativity? To answer, we should first realize that the question is about semantics, not about physics. It is about the naming of a construct, not about the meaning of the construct, which is well defined. However, we do want to link technical speech with common speech, and we can defend the use of the term "mass" on the basis that this use is consistent with a primitive meaning of the word. Primitively, mass is most commonly understood as a weight; later, we will relate weight to the inertial resistance to force. In particular, if we apply a force F to an object, perhaps attached by a spring to something very massive such as the earth, the object will undergo an acceleration a proportional to the force and inversely proportional to a property of the body we call mass (which, in turn, is proportional to the weight of the object or the force of gravity on the body). In short, we write the familiar relation

$$F = ma \qquad \text{or} \qquad m = F/a$$

as a useful definition of m. (Because force is best defined in terms of momentum change and mass, there is a certain circular character about this argument that can be removed by more intricate analyses that need not concern us here.)

Then, if we place a particle of nominal mass m_0 (rest mass) in a box (taken as having negligible mass for convenience) and apply a force F on the box, the box will accelerate with a certain acceleration a_0 such that

$$F = m_0 a_0 \qquad \text{or} \qquad m_0 = \frac{F}{a_0}$$

Now if we arrange that the particle is moving rapidly about in the box, bouncing off the walls with a velocity v, and we apply the same force, using the same spring, we will find the acceleration reduced as if the mass of the box had increased. The new acceleration a_1 will have a value such that

$$F = \frac{m_0}{\sqrt{1 - v^2/c^2}} a_1 \quad \text{or} \quad m = \frac{m_0}{\sqrt{1 - v^2/c^2}} = \frac{F}{a_1}$$

We find in general that all energy has an inertia of the sort we have associated with mass—all energy has mass. We might therefore consider that all mass, in the ordinary sense of the mass of a particle, is the mass associated with a condensation of energy. Perhaps particles are essentially "boxes" that hold energy. Indeed, modern views of the origin of the mass of elementary particles have very much this character. Perhaps there is no meaning to the concept of mass separate from energy!

Notes

1. For a car moving with a velocity v, the round-trip time is

$$t' = \frac{L}{c + v} + \frac{L}{c - v}$$

A little algebra puts this in the form

$$t' = \frac{2L}{c} \left(\frac{1}{1 - v^2/c^2} \right)$$

which is our basic relation. For small values of v/c, one can use the binomial expansion $(1 + a)^n \approx 1 + na$ to get

$$t' \approx \frac{2L}{c} [1 + v^2/c^2]$$

and the proportional extra time is

$$\frac{\Delta t}{t} = \frac{t' - t}{t} = \frac{v^2}{c^2}$$

2. Assuming the flatcar is moving through the air, if the pigeon aims at a goal transverse to the motion of the car, closes its eyes, and starts flying toward the goal, it will miss the goal by a large margin. From Oliver's view, the goal is carried upstream by the motion of the car while the pigeon is in flight: from Olivia's view, the wind sweeps the pigeon downstream. Figure 6.3 shows the intended and actual path of the pigeon as inscribed on the map Olivia draws. If the pigeon is to actually reach the goal, the bird must aim at a point upstream before closing its eyes and beginning the flight. How far upstream must the bird aim? And then, how far must the bird fly and how long will it take for the bird to fly to the target? This is most easily determined by taking the view of Oliver, who is stationary with respect to the medium in which the bird is flying—that is, the air. From Oliver's view both the pigeon and goal are in motion. If the pigeon's course is correct, the pigeon will reach the goal after a flight of t seconds. During this time the goal will have moved upstream a distance of vt meters, where v is the velocity of the car. If the bird's flight speed is c meters per

second, the bird will have flown a distance of ct meters, and this distance will represent the hypotenuse of a triangle where L, the transverse distance from the starting point to the goal, is the base, and vt is the altitude. This is suggested by Figure 6.3b. From the Pythagorean theorem,

$$L^2 = (ct)^2 + (vt)^2$$

Then

$$t^2 = \frac{L^2}{c^2 + v^2} \quad \text{or} \quad t = \frac{L}{c} \frac{1}{\sqrt{1 - v^2/c^2}}$$

and the round trip takes twice as long:

$$t'' = \frac{2L}{c} \frac{1}{\sqrt{1 - v^2/c^2}}$$

And the pigeon's round-trip flight takes longer than the time, $2L/c$, it would if the railroad car were stationary—or the wind were not blowing—by the factor

$$\frac{\Delta t''}{t} = \frac{t'' - 2L/c}{2L/c} = \frac{1}{\sqrt{1 - v^2/c^2}} - 1 \approx \frac{1}{2} \frac{v^2}{c^2}$$

3. Historians of science have debated the question as to whether the Michelson-Morley experiment was considered by Einstein in his development of the Special Theory of Relativity. To this working physicist, it seems clear that the answer is yes. In his original paper, published in 1905, Einstein referred to the existence of null experiments. However, other considerations played a very large part in Einstein's development of the Theory of Special Relativity, and it is quite conceivable that Einstein knew little of the details of an experiment conducted when he was seven years old; moreover, he may not have been familiar with the name of Michelson in 1905.

4. The time required will be the distance traveled divided by the speed of light, c. For one leg of the trip, that distance can be calculated from the length of the course, L_y, and the distance the course traveled during the time t required for the passage of the light over that part of the course, according to the Pythagorean theorem. Then

$$t = \frac{\sqrt{L_y^2 + v^2 t^2}}{c} \qquad c^2 t^2 = L_y^2 + v^2 t^2 \qquad \text{and} \qquad t = \frac{L_y}{c} \frac{1}{\sqrt{1 - v^2/c^2}}$$

The round trip takes twice as long, so

$$t_y = 2t = \frac{2L}{c} \frac{1}{\sqrt{1 - v^2/c^2}}$$

With $L/c = 1$ second and $v/c = \frac{1}{10}$, the calculated elapsed time is indeed 2.01 seconds. Olivia and Oliver do find that their measurements of the length of the course agree: $L_y = L_y' = 300,000$ kilometers.

5. Like Oliver, Olivia finds the time by simply dividing the distance the light travels by the speed of light, c. But there is no correction, from her view, for motion of train or platform; the distance is simply $2L_y'$. Hence

$$t_y' = \frac{2L_y'}{c}$$

6. Olivia reads $(1 - v^2/c^2)^{1/2}$ seconds on her clock in timing the event, whereas Oliver records 1 second on his.

7. The first leg of the trip should take less than 1 second because the mirror reflecting the light at the end of the course has moved during the time the light takes to reach it. The light and clock are moving together, then, with their combined velocities, $c + v$; and the time required is just the distance along the course divided by the sum of the velocities. Taking the distance as the value, $L_x = 300,000$ kilometers, which Oliver measured when the train was halted beside the platform, he finds that

$$t_1 = \frac{L_x}{c + v}$$

On the other leg, the time required is longer than 1 second because the goal is moving away as the light moves toward it. That time is then

$$t_2 = \frac{L_x}{c - v}$$

With a little algebra, we find the total time

$$t_x = t_1 + t_2 = \frac{L_x}{c + v} + \frac{L_x}{c - v} = \frac{2L_x}{c} \frac{1}{1 - v^2/c^2}$$

8. More precisely and more generally, in order to reconcile the times Oliver recorded with his sensible calculations, the relation between the length Oliver observes as the train moves, L_x, and the length he and Olivia measured on the train, L'_x, must be

$$L'_x = L_x\sqrt{1 - v^2/c^2}$$

9. That is,

$$t'_x = \frac{2L'_x}{c}$$

10. Manipulating the second relation

$$x = \frac{x' + vt'}{\sqrt{1 - v^2/c^2}} \rightarrow vt' = x\sqrt{1 - v^2/c^2} - x'$$

then substituting for x' from the first equation

$$vt' = x\sqrt{1 - v^2/c^2} - \frac{x}{\sqrt{1 - v^2/c^2}} + \frac{vt}{\sqrt{1 - v^2/c^2}} \rightarrow \frac{-x \cdot v^2/c^2 + vt}{\sqrt{1 - v^2/c^2}}$$

and dividing through by v we have the result

$$t' = \frac{t - xv/c^2}{\sqrt{1 - v^2/c^2}}$$

The relation for t follows from a similar procedure.

11. The invariant character of the proper time s follows from the Lorentz equations. We write the equations in the form

$$x' = x \cos \theta + ict \sin \theta \quad \text{and} \quad x = x' \cos \theta - ict' \sin \theta$$
$$ict'' = -x \sin \theta + ict \cos \theta \quad \text{and} \quad ict = x' \sin \theta + ict' \cos \theta$$

where i is the square root of -1 [$i^2 = -1$]. The relation

$$c^2t^2 - x^2 = c^2t'^2 - x'^2$$

follows from direct calculation.

12. We define the start of the course as the origin of the coordinate system in both train and platform coordinate systems and begin the measurement of time in both systems at the start of the run. After 1 second goes by on her watch, $t' = 1$, Olivia notes that her dog has traveled u' meters and concludes that the dog was running at a speed of u' meters per second. From the equations of the Lorentz transformation, we find that the point the dog reached at the end of the 1-second run on the train corresponds to a point on the platform

$$x = \gamma u' + \beta \gamma c \quad \text{where } \gamma = 1/\sqrt{1 - \beta^2} \quad \text{and} \quad \beta = v/c$$

where v is the velocity of the train. And the dog reached that point at a time t measured by the clock on the platform, where

$$ct = \gamma c + \beta \gamma u' \quad \text{or} \quad t = \gamma + \beta \gamma \frac{u'}{c}$$

Oliver, on the platform, measures the velocity of the dog as $u = x/t$:

$$u = \frac{x}{t} = \frac{\gamma u' + \beta \gamma c}{\gamma + \beta \gamma u'/c} = \frac{u' + v}{1 + u'v/c^2}$$

which is the general formula for the relativistic addition of velocities. For $u'v \ll c^2$, the relation reduces to the classical formula $u = u' + v$.

13. After the spring sends the two masses apart on the train with velocities u' and $-u'$, their classical momentum in the platform system can be calculated as

$$p = m \frac{v + u}{1 + uv/c^2} + m \frac{v - u}{1 - uv/c^2} = \frac{2mv(1 - u^2/c^2)}{1 - u^2v^2/c^4} \neq 2mv$$

14. Although Oliver's measurements of the distance the carts travel in the y-direction to the collision are the same as Olivia's measurements, he finds that they are moving more slowly in that direction as he notes that Olivia's clock is running slow. His value for the y-velocity is $u'/\sqrt{1 - v^2/c^2}$—rather than u' measured by Olivia and he finds the initial velocities of the carts in the x-direction to be v, the velocity of the train. He then adds the squares of the velocity components to find the kinetic energies of the carts before the collision:

$$2 \cdot [\tfrac{1}{2}m(3u'^2/4 + v^2)] = m(3u'^2/4 + v^2) = \tfrac{7}{16}mc^2$$

Oliver uses the relativistic formula for the addition of velocities after the collision to find a value of the classical kinetic energy:

$$\frac{1}{2} m \left(\frac{v + u'}{1 + u'v/c^2} \right)^2 + \frac{1}{2} m \left(\frac{v - u'}{1 - u'v/c^2} \right)^2 = \frac{8}{25} mc^2$$

15. Using the binomial expansion valid for small values of v^2/c^2,

$$(1 - v^2/c^2)^{-1/2} = 1 + \tfrac{1}{2}v^2/c^2$$

we have

$$mc^2 \left(\frac{1}{\sqrt{1 - v^2/c^2}} - 1 \right) = mc^2 \left[1 + \frac{1}{2} v^2/c^2 \right] - mc^2 = \frac{1}{2} mv^2$$

7

The Equivalence
Principle and
the General Theory
of Relativity

God ever geometrizes.

Plato

Acceleration and Gravity—The Equivalence Principle

Almost aside from its value as a description of nature, Einstein's General Theory of Relativity is extraordinarily interesting in that the formulation of the description arose less from complex analyses of experimental measurements of the character of nature than from Einstein's insight into the necessary beauty of nature. Moreover, he saw that this beauty is the beauty of geometry, where the straight lines of geometry are beams of light.

If the General Theory did not evolve from the necessity of bringing some arcane set of facts into the general map of nature, the theory as a description of nature did follow from early observations of nature and, in particular, from analyses of the character of acceleration. As we have found, only relative position is important; there is no absolute position. And only relative velocity, the rate of change of position, is important; there is no absolute velocity. We can well ask whether only relative acceleration, the rate of change of velocity, is important: is there an absolute acceleration? The question of the absolute or relative character of acceleration is not a question to be decided from any logical or mathematical procedure; one must decide this by observation or by the controlled observations we call experiments. Here our accumulated experience suggests that there *is* such a thing as absolute acceleration.

If we are sitting in a train next to another train in the station and the only view we have from our window is of that train, we *can* determine whether our train, rather than the train next to us, is starting to move from the station, if our train begins to move with an appreciable acceleration. If we restrict our observations to measurements of space and time alone, we cannot differen-

103

tiate between the two possibilities (1) that our train is moving forward out of the station and (2) that the other train is moving backward out of the station. The distance between the two trains will change with time in the same manner for the two possible situations. However, the behavior of a simple accelerometer, a weight hanging from the end of a string, will differentiate between the two possibilities, as suggested by Figure 7.1.

Consider the two observers, Oliver and Olivia, on adjacent tracks in the station. Each observes the other train and notes that one of the trains is beginning to move (accelerate) in the course of leaving the station. Moreover, it is clear to each that either Olivia's train is moving out in the forward direction or Oliver's train is beginning to back out of the station. Using only measurements of the different relative positions of the two trains, neither Oliver nor Olivia can tell which is leaving. However, Oliver and Olivia may both measure their acceleration using a simple accelerometer. Each can tell if his or her train is accelerating by the position of the weight on the end of the string. If Olivia's train is accelerating, for instance, the weight on her string will swing back opposite to the direction of acceleration; if her train is stationary and Oliver's train is beginning to leave the station, the weight will not move but will continue to hang straight down toward the floor.

An observer in an enclosed laboratory can then determine the absolute acceleration of the laboratory by using such simple accelerometers. If absolute acceleration exists, the state of zero acceleration must have some absolute meaning in terms of a reference system. What is that preferred frame of reference which has no acceleration? Again, we must defer to observation or experiment, and the most meaningful thing we can say is that our zero of acceleration appears to be the general frame of the fixed stars. The acceleration of the entire mass of the universe, defined empirically as the acceleration of the fixed stars, seems to establish a zero for measurements of acceleration.

Oliver, in an enclosed laboratory on earth, can measure the acceleration of the laboratory using simple techniques. He might determine the acceleration by holding a ball in his hand, at rest with respect to the laboratory, and releasing the ball. If he drops the ball from a height of 5 meters and the ball meets the floor 1 second later, he can use the equations of motion to deduce that either the laboratory is accelerating at a rate of 10 m/sec^2 upward or the ball is accelerating downward at the same rate. Of course, he would probably

Figure 7.1 Oliver and Olivia in different coordinate systems with simple accelerometers. Oliver's car is accelerating to the right; Olivia's is moving with constant velocity (or is stationary).

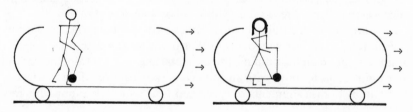

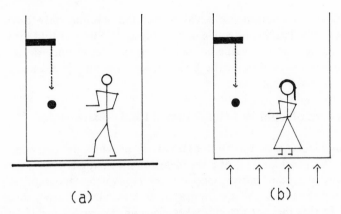

Figure 7.2 Oliver in his laboratory in the gravitational field of the earth (*a*) and Olivia in her accelerating laboratory in deep space (*b*) considering the behavior of a ball released in the system.

state that the ball was accelerating downward under the force of gravity at a rate of 10 m/sec^2.

Now consider Olivia's laboratory, furnished in the same manner, situated very far from the earth in interstellar space where gravity is negligible. The two laboratories are shown in Figure 7.2. Assume that some sort of rocket drive is attached to her laboratory, and the laboratory is accelerating upward at a rate of 10 m/sec^2. Olivia, in that laboratory can determine the acceleration of the laboratory in the same manner as that used by Oliver on earth: she can release a ball and observe the motion of the ball. If she does so, she will see the ball strike the floor (or the floor strike the ball) 1 second later, and she will deduce that the relative acceleration of the ball and laboratory is, again, 10 m/sec^2. The observations of the observers in the two laboratories will thus be similar.

Now we might ask the same kind of questions we asked in our investigation of possible differences between laboratories at different places or moving with different velocities: can Oliver or Olivia make *any* measurements in their laboratories that will differentiate between the stationary laboratory on earth, subject to the earth's gravitational field, and the accelerating laboratory on the spaceship in gravity-free space? This is an experimental question, and the answer appears to be no. The effects of the gravitational field appear to be the same in every respect as the effects induced by the acceleration of the system. It appears to be impossible for the observers to tell which laboratory they are in by measurements confined to that laboratory. This is the *Equivalence Principle,* first formulated clearly by Einstein, and it is one of the most important clues to our understanding of the basic relations between space and matter and of the general nature of forces. The effects that result from an acceleration of the reference system are indistinguishable from the effects of gravity. Then gravity and acceleration must be the same!

An apocryphal story is told of Einstein. It seems that he was supposed to have seen a construction worker fall from a tall building, only to be miraculously saved by landing on a pile of soft refuse. The worker told Einstein that he seemed to be at rest while he was falling and the ground was moving up to strike him. According to the Equivalence Principle, it did not just *seem* that way—it *was* that way.

The Bending of Light in a Gravitational Field—Black Holes

The equivalence of acceleration and gravity provides us with some further insights. We can look first at a problem of an essentially trivial nature that introduces important consequences of the equivalence theorem. Suppose a gun is fired at a ball at the instant the ball is released to fall from some height. Assume further that the activity takes place on the earth's surface at normal gravity and that air resistance can be neglected. Figure 7.3 is an illustration of the scene. Both ball and bullet will undergo an acceleration downward as a consequence of gravity. The bullet will follow some kind of curved path. Our question is: how should the gun be aimed so that the bullet will strike the ball?

Without knowing the original height of the ball, the distance of the ball from the gun, the muzzle velocity of the bullet, or even the value of the acceleration of gravity, we can state that the gun should be aimed directly at the ball just as if there were no gravity at all.

Although one can prove the validity of this conclusion without using the Equivalence Principle explicitly, the result is transparent when the principle is applied. Consider that there is no (gravitational) force on the ball, which then remains stationary, but the earth is accelerating upward. Because the ball is stationary and there are no forces on ball or bullet, the gun should be aimed precisely at the stationary ball. We need not consider the earth at all, except insofar as the earth, moving upward, might strike the ball before the bullet hits it.

We can use much the same kind of reasoning to consider the behavior of

Figure 7.3   A gun firing at an object that is released to fall at the time the gun is fired.

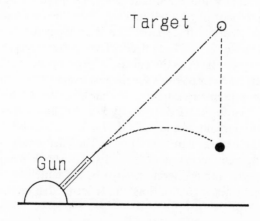

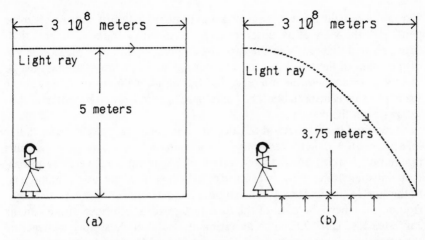

Figure 7.4 A light beam passing through (*a*) a nonaccelerating system and (*b*) an accelerating system.

a light beam in a gravitational field. Assume that we have available a laboratory in interstellar space far from the effective gravitational fields of any star or planet. Figure 7.4 shows such a laboratory; it is 300,000 kilometers long. This ridiculously large distance is chosen to put the deviation of the beam from the acceleration (or gravity) on a human scale: the speed of light is 300,000 kilometers per second, and a beam of light will cross the room in just 1 second. The laboratory is at rest with respect to some far-off source of light, perhaps a distant star, and the light from the star enters the laboratory parallel to the floor through a hole in the wall that is 5 meters above the floor. Olivia, working in this laboratory in interstellar space where gravity is negligible, measures the height of the beam above the floor along the length of the laboratory room and finds that the beam is 5 meters above the floor at every point and strikes the far wall just 5 meters above the floor.

Then rocket motors are started that give the vehicle containing the laboratory an upward acceleration of 10 m/sec². After 0.5 seconds, the room will move upward[1] 1.25 meters. The light that enters the room at the time the motors start and the acceleration begins will reach the center of the room by that time, and the light beam will be only 3.75 meters above the floor: the floor moves upward 1.25 meters. One second after the motors start, the light that enters the room at that time will reach the far wall. By this time the room, and floor, will have moved up 5 meters, and the ray of light will hit the far wall at floor level. The sketch at the right of Figure 7.4 shows the path of the light from the view of Olivia in her laboratory. From her view, the light has curved!

Assuming that she has no source of information beyond her own measurements, Olivia will not know whether the laboratory has accelerated as the result of the ignition of rocket motors attached to it or whether the laboratory has suddenly come under the influence of a gravitational field. If the Equivalence Principle is valid, it is *impossible* for her to tell the difference. Therefore, a ray of light must be bent in a gravitational field just as it *seems* to be bent from the view of an observer in an accelerating frame of reference. If light behaved differently in the two cases, Olivia could make measurements of the beam of light and deduce whether she was in a gravitational field or merely in an accelerating reference frame; but according to the Equivalence Principle, this is impossible. The bending of light in a gravitational field thus follows from the principle.

According to this method of calculating the bending of light, light "falls" in a gravitational field very much as a material object. We can use this method of thinking about the curvature of a light ray under gravity to estimate the magnitude of the curvature of light caused by the gravitational field of the sun. In particular, we can estimate the magnitude of curvature of light from a star where the path of the light just grazes the surface of the sun, as suggested by Figure 7.5. Such an estimate can be made simply[2] by treating the sun as an elevator with a length equal to the diameter of the sun, where the elevator undergoes the acceleration of gravity at the surface of the sun (which is about 270 m/sec^2, 27 times that of the earth at the earth's surface). The result of such a calculation gives the curvature as 0.87 seconds of arc (1 second, also written as 1″, is equal to $\frac{1}{3600}$ degrees). This is a deflection of about ¼ inch in a mile. Mathematically our calculation is correct, but there are conceptual or physical flaws in our handling of the problem. Although an observer stationed on the sun's surface would have seen a deflection of the magnitude we calculated, the relation between his observation and the observation of a man on earth is not so simple and depends on details of the structure of space that are considered properly by Einstein's General Theory of Relativity. According to the General Theory, the deflection of starlight by the sun at the edge of the sun as seen by an observer on earth should be about 1.75 seconds of arc. This is just twice that which we calculated using a very simple model.

Values of the deflection of light by the gravitational field of the sun which are in agreement with the calculated value have been determined by measuring the change in the apparent position of stars, so placed that the light of the stars passes very close to the sun. Such measurements are usually possible only during a total eclipse of the sun.

Figure 7.5   The geometry of a schematic description of the bending of light that passes very near the sun.

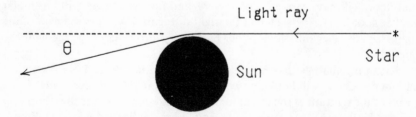

Black Holes

From the foregoing arguments we can calculate the bending of a ray of light passing over the earth's surface. The deflection is extremely small and has not been detected. However, if the acceleration of gravity at the earth's surface were enormously greater, a ray of light emitted parallel to the surface would simply circumnavigate the earth if it were high enough so that it would not be absorbed by the atmosphere. The light would never leave the earth. If the gravitational effect were only a little larger, no ray of light could leave the earth, and rays from outside the earth—from stars, for example—would be trapped and could not get out. If the earth were so massive, so dense, that it would generate such a gravitational field, it would absorb all light incident on it and let none escape.[3] The earth would be completely black: a *black hole*.

Late in the lifetime of a star, as the energy from nuclear processes runs low, the star collapses under the immense gravitational forces generated by the mass of the star to a small sphere with a density about that of atomic nuclei, that is, about $2 \cdot 10^{15}$ grams per cubic centimeter (about 2 billion tons per cubic centimeter). After our sun undergoes such a collapse—to a neutron star—its radius will be about 5 kilometers. Because this is larger than the critical radius of about 1.5 kilometers for a black hole the mass of the sun, the sun will never become a black hole. The critical radius for black hole formation is proportional to the total mass; therefore, that radius for a star eight times the solar mass would be $8 \cdot 1.5$ or 12 kilometers. At the density of nuclear matter, the radius after collapse of such a star would be only 10 kilometers; hence, such a more massive star would become a black hole. It is quite possible that much of the mass of the universe is held in such black holes, the burned-out remnants of large stars common in the youth of the universe.

It now seems probable that large black holes with masses a million times that of our sun occupy the center of galaxies such as ours, and much larger black holes, with masses on the order of 1 billion stellar masses, form the center of those enormously luminous galaxies identified as quasars.

Gravity and the Curvature of Space

In many respects the gravitational force is very much like the electrostatic force; in certain fundamental ways, it is quite different. For moderate values of masses, that is, outside the black hole regime, Newton taught us that the gravitational force between a mass m_1 and a mass m_2, separated by a distance R, can be described by the relation

$$F = \frac{m_1 m_2}{R^2}$$

where the masses are measured in "natural" units.[4]

This relation is analogous to Coulomb's Law for electrostatic fields. Like electrostatic force, gravity is a long-range force. In further analogy with the description of electrostatics, we can define a gravitational field g such that

$$g = \frac{F}{m_0}$$

where m_0 is a unit mass. In such a field the force on a mass m_1 is

$$F = g m_1$$

and we can define lines of gravitational force that have the direction of g and a density equal to the value of g. Again, for moderate masses (for instance, the masses of small stars such as our sun) and small masses, we have analogues to the divergence law for electrostatics in that we can state that the number of field lines that enter a volume is equal to $4\pi m$, where m is the mass held in the volume. Also, the circulation or curl of g is zero.

But there are also important fundamental differences between gravitation and electromagnetism. Gravitational forces are always attractive and never negative, and the superposition theorem of electrostatics does not hold for gravity. We can gain some insight into the breakdown of the superposition principle by considering the energy stored in a gravitational field. As with energy stored in an electrostatic field, energy is stored in a gravitational field. And according to the mass-energy relation $E = mc^2$ of Special Relativity, a mass is associated with this energy. But that field mass itself is then a source of a gravitational field! There are thus forces between fields. The existence of such forces modify the divergence theorem and the superposition theorem; the field generated by a mass m_1 cannot be simply added vectorially to the field from another mass m_2, as the fields affect each other through their mutual attraction. (Later, when we discuss gravity as a curvature of space, we will note that gravity is a tensor force whereas the electrostatic field is a vector force, and we will clarify the meaning of those terms).

For the most part these differences between gravity and electrostatics are important only for very large gravitational fields. In some fundamental sense, gravity is a very weak force: the electrostatic force between two protons (for example) is about 10^{38} times greater than the gravitational force. But however weak they are fundamentally, gravitational forces are dominant for macroscopic affairs, as the forces from different masses always add. Unlike electrical charges, there is no negative mass and there are no repulsive combinations; hence, there is no gravitational shielding, and enormous amounts of material add their effect cooperatively.

Gravitation is especially interesting because it is possible to express the effects of gravity in terms of the geometry of space-time. Operationally, space can be defined only in terms of observations and measurements. It is especially useful to define the character of a space in terms of the properties of the geodesics of the space, that is, the character of the shortest paths that connect points. From Special Relativity the shortest path between two points is the path of a light ray connecting the points. (Here we are considering light in the absense of matter, or the path in a vacuum, to simplify our discussion.) Then, if the path of light is distorted by the gravitational field associated with matter, we must consider that space itself is distorted by the presence of matter.

In order to better understand the character of a distortion or curvature of a three-dimensional space in a manifold of higher dimensionality, it is instructive to consider distortions of a two-dimensional space in a three-dimensional manifold. Assume a race of observers, perhaps flatworms, who are cognizant only of the surface on which they crawl and cannot even conceive of a third dimension—up or down. On this surface they will define a straight line as the shortest distance between two points and call these shortest paths geodesics. If they live on a plane surface, a theorem of their geometry is that the sum of the internal angles of a triangle will be 180°. Now if this surface is not plane but curved as the surface of a sphere, the flatworms will find that the sum of the angles of large triangles—constructed, of course, from three geodesics—will be greater than 180°. (On saddle-like surfaces, the sum will be less than 180°). Figure 7.6 illustrates the different geometries.

This story of the flatworms suggests that we can investigate any possible curvature of our universe by examining the geometry. In order to study the geometry, we must use straight lines. Like any other concept in science, straight lines, the geodesics of our universe, must be operationally defined in terms of some particular technique of measurement. Our experience, codified in the Special Theory of Relativity, suggests to us that nothing travels faster than the speed of light, and the path of a light ray in a vacuum between two points therefore establishes the shortest distance between these points and

Figure 7.6 Flatworms determining the geometry of their universes: to the left, a satisfied flatworm geometer on a plane surface; to the right, a puzzled flatworm geometer on a spherical surface.

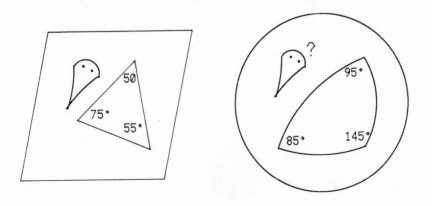

defines the geodesics of our space. We can then examine the curvature of our
space by examining geometric figures constructed from these geodesics.

It was early in the last century that it was recognized, particularly by the
great mathematician C. F. Gauss, that geometry was an experimental science
and that the character of the geometry of our universe must be determined
experimentally. Gauss then performed the flatworm experiment in 1823
using beams of light from alpine peaks as geodesics. He found, within the
accuracy of his measurements, that the angles of a triangle did add to 180°
for a triangle with sides on the order of 100 kilometers. With larger triangles
encompassing the sun, we have now found the deviation Gauss suspected.

The curvature of space in the presence of matter has been measured by
an analysis of the curvature of light near the most massive body in our vicin-
ity, the sun. Figure 7.7a shows a triangle formed by two distant stars and a
point on the earth in the absence of the sun. The sides of the triangle are the
geodesics defined by the paths of light from the stars to the observers on earth.
Figure 7.7b shows the changed geodesics generated by the same configuration
at a time when the sun is between the earth and the two stars. Observations
show that the angle θ' separating the stars is greater at a time when the sun is
between the light paths than the separation θ measured when the sun is
absent. If the angles of the triangle shown in Figure 7.7a add to 180°, the sum
of the angles of the triangle of 7.7b surely add to a greater sum.

Hence, if light is bent by a gravitational field, we must either give up the
idea that a beam of light represents a geodesic or conclude that space itself is
warped or curved by the source of the gravity. Because the concept of light as
a geodesic (nothing travels faster between two points) is fundamental to Special Relativity, which leads to so many experimentally confirmed conclusions, it is very difficult to postulate a better straight line than that defined by
light. Consequently, by far our most compelling explanation of gravitational

Figure 7.7   Paths of light rays to the earth from different stars (a) when the sun is far
from the paths of the light from the distant stars and (b) when the sun is near the light
paths.

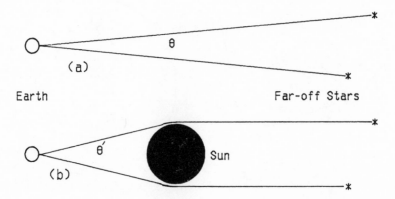

curvature of light is made by defining space itself in a manner that leads to a curvature of that space.

Space is then distorted or warped by the presence of mass. Our three-dimensional space is curved in a space of higher order—with four or more dimensions. In the General Theory of Relativity, Einstein explains the acceleration of gravity by relating it to the geometry of this curved space.

The Distortion of Time in a Gravitational Field

Just as the Equivalence Principle leads naturally to the conclusion that space is curved or distorted by a gravitational field, the postulated equality of an accelerating reference frame and a similar system in a gravitational field shows that the flow of time is changed by gravity. In particular, an observer in a gravitational field such as the field at the surface of the earth will find that a clock below him will run slow and a clock held above him will run fast: time will pass more slowly below and more quickly above him.

To demonstrate the origin of this conclusion, we again assume the equivalence of gravity and acceleration and examine the observations of a man in an elevator car where the elevator is on some planetoid where gravity is negligible. Figure 7.8 shows the position of such an elevator car at four different times spaced at intervals of a second. We assume that the elevator is stationary at the time $t = 0$ but is accelerating upward at a rate of 10 m/sec^2. The height of the elevator car is taken conveniently (for us, if not for the elevator manufacturer) as $3 \cdot 10^8$ meters. Light, which travels at a speed of $3 \cdot 10^8$ meters per second, will then pass from the floor to the ceiling of the car in 1 second if the car is not accelerating and can be considered an inertial frame.

The elevator is furnished with a clock at the floor of the car and a comfortable chair at the top, where the observer sits by another clock. The observer also has a telescope so he can read the clock on the floor. Of course, the clocks were compared before their installation and were found to run at the same rate. To complete the setting, we anticipate the observations and reasoning that make up the Special Theory of Relativity, and we presume that no signal can be sent faster than light and that the velocity of light is the same in any inertial system.

In the discussion that follows, from an inertial system at rest we track the position of an accelerating elevator car as flashes of light emitted every second by a clock on the floor are seen at the ceiling—that is, as light beams start from the floor and arrive at the ceiling. Figure 7.8 shows the position of the laboratory at different times as it moves under the constant acceleration. If a light beam leaves the floor of the laboratory at time $t_0 = 0$, 1 second later it will reach the point where the elevator ceiling was when the light left, but in the second during the travel of the light, the elevator will move 5 meters. Hence, the light has further to go and will take a little more than 1 second—

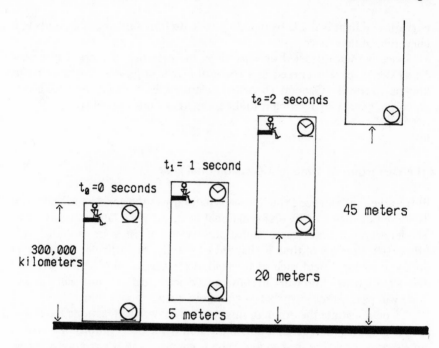

Figure 7.8 An observer in an accelerating elevator car reading a clock near him and a clock far below him.

one second plus (5 meters)/c, where c is the speed of light—to reach the ceiling.

The second light flash, which leaves the elevator floor 1 second after the elevator begins to move, will have an extra 15 meters to go to reach the elevator ceiling. Although it will start from a position 5 meters above the ground, the ceiling will be 20 meters above its original position 1 second later. Then the second light beam will take even longer to reach the ceiling than the first, by an incremental time of (15 meters)/c. Similarly, the third light beam must travel even further than the second and will take commensurately longer to reach the ceiling.

The observer at the ceiling will receive signals at intervals that are larger than 1 second. If he is watching the hands of the clock instead of the flashes, he will see the clock on the floor count out the seconds as if the clock were running too slowly. Time can have meaning only in terms of measurements; so, from the view of the man at the ceiling, time will be passing more slowly on the floor of the elevator car.[5] If the Equivalence Principle is valid, he cannot determine in any way whether he is in an elevator accelerating upward or in a stationary elevator in a gravitational field. So, time must run more slowly at the bottom of a gravitational field. That is, as we stand on earth, time must run more slowly at our feet than at our head.

We could examine the converse situation where the observer is at the floor of the elevator and the clock is at the ceiling. We would find then that the clock appears to run fast. Time proceeds more quickly at the top of the elevator or over our head in a gravitational field.

Are these effects "real," or are they illusive effects of the particular observations? If the two clocks are brought together again, will they agree, or will the lower clock have fallen behind permanently? If the acceleration were carried on for a sufficiently long time, the lower clock would fall behind the upper clock by a certain amount—say, an hour for definiteness. Then, if the clocks were brought slowly together again—perhaps the upper clock would be lowered to the floor—the upper clock would still be ahead of the lower clock by an hour (with a correction of a second connected with the process of lowering the upper clock). If two twins were stationed near the clocks, the lower twin would have aged 1 hour less than the upper twin: aging is a biological clock and can not proceed differently than a mechanical clock if the Equivalence Principle is to be generally valid. (If the biological clock read differently than the mechanical clock, that difference could be used to differentiate gravitation and acceleration.)

In summary, in an accelerating frame of reference, whatever the source of the acceleration—gravity or rocket motors—as 1 second elapses as measured by the watch on the wrist of an observer,

$$1 - \frac{ah}{c^2} \text{ seconds}$$

passes on the clock h meters below him where a is the acceleration and

$$1 + \frac{ah}{c^2} \text{ seconds}$$

passes on the clock at a distance h above him.[6]

Although the effects on the earth's surface are small—over a distance of 10 meters the dilatation factor is only about 1 part in 10^{15}—this change in the flow of time has been accurately measured and is in accord with these calculations.

Time Dilatation and the Twin Paradox

Although the slowing of clocks follows directly from a relativist interpretation of the Michelson-Morley experiment and is therefore a starting point, rather than a conclusion, of our logical discussions, this time dilatation leads to such intriguing results that we will consider some of them in detail. Furthermore, there is ample experimental evidence concerning most aspects of time dilatation.

Nature is well supplied with minute natural clocks. The vibrations of atoms or molecules or the rates of decay of unstable nuclei or elementary particles allow these atoms, nuclei, and particles to serve as clocks themselves. We can easily accelerate these clocks to very high velocities and detect the time dilatation factor of

$$\sqrt{1 - \beta^2} \qquad \text{where } \beta = \frac{v}{c}$$

From the view of an observer of such a clock moving at a velocity v, 1 second of time in the system of the clock corresponds to

$$\frac{1}{\sqrt{1 - \beta^2}}$$

seconds by the clock of the observer. If the clock were moving at a velocity three-fourths the speed of light, $\beta = \sqrt{\frac{3}{4}} = 0.866$, 1 second of the moving clock would be equivalent to 2 seconds of the stationary clock as both are read by the stationary observer.

Time dilatation as such was first detected by Herbert Ives, an American physicist who used radiating atoms as his clock. When he accelerated the atoms to high velocities, he noted that their frequency of vibration, as determined by the frequency of light they emitted, was slightly reduced by the degree predicted by the Lorentz relations. More recently, it has been possible to make measurements on unstable elementary particles such as pions and muons, which can be produced with velocities very near the speed of light, so that the dilatation factor is quite large. Dilatation factors exceeding 100 have been measured.

It is possible to question the universality of the clocks that are used; perhaps those clocks are slowed by the velocity (the effects of the ether wind), but other clocks are not. In particular, does time dilatation hold for biological systems? If any clock did not follow the time dilatation relations, that clock together with a standard clock could be used to establish an absolute velocity. The observer in the closed car could tell whether he was moving by comparing his clocks. If they told the same time he would be motionless; if they ran at quite different rates, he would be moving at high velocity. If, as we believe, there is no absolute velocity, then all clocks, biological or otherwise, must run at the same rate.

Some of the results of time dilatation and length contraction can be illustrated in a somewhat gaudy manner by considering a fanciful trip from the earth to a nearby star—perhaps Alpha Centauri, which is about 4 light-years from the earth. (That is, light from Alpha Centauri takes about 4 years to reach the earth.) We will use poetic license to disregard constraints on the journey that might result from the physiological limits of the crew (that is, the acceleration they can tolerate) and limits resulting from the conservation of energy on the velocities of a spaceship that carries its own fuel. Two records of the trip are kept, one from the view of Oliver on the ground and one from the view of his twin, Roland, on the ship.

From the view of Oliver on earth, the ship leaves the earth on January 1, 2101, and accelerates quickly to a velocity very near that of light. After 4 years the ship (he assumes) reaches Alpha Centauri. The crew explores the system for a year (according to plan, Oliver has no direct communication with them) and then returns to earth, landing on January 1, 2110. From the television shots of the debarkation, although the ship has been away 9 years, the crew seems hardly to have aged.

Roland's diary reads as follows:

> As we begin our journey, Alpha Centauri is 4 light-years away according to the captain. We quickly accelerate to a velocity $v = 0.866c$. Although this takes only a few days and we have hardly left the solar system, the captain announces that our destination is now 2 light-years away. We are already half-way there! [The ship has covered very little distance, but the space-contraction factor is now ½. From the view of the crew on the ship, the distance has been shortened by that factor to 2 light-years.] A few days later, we reach a velocity $v = 0.99c$. The captain says that we are now only 0.4 light-years away. [They are still near the earth, but the foreshortening has increased.] After a few more days, the captain announces that we have reached our cruising velocity of $0.9999c$, that we are now 20 light-days from Alpha Centauri, and that we will get there in 20 days.

Oliver knows from the flight plans that Roland and the rest of the crew have barely left earth, have reached their cruising velocity of $v = 0.9999c$, and will now travel for 4 years to reach Alpha Centauri. In Oliver's view, Roland thinks he has only a little ways to go, about ⅟₇₀ of the actual distance, because of the contraction in length he sees. And he thinks the trip takes 20 days rather than 4 years because of the time dilatation he experiences. Roland's clocks, and of course all physical and physiological processes, are slowed by a factor of 70.

The same process takes place on the trip back. So Roland, who feels he has been gone 1 year and 2 months—a year exploring and 2 months traveling—finds that 9 years have elapsed on earth. Oliver, who remained on earth, is now physiologically 8 years older than his brother. Of course he has eaten 8 more years of meals, slept more nights, and so on.

There seems to be a defect in this result. Can't we consider that it was the earth and Oliver that moved away from the spaceship and that Roland traveled at a velocity of $v = 0.9999$ away from the ship and returned? Why did not time pass more slowly for Oliver on the earth than for Roland on the ship? Do we not have a postulate that velocity is relative? Are the relations not symmetric between the twin brothers?

This is the famous twin paradox. The paradox—that the twins age differently as a result of the velocity of one of the twins when we have postulated that velocity has only a relative meaning—is removed when we observe that Roland undergoes an acceleration while Oliver does not. The situations are not symmetric. If we had put both twins in enclosed rooms without access to the outside, Roland would notice accelerations when the ship starts out, turns around, and finally stops upon returning to earth. Oliver would notice no such acceleration. There is an asymmetry.

We can understand the difference in elapsed time, which must be related to the fact that an acceleration exists, by using the results of our consideration of time dilatation in an accelerating reference system. If one stands in an accelerating system, time passes more slowly below and more quickly above—and the difference is larger if the height differences are large. Roland is accelerating. And when that acceleration takes place near Alpha Centauri, the "height" difference between Roland and Oliver on earth is very large— about 4 light-years.

If Roland is a well-informed physicist, he should be able to calculate the difference in age between his brother Oliver and himself from his view-point—from the reference system of the ship. Assume, then, that Roland knows the character of the theory of relativity and records his musings about the relative ages of his brother and himself in his diary. As Roland is accelerating upon leaving the earth, he knows that Oliver is below him, as seen in Figure 7.9. But Oliver is not very far below him, so the fact that Oliver is then aging more slowly is not too important. On the month-long trip to Alpha Centauri, Roland says that his brother is moving away from him at a very high velocity so that during that month, Oliver ages less than a day. However, upon deaccelerating at Alpha Centauri, he notices that Oliver is very far "above" him—hence, that time must be traveling very quickly where Oliver is and that Oliver must be aging very much (about 4 years, as he calculates

Figure 7.9 Four periods of acceleration of Roland, the twin who travels to Alpha Centauri. (*a*) Roland accelerates upward (in the diagram), leaving earth. (*b*) He slows down to stop at Alpha Centauri, accelerating downward. (*c*) He speeds up taking off from Alpha Centauri on the return trip, accelerating downward. (*d*) He slows down to land on earth, accelerating upward.

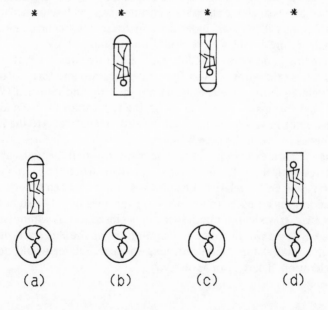

it). The same effect occurs as the crew leaves Alpha Centauri a year later (during which time, Roland calculates that Oliver also ages 1 year): his brother ages another 4 years during the brief time they take in their acceleration. During the month-long trip back home, Roland notes (as a good relativist) that Oliver is traveling with very high velocity with respect to him and again will age less than a day. And during the brief deacceleration, Oliver ages even less than he. But the damage was done during the turn-around at Alpha Centauri; when they meet, Roland finds that Oliver is clearly about 8 years older than he.

One should not conclude from the diary, valid though it may be, that the "cause" of the discrepency is to be found wholly in the act of turning around—there is no operational significance to such a statement. The only observations made are the comparison of clocks by the twins at the beginning and end of their journeys. If Roland does not turn around, he will never get back to earth, and clocks and ages can never be compared.

The time dilatation of twins has been seen and measured experimentally, although the twins were twin nuclear clocks rather than men. It has been found that a nucleus in a hot material will decay more slowly than a similar nucleus in a cold material. The atoms in the hot material are moving back and forth (Roland's journey) more rapidly that are those in the cold material, and as a consequence of their motion, their clocks are slowed down. The difference in lifetime is extremely small, but it has been measured and is in accord with calculations from relativity.

Much larger effects are obtained from less transparent but logically equivalent measurements using twin sets of beams of short-lived elementary particles such as pi-mesons. Here the round trip is provided by the back-scattering of one set of particles in the course of collisions with nuclei, and the aging is noted by a determination of the portion of particles that have died—that is, decayed. In summary, it is important to note that the paradoxical result is *observed* as a property of nature and is not just an element of a possibly erroneous theory.

The General Theory of Relativity

Special Relativity is derived from the necessity of finding a conceptual framework in which one can fit experimental information concerning the impossibility of defining an absolute coordinate system. The Michelson-Morley experiment, together with the results of classical experiments on electromagnetism, require us to adopt a description of the universe in terms of a new geometry—a geometry in which time and space are considered together.

The success of this use of a new geometry to solve serious problems, or to classify otherwise unrelated phenomena, suggested to many physicists—Einstein in particular—that other fundamental problems of physics might be understood through further extensions of the geometric description of spacetime. In particular, the existence of the Equivalence Principle, the exact equivalence of gravitational and inertial mass and the exact equivalence of acceleration and the gravitational field, suggested that gravitation might be another aspect of acceleration and therefore accountable as a kinematic effect of the character of space-time. Einstein's General Theory of Relativity encompasses this aim; gravitation is expressed as a part of the geometry of space-time.

The gravitational force is unique at present in that we have been able to devise a view of the universe in which gravity does not exist as a force but only as a geometric consequence. The General Theory of Relativity has the result of reducing the problem of the description of the motion of several bodies from dynamics to kinematics, where dynamics is the study of the effects of forces on bodies and kinematics is the study of the motion of bodies without reference to forces. If we consider only gravitational forces and adopt an appropriate space-time geometry, *bodies travel only in straight lines.* The presence of other masses distorts the shape of space in such a way as to change the character of a straight line.

We can construct two loose analogies to provide an intuitional framework for a qualitative understanding of the effects of curvatures of space. For simplicity, we will discuss curvatures of two-dimensional spacelike manifolds rather than the four-dimensional universe we live in, where the coordinates of events are described by three space dimensions and one time dimension.

We first view motion on the surface of a globe or another curved surface. As on a flat surface, a straight line or geodesic is defined as the shortest distance between two points. On the surface of a globe, such a geodesic or straight line will be an arc of a great circle.

Consider two objects, a and a', moving in parallel straight lines at some time on the surface of a globe, as suggested by Figure 7.10. As time passes the two objects will draw closer and may even collide eventually at point c. A two-dimensional flatworm inhabiting this global universe, who found it difficult to conceive of a third dimension, might construct a dynamics to explain the phenomenon. The flatworm could assume that space is flat (indeed, the mathematical beauty of plane geometry might convince the flatworm that space must be flat) and invent a force called "gravity" that would act on the two bodies, forcing them together. Of course a and a' would not "feel" a force (if the bodies were sentient flatworms), but this is also true of a man in free fall in our universe.

The "force" on a' and the path of a' are independent of the presence of a in the global system we have just described, and the analogy is still far from our own experience. We can modify this simple picture by considering a two-dimensional system such that the space is flat if no bodies (no matter) exist at all, but the existence of matter distorts the space. A two-dimensional space defined as the surface of a thin sheet of rubber stretched over a horizontal frame will have such a character. A stone placed on the rubber will now distort the space in a manner such that the distortion will be greater for a massive stone than for a pebble, and the geodesics or shortest paths between points on the rubber sheet near the stone will depend on the weight of the

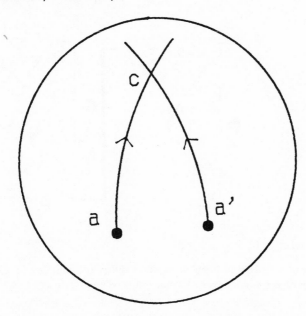

Figure 7.10 Paths of objects moving in straight lines on a spherical surface.

stone. In such a space the presence of mass will induce a curvature in the space, and that curvature will depend on the magnitude of the mass. The path of a' will depend on the presence of a, because the geometry near a' will be determined by the distance of a' from a and the mass of a. In general, the geometry of that rubber-sheet universe depends on the distribution of mass in that universe.

Einstein's General Theory of Relativity is just such a geometric description of our four-dimensional universe. According to this view, gravity does not exist as a force at a distance but as a manifestation of geometry. A body falls to earth in a path that is a "straight line" in space-time near the earth. The meaning of this is made somewhat clearer by Figure 7.11, which shows the paths of a particle in a two-dimensional Cartesian space where one dimension corresponds to a space dimension and one to time. (We are still discussing illustrations that suggest only the character of space-time, as this coordinate system leads to an incorrect invariant description of proper time.) Figure 7.11a shows the path in this space-time of a body at rest in the system of reference; Figure 7.11b shows the path of the body moving with constant velocity in the (negative) x-direction; and Figure 7.11c shows the body moving under acceleration in the x-direction. In the geometry we have adopted, the path of Figure 7.11c is a curved path; if we construct an appropriate geometry, as Einstein did, we might better consider that the path is a straight line in a curved space. The curvature of the space is caused by the presence of the mass of the body we normally consider as the source of the gravitational force.

According to the General Theory, the path of Figure 7.11c appears curved

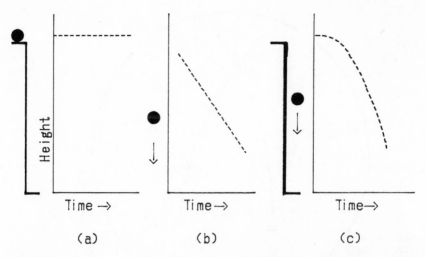

Figure 7.11 Paths in space-time of an object under different conditions: (*a*) a stationary object; (*b*) an object moving with constant velocity; and (*c*) an object falling in a gravitational field.

for much the same reason that the path of an airplane flying in a straight line (the great circle route) over the surface of the earth appears as a curved line on the "flat" map of the earth shown in Figure 7.12. The map is not an adequate representation of the curved space, just as it is not an adequate map of the spherical surface of the earth. Newton's First Law—which can be stated as "a body that is not under the influence of an unbalanced force will move in a straight line in space-time"—is thus extended, so that the effects of gravity are considered not as a force but as the result of the curvature of space-time induced by the presence of a perturbing mass, and therefore a change in the character of the "straight line." Dynamics is changed to geometry.

The Equivalence Principle is an axiom of any general theory of relativity rather than a result, so that the equivalence of acceleration and gravity and any consequences of this equivalence are not tests of a particular theory. In

Figure 7.12 The great circle route (really a straight line) from Chicago to Paris shown on a flat map (dashed line) of the curved surface of the world.

that sense the change in the flow of time in a gravitational field is not a strong test of general relativity.

But there are two consequences of Einstein's geometric theory that do not arise simply from the Equivalence Principle. The exact value of the curvature of light by the sun is one, and the rate of precession of the orbit of an object in a gravitational field is the second.

Although the curvature of light in a gravitational field measured by an observer in that field must be the same as if the observer and reference system were accelerating, the magnitude of the curvature of light by a gravitational field, as measured by an observer who is in another frame of reference, does not follow simply from the Equivalence Principle. That relation between observers in different accelerating or gravitational reference frames is determined by Einstein's geometric theory, and the curvature of light passing near the edge of the sun, as observed by someone on earth, is calculated by that theory to be 1.75 seconds of arc. It is difficult to make precise measurements of this deflection, but the results of many measurements made during total eclipses of the sun are in agreement with the geometric theory with an accuracy of about 10%.

The other striking consequence of the General Theory that does not arise immediately from the Equivalence Principle and differs from the classical theory of gravitation is found in the prediction of the rate of *precession* of the orbit of an object in a central gravitational field. The orbits of two bodies in their mutual gravitational field precess as a consequence of relativistic effects.

Precession can be illustrated somewhat more simply if we consider the orbit of a light body about a very massive body. Classically, the orbit of such a light body about a massive body, such as the orbit of a planet in the gravitational field of the sun, takes the form of a conic section. Stable orbits have the form of ellipses such as that shown in Figure 7.13, where the planet moves faster at the perihelion *a* and slower at the aphelion *c*. If we include the effects of the Special Theory of Relativity, the planet will be heavier at the perihelion than at the aphelion. It will move as shown in the figure (all effects are exaggerated in the figure for clarity) and will change the position of the axis of the ellipse. This change in position of the ellipse is called a precession; for Mercury, the innermost planet, the precession can be observed and measured with some accuracy to be 574 seconds of arc per century with an uncertainty

Figure 7.13 An elliptical orbit of a planet about the sun showing the relativistic precession of the orbit.

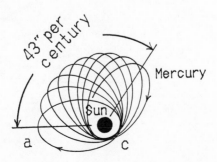

that is now believed to be less than 0.5 seconds (a revolution every 225,000 years!). The shift from the perturbations of other planets is 513 seconds, leaving 43 seconds of arc unaccounted for, classically. The mass increase at the perihelion, which follows from the Special Theory, accounts for 7 seconds per century; but Einstein's General Theory, which considers all motion under gravity as motion in a straight line, predicts 42.9 seconds per century—just what is observed.

Previously we compared gravity and electrostatics. An electrostatic field is a vector field requiring three numbers to describe it completely, that is, the values of the components of the field in three dimensions. In a Cartesian coordinate system (with x, y, and z-directions), the field **E** is defined completely by the values of the components E_x, E_y, and E_z. The gravitational field, now associated with distortions or curvatures of space, requires a more complicated description in terms of the components of a tensor rather than a vector.

The nature of the formal reasoning behind the description of space in the General Theory and the use of tensors in that description becomes clearer when we examine analogous situations in spaces of lower dimensionality. Let us start by examining the problems of a one-dimensional physicist, a lineworm, shown in Figure 7.14a, as he contemplates the possible curvature of his line universe.[7] Perhaps, in the course of analyzing some natural phenomena in his linear universe, he concludes that his space has a slope in a (hypothetical) second dimension. That slope is shown as a tangent in the figure. Moreover, he finds that the slope changes as he moves in his universe. He gives the concept of a changing slope the technical name "curvature" (technical, because he has no intrinsic concept of a second dimension or curvature as we see it), where he defines curvature as the rate of change of the slope. As a physicist, the lineworm is also a competent applied mathematician, so he writes the rate of change of slope formally as d^2z/dx^2, where z is the (imaginary to lineworms) position of his space in the hypothetical second dimen-

Figure 7.14 (a) A lineworm physicist in Lineland examining the curvature of his space; (b) a flatworm physicist in Flatland examining the curvature of his space.

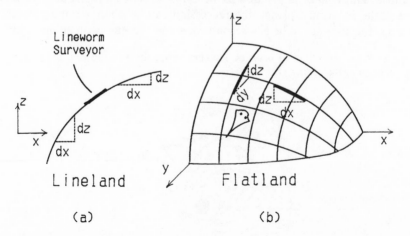

sion and x is the coordinate in his map of his world.[8] Of course, as we can see on our figure of his universe, that curvature varies with x.

For the lineworm in one-dimensional line-space, the curvature at any point in his space is defined simply by one number, d^2z/dx^2. The flatworm physicist, whom we will call Andrew, in the two-dimensional surface-space shown in Figure 7.14b has a more complicated problem. In the course of his measurements at any point in flatland (which is not really flat), he can define slope in two directions: he can identify a slope in the x-direction (shown in the figure as a tangent) and a slope in the y-direction (also shown as a tangent). Moreover, the slopes vary from point to point, and Andrew can define the rate of change of the slope as a "curvature" (which is also a technical name for him, as he cannot visualize the third dimension in which the curvature takes place). But now he has four different rates of change of slope or curvatures, rather than the lineworm's one. He can describe the rate of change of the y-slope as one moves in the y-direction, which he writes formally as d^2z/dy^2; the rate of change of the x-slope as one moves in the x-direction, d^2z/dx^2; the rate of change of the y-slope as one moves in the x-direction, $d^2z/dy\,dx$; and the rate of change of the x-slope as one moves in the y-direction, $d^2z/dx\,dy$. Again, the z-coordinate is to him a mathematical fiction beyond anything he can imagine physically.

Now in Flatland, as well as near Boston, one can use different maps. An associate of our first flatworm, Andrea conducts the same survey using a different orientation for her choice of the x- and y-coordinates. However, Andrea's four numbers defining the curvature of flat-space for any point must define the same physical reality as Andrew's numbers and must then be related to his numbers in some rational way through transformation equations. Flatland physicists define a set of such numbers as tensors just as we do in our universe of three space dimensions plus one time dimension.

In a space of two dimensions, a curvature tensor is defined by four numbers, a vector by two. In three dimensions, such a tensor requires nine numbers, a vector three. Then, for the description of the effects of gravity in four-dimensional space-time, 16 components must be specified in general. For static gravitational fields, nine components define the field, whereas the vector electrostatic field is completely described by the values of the three vector components. The reason for the close analogy between gravity and electromagnetism when the forces between two bodies are considered is that in such situations, the gravitational field tensor can be defined such that only three of the 16 numbers are not zero, and those three numbers transform in the same manner as the three numbers defining a vector. The gravitational field tensor then reduces to a vector when curvature (or gravity) is small.

Notes

1. The distance s traversed at a time t after a body starts moving from rest with an acceleration a is simply

$$s = \tfrac{1}{2}at^2$$

2. The result follows, assuming that the acceleration of the light beam toward the sun, a_d, is just that at the sun's surface (about 270 m/sec^2) over a region equal to the sun's diameter D of about $1.3 \cdot 10^6$ kilometers. Then, in the time D/c of 4.3 seconds the light travels through this field, the light gains a velocity, $v = a_d t = 1160$ meters per second, toward the sun, which is equivalent to an angle of deflection of

$$\theta = \frac{v_d}{c} = \frac{1160}{3 \cdot 10^8} = 3.9 \cdot 10^{-6} \text{ radians}$$

which equals 0.87 seconds of arc.

3. Knowing the gravity at the surface of a spherical mass, one can estimate the size reduction required so that the mass would act as a black hole. The acceleration downward, g_b, required for anything moving at the speed of light, c, to follow the curvature of the surface of a sphere of radius r is

$$g_b = c^2/r$$

If the gravitation at the surface of a mass of radius r_0 is g_0, the gravitational acceleration that would occur if the mass were reduced in size to a radius r is

$$g = g_0 \frac{r_0^2}{r^2}$$

Equating g_b with g,

$$r = \frac{g_0 r_0^2}{c^2}$$

4. The equation is valid numerically if F is measured in dynes, r in centimeters, and m in units of 3870 grams.

5. A more detailed calculation can be constructed if we note that the distance s the elevator moves as a function of time t follows from the formula for the motion of bodies accelerated from rest,

$$s = \tfrac{1}{2}at^2$$

where a is the acceleration.

A first flash from the lower clock is emitted at a time $t = 0$ as the elevator leaves from its ground position in Figure 7.8. If the elevator were not to move, the light would arrive at the ceiling of the elevator car at a time

$$t_0' = t_0 + \frac{h}{c} = 1 \text{ second}$$

where h is the height of the car. However, because the car is accelerating upward, the ceiling will move up during the time the light is in transit, and the signal will take longer to get to the ceiling because it has farther to go. The extra time will be equal to the extra distance divided by the speed of light. Then the time of arrival will be

$$t_0' = t_0 + \frac{h}{c} + \frac{1}{c} [\tfrac{1}{2}a(h/c)^2] = 0 + 1 \text{ sec} + \frac{5 \text{ m}}{c} \text{ sec}$$

The second flash will start at a time $t_1 = 1$ second, when the floor of the car will be 5 meters above ground. By the time the light reaches the ceiling about 1 second later, the ceiling will have moved a total of 20 meters from its original position, and the signal will have to travel an extra distance equal to 20 meters minus 5 meters. It will reach the observer at the ceiling at a time

$$t_1' = t_1 + \frac{h}{c} + \frac{20 \text{ m} - 5 \text{ m}}{c} = 1 \text{ sec} + 1 \text{ sec} + \frac{15 \text{ m}}{c}$$

Similarly,

$$t_2' = t_2 + \frac{h}{c} + \frac{45 \text{ m} - 20 \text{ m}}{c}$$

The time intervals between the reception of the signals by the observer near the ceiling will be

$$t_1' - t_0' = 1 \text{ sec} + \frac{10 \text{ m}}{c}$$

$$t_2' - t_1' = 1 \text{ sec} + \frac{10 \text{ m}}{c}$$

$$t_3' - t_2' = 1 \text{ sec} + \frac{10 \text{ m}}{c}$$

and in general

$$t_{n+1}' - t_n' = 1 \text{ sec} + \frac{10 \text{ m}}{c}$$

where n is any integer. Going further,

$$\Delta t' = \Delta t \left[1 + \frac{ha}{c^2} \right] \quad \text{for } ah \ll c^2$$

6. These relations hold for uniform gravitational fields and for $ah \ll c^2$.

7. Flatland and Lineland are described ingeniously in *Flatland* by A. Square (Edwin A. Abbot), first published 100 years ago. Olivia says that the views of A. Square—male chauvinist nonpareil—concerning women are similar to those of many men she knows.

8. The lineworm determines (somehow, since he cannot imagine the z-direction) the slope dz/dx at two different points (1) and (2), a distance x_{12} apart, and defines a "curvature" as

$$d^2z/dx^2 = [(dz/dx)_2 - (dx/dx)_1]/x_{12}$$

which is the (average) rate of change of the slope dz/dx over the distance x_{12}.

The flatworm procedes similarly, but he has two slopes to consider: dz/dx, the slope in the x-direction, and dx/dy, the slope in the y-direction. As he moves in the x-direction, both slopes change, and he defines a rate of change of the x-slope with a change in the x-position as d^2z/dx^2, and a rate of change of the y-slope with the x-position as $d^2z/dydx$. Similarly, if he examines the changes in the two slopes as he moves a distance y in the y-direction, he defines the other two components of a four-component curvature tensor: the rate of change of the x-slope with respect to the y-position as $d^2z/dxdy$, and the rate of change of the y-slope with respect to a change in y as d^2z/dy^2.

8

The Electromagnetic Field—
The First Unified
Field Theory

> Then God said, *Let there be light, and there
> was[e] light.
>
> [e]The light was made before ether sunne or
> moone was created therefore we must not
> attribute that to ye creatures that are Gods
> instruments.
>
> *The Geneva Bible, 1560*

Forces on Charges in Different Reference Frames

Electromagnetic forces in all their complexity are better understood than any
of the other forces. Moreover, electromagnetic forces are similar to the weak
and strong nuclear forces in fundamental ways. Hence, we must study electromagnetism if we are to gain insight into its more complicated sibling, the
weak nuclear interaction, and its still more complex cousin, the strong
nuclear interaction. In particular, the constraints and enrichment of the electric field required to satisfy the invariances of the Special Theory of Relativity
can be considered a model of relativistic effects for the other vector fields,
and the classical view of the radiation of electromagnetic waves—the emission of light—leads to insights into the radiation of particles by the other
vector forces.

Until the nineteenth century, electric and magnetic forces were considered
separately; no connection between them was known. Then, early work by
such scientists as Hans Christian Oersted and André Ampère led to the developments by Faraday and Maxwell of the first unification of fields—the first
Unified Field Theory, the description of the electromagnetic field. And then
Maxwell's electromagnetic theory led directly to Einstein's formulation of the
Special Theory of Relativity.

Our interest here in fundamental descriptions of fields leads us to proceed
anhistorically and to use the Special Theory to develop the relations between

electricity and magnetism. We begin by considering the character of the extension of electrostatics to moving systems.

The Magnetic Field as a Moving Electric Field—Ampere's Law

The force **F** on a stationary electric charge q which results from any assembly of electric charges or currents is equal to **E**q. This statement stands as a definition of the electric field **E**. However, if there are no privileged frames of reference, the definitions must be incomplete, just as "stationary" can have no fundamental meaning. A complete description of electric phenomena must consider moving charges, and that description must not differentiate among different inertial reference frames. To consider relations between observers in different inertial systems, we again enlist the assistance of Oliver and Olivia, who gained experience in such matters in their work described in Chapter 6.

Properly sensitive to these concerns, Oliver and Olivia decide they want to extend their description of electrical force on stationary charges to include moving charges, and to make some relevant measurements. To this end they work together to construct a pair of long, broad, parallel conducting plates, which they attach to a flatcar on Olivia's train while it sits in the station near Oliver's laboratory on the platform. They place a positive electric charge density $+\sigma$ on the lower plate and an equal negative charge density $-\sigma$ on the upper plate, thus generating a uniform electric field between the plates directed upward that is proportional to the charge density σ ($E = 4\pi\sigma$). They also prepare two small balls of equal mass m upon which they place equal positive charges $+q$.

Having finished preparing this equipment, they arrange for the fast train to pass by the platform at a velocity v of one-half the speed of light c, or $v = c/2$, while Oliver on the platform and Olivia on the train observe the motion of the two charged balls acted on by the electric field between the plates. This scene is shown in Figure 8.1. Our two investigators arrange that one of the balls is (initially) stationary with respect to Oliver on the platform, and one is stationary with respect to Olivia on the train. They then agree to measure the forces on the balls by determining the time required for the balls to move

Figure 8.1 Oliver and Olivia in different inertial systems considering the forces on two charged balls, one stationary with respect to the platform, one stationary with respect to the train. Each determines the force on the balls by measuring the time required for them to move a small distance upward Δz.

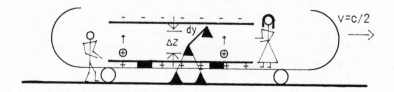

upward a distance Δz from rest—a large force will move the balls more quickly than a small force.[1]

Oliver finds that the ball he saw as stationary moves upward the distance Δz in 0.93 second, a little shorter than the time of 1.0 second he and Olivia measured while testing their apparatus at the construction site in the station. But the moving ball takes a longer time, 1.155 seconds, to travel the same vertical distance. Olivia, on the train, finds that the ball, which is stationary with respect to her, travels the distance in 1.0 second by her clock, whereas the ball initially at rest with respect to the platform—which she sees moving backward—takes 1.074 seconds to rise the distance Δz.

Later, Olivia and Oliver discuss the interpretation of their measurements. Olivia says, "The results are completely explained by Special Relativity. You saw your stationary ball move upward a little faster than the ball we measured in the station because the length of the plates was contracted by their motion and the charge density was then a little larger; hence, the electric field—and force on the charged balls—was a little larger. As for the ball you saw moving along with the train, I found that it moved upward in 1.0 second as in the station, but from your view the elapsed time was greater because you saw my watch running slow as a consequence of relativistic time dilatation. I worried about the possibility that the magnitude of the charge might change with velocity, but the internal velocities of the charges held in atoms are large and vary widely, and the atoms are accurately neutral. Thus there can be no effect of motion on charge." [In Olivia's and Oliver's design the electrical forces are much greater than the gravitational forces, so they need make only small corrections for gravity, which we ignore here.]

Oliver answers, "What you say is correct, but such a formulation is often difficult to apply. I need a way to describe matters in my reference system. From my view the moving ball travels upward more slowly than the stationary ball for two reasons: one, it is heavier—there is a relativistic increase in the mass of the ball—and two, there is a magnetic force acting on the moving body acting opposite to the electric force. In my formulation, using Figure 8.1, I define a magnetic field B directed out of the paper with a strength proportional to the magnitude and velocity v_E of the moving electric field, $B = Ev_E/c$. With this definition of magnetic field, there is a magnetic force F on the moving charged ball such that $F = qBv_q/c$ in a direction perpendicular to the field and to the velocity v_q of the charge—and directed downward in the experiment we conducted. With this formulation, I account for my observations as well as with your logic.[2] Moreover, if I apply these same rules to the observations you made, it correctly describes your observations."

Oliver continues, "In my formulation of electric and magnetic fields, the total force is described as the sum of two forces: there is a force on a charge Eq that is independent of the velocity of the charge, and there is a force Bqv/c that is proportional to the current—that is, the charge multiplied by its velocity. Moreover, the field E depends only on the charge distribution in the environment, whereas B depends only on the currents present—that is, the motion of the charges."

Olivia answers thoughtfully, "Then you have *invented* a 'magnetic field' to describe relativistic effects. I agree that your scheme works nicely, but isn't it a bit artificial?"

Oliver shakes his head. "More artificial than the concept of the electric field? I don't think so. I would prefer to think that this formulation in terms of two fields is an elegant way to *describe* nature."

We can consider the definition of the magnetic field **B** in a somewhat different manner that facilitates the determination of B in a broader set of circumstances. Figure 8.1 shows a line in the y-direction, attached to the platform, which has a length dy. As a consequence of the motion of the train carrying the capacitor, lines of electric field cut the line dy. The direction of **B**, according to Oliver's definition, is along this line, and Oliver finds that the magnitude of B multiplied by the length of the line segment dy is proportional to the number of lines of electric field that cut dy per second.[3]

In general, the magnetic field in a given direction is equal to the number of electric lines cutting a unit length of an imaginary line in that direction divided by c, the speed of light. This is a formulation of Ampere's law.

The Electric Field as a Moving Magnetic Field—Faraday's Law

As described, the magnetic field is a somewhat artificial device designed to extend electrostatics to moving charges. We have considered the magnetic field as generated by a changing electric field. Faraday's discovery (made independently by Joseph Henry in the United States) that a changing magnetic field also generates an electric field led to the understanding—especially by Maxwell—that one can equally consider that the electric field is just the addition necessary to extend magnetostatics (the description of magnetic fields in systems at rest) to moving systems. Moreover, this reciprocal generation of electric fields by magnetic fields and magnetic fields by electric fields leads to the propagation of the radiation we call light.

To examine the relationship of moving magnetic fields and electric fields, we again use the Special Theory of Relativity. And once more, Oliver is on the station platform and Olivia is on the train, ready to make some relevant electrical measurements and report to us.

They decide to modify the apparatus they used previously by adding an electric field from extra plates mounted on the platform to cancel the electric field generated by the moving plates. There will then be no electric field on the platform, only a magnetic field generated by the currents carried by the moving plates. Figure 8.2 shows this electromagnet mounted on the platform such that the railroad car, carrying a charged ball at rest in the car, passes through the magnetic field.

With the new apparatus in place, Oliver finds that there is no force on the stationary ball, but he notices that the ball moving along with the train is forced downward. He ascribes this to the effects of the magnetic field he measured previously (which should not be affected by the stationary charge distribution that nullifies the electric field he measured previously.) As he expected, he finds that the ball travels a distance Δz (the same distance as in

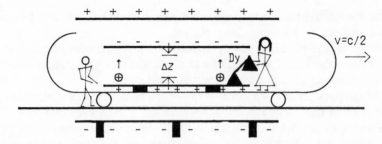

Figure 8.2   Olivia, on a train, considering the force on a charge that results from a magnetic field moving in her system. Oliver, on the platform, considers the force from his viewpoint.

the first experiment, but directed downward) in a time he determines to be 2.00 seconds.[4]

Olivia, of course, also sees the ball move downward, and she determines that it moves a distance Δz in 1.73 seconds. (This is what Oliver expects, as he knows that Olivia's clock runs more slowly than his as a consequence of time dilatation.) Because the ball is initially at rest in Olivia's train, she considers that she has measured the electric field in the train system, and that field is just what she expected—knowing as she does the character of the charged plates and calculating the fields she expects from the charge distributions on the plates.[5]

Oliver agrees that Olivia calculated the force on the ball correctly from relativistic principles, but he suggests that it might have been better if she had considered only the magnetic field. He says the moving magnetic field *generated* an electric field according to the relation

$$E = Bv/c$$

where B is the field on the train and v is the velocity of the field. Oliver calculates the force that way and gets the same result as does Olivia.[6]

The moving magnetic field, as shown in Figure 8.2, generates an electric field in the y-direction. If Olivia mounts a line segment Dy' on the railroad car, that line will cut the lines of magnetic field from the magnet mounted on the platform as the train moves by the platform. Olivia notes that the product of the electric field and the length of the line Dy' fastened to the railroad car is equal to the number of lines of the magnetic field $B'vDy'$ that cut Dy' per second. In general, an electric field is produced in the direction of a line segment which is proportional to the number of lines of magnetic field that cut a unit length of the segment per second. This is *Faraday's law;* it is similar to Ampere's law with the roles of Bc and E reversed.

Maxwell's Equations

The essential results of the study of the electromagnetic field can be summarized in the four Maxwell's laws.

1. *Gauss' law applied to the electric field:*

   The net number of **E** lines leaving a volume = $4\pi q$

   where q is the total charge within the volume.

2. *Gauss' law applied to the magnetic field:*

   The net number of **B** lines leaving a volume = 0

3. *Ampere's law:* The magnetic field in a given direction is equal to $1/c$ multiplied by the number of lines of electric field cutting a unit length of an imaginary line in that direction.[7]

4. *Faraday's law:* In addition to the electric field from stationary electric charges, there is an electric field generated by any changing magnetic field. The component of that field in a given direction is equal to $1/c$ multiplied by the number of lines of magnetic field cutting a unit length of an imaginary line in that direction.[8]

Together with the ponderamotive equation, which relates the fields to the forces on charges, Maxwell's equations summarize the classical (where quantum effects are not important) properties of the electromagnetic field. The ponderamotive equation takes the simple form

$$F = Eq + Bq\,\frac{v}{c}$$

for the case where the two vector fields and the velocity are orthogonal and the sign conventions of Figure 8.2 are followed.

The four Maxwell's equations are valid in any inertial reference system. We did not follow history in our discussion of those relations where we relied on Special Relativity to develop the equations from Coulomb's law. Originally, Maxwell constructed these relations to summarize and describe experimental observations concerning the forces among charges. In particular, the constant c that enters the equations and has the dimensions of a velocity was not logically associated with the velocity of light in the original construction of the four relations. That connection was made later by Maxwell. Lorentz, understanding the importance of the fact that the equations had the same form in different inertial systems, then used Maxwell's equations to develop the transformation equations that bear his name. In this way the study of electromagnetism led to the theory of relativity.

In the domain of the very small, such that units with the dimensions of energy multiplied by time are on the order of magnitude of Planck's constant $\hbar$, there are some modifications to descriptions of phenomena based wholly on Maxwell's equations. Conversely, there are extremely small deviations from Maxwell's equations in descriptions of classical phenomena. As an example, the linearity of fields we have emphasized forbids the scattering of light by an electromagnetic field, or by other light. In fact, there are very small nonlinear effects, which we will touch upon later, that do lead to some scattering of light by very strong electric fields such as the fields near atomic nuclei and some scattering of light by light.

The Symmetry Between **E** and **B**

The symmetry between **E** and **B** exhibited in Maxwell's equations is quite striking. Indeed, the equations are completely symmetric, with the exception that electric charges and currents have been observed but not magnetic charges (or poles, as they are usually named) or, of course, currents of magnetic poles.

Magnetic poles would have properties analogous to those of electric charges. Each pole would emit $4\pi g$ lines of magnetic field **B**, where g is the strength of the pole (corresponding to the charge q). The force on a pole in an electromagnetic field would be

$$F = Bg + Eg\frac{v}{c}$$

where, again, we choose to describe the simple configuration in which the field vectors and velocity are orthogonal. In short, the force on a stationary magnetic pole would be a measure of the magnetic field B, and the force on a moving pole would be proportional to the velocity of the pole and the magnitude of the field E. The symmetry of pole and charge seems to be exact.

If charges and poles are so similar, why hasn't nature provided us with poles? (Of course, perhaps she has, but we have not seen poles in spite of careful searches.) However, as we shall show later, if poles are found they must have much larger charges than the unit electrical charges found on elementary particles such as the electron. So this universe cannot be completely symmetric between pole and charge on the microscopic level.

We might still ask why, if poles and charges are symmetric in principle, we have charges and not poles. If this universe or some other universe were constructed so that there were no electric charge but only magnetic poles with the same value of pole strength as the fundamental charge strength, we believe that that universe would be indistinguishable from ours. If we could communicate with the inhabitants of that universe (who are made of protons, which have no electric charge but hold a unit magnetic pole strength, and electrons, with no charge but with an opposite magnetic pole strength), we could not determine whether they live in a magnetic universe or an electric universe as we do. Do we then live in an electric or magnetic universe? The question has no operational meaning; we have named our charge the electric charge.

Electromagnetic Waves

Changing charge and current distributions generate changing electromagnetic fields. In space, away from physical currents or charges, we see from Maxwell's equations that a changing electric field causes changes in the magnetic field (Ampere's law) and that changing magnetic fields cause changes in electric fields (Faraday's law). Maxwell was able to show that this coupling of changes of electric and magnetic fields will propagate in space with the speed of light, as electromagnetic radiation. Moreover, because the equations are valid in any inertial system, it is immediately evident that the speed of light must then be the same in any inertial system. Einstein's remarkable insights were extracted from a jungle of ideas present at the time, not from a desert.

We have emphasized the generation of electric fields from sets of charges. In particular, a field of characteristic shape, as suggested by Figure 8.3, is produced by a *electric dipole* consisting of equal charges of opposite sign. A similar *magnetic dipole* field is generated by a simple current loop, also shown in Figure 8.3. Both the electric dipoles and the magnetic dipoles can change in time, and such changes must result in changing fields. Figure 8.3 suggests how electric fields at a distance from a changing electric dipole, and magnetic fields at a distance from a changing magnetic current loop (or magnetic dipole), must change with time just as the sources change. The diagrams are incomplete, however, inasmuch as the changing electric fields must generate magnetic fields (which must also change), and the changing magnetic fields from the changing current loop must generate (changing) electric fields.

To discuss the generation of changing magnetic fields by changing electric fields and the induction of changing electric fields by the changes in magnetic fields, we are implicitly asking for an explanation of the propagation of an electromagnetic disturbance and then for an explanation for the propagation of light. Figure 8.4 shows the electric and magnetic field configuration of a wave traveling in the x-direction. The vectors show the field strengths at each point along the x-axis at a particular time for a wave traveling in the x-direction with a velocity v. The wave is polarized so that the electric field vectors lie in the z-direction and the magnetic field is directed in the y-direction.

In the path of the wave, we construct two lines L centimeters long: one in the y-direction, arranged so that the moving electric field lines will cut the line, and one in the z-direction, arranged so that the moving magnetic field lines will cut that line. An observer, perhaps Albert Einstein at the age of 16,

Figure 8.3 Above, the electric field from a changing electric dipole, and below, the magnetic fields from a changing current loop.

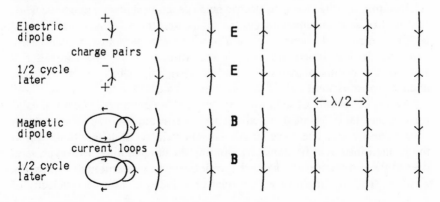

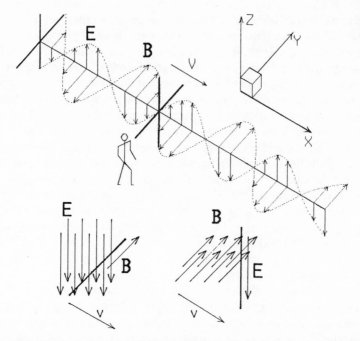

Figure 8.4 The variation with distance of the electric and magnetic fields of a polarized electromagnetic wave, together with elements used to deduce the velocity of the waves.

sometimes runs along the path of the wave carrying the y-z-lines with him. The lower diagrams of Figure 8.4 show the moving fields cutting the lines. (The diagrams are separate for visual ease, but they represent the same region of space.) From Ampere's law, the electric field lines cutting the y-lines generate a magnetic field along that line, as shown in the figure such that $B = Ev/c$, where v is the velocity of the E-field—and of the wave. On the other hand, from Faraday's law the magnetic field lines cutting the z-line generate an electric field along that line such that $E = Bv/c$. But if $E = Bv/c$ and $B = Ev/c$, v must be equal to c; the waves must travel at the speed of light.

In summary, the changing electric fields generate changing magnetic fields that in turn generate changing electric fields, and the wave is driven onward by these reciprocal changes. Naturally, these changes must be consistent. From our analysis based on Faraday's and Ampere's laws, which are descriptions of the results of measurements on changing fields, we find that the changes are consistent only if the wave moves with the speed of light.[9]

Again, the results depend only on Maxwell's equations, which are valid for any inertial coordinate system; hence, our results are valid for any such coordinate system. The wave moves with the velocity of light with respect to the young Einstein as he stands by, holding the crossed lines. However, if he runs in the x-direction at whatever speed, the equations still hold; hence, the wave must be passing him with a relative velocity equal to the velocity of light. The velocity of light must be the same for any inertial coordinate system if Maxwell's equations are valid. Such results violate Galilean relativity and classical physics.

This point, that the speed of light must be the same in any coordinate system, is made neatly by assuming the converse—that the velocity of light is absolute. In the view of the fields from a moving observer (perhaps Einstein, running alongside the wave at the speed of light), neither the magnetic nor the electric fields change. So how could they generate each other? There could be no propagation and no light. In fact, this crucial argument was made by Einstein, at the age of 16, in a letter to his uncle.

Notes

1. Oliver and Olivia use the relation $F = ma$, where F is the force and m the mass of the ball, and the acceleration is related to Δx and the elapsed time t through the equation

$$\Delta z = \tfrac{1}{2} at^2 \quad \text{hence } t = \sqrt{2a\Delta z}$$

2. Olivia calculates the drift times using the relations of the Special Theory of Relativity. When Olivia and Oliver tested their apparatus while the train was in the station, they found that the transit time was 1.0 second for a force $F_y = E_y q$ and a mass m, where $E_y = 4\pi\sigma$ and σ was the charge density. From the platform system, the length of the plates is reduced by the relativistic contraction factor $1/\gamma$, where

$$\gamma = \frac{1}{\sqrt{1 - v^2/c^2}} \quad \text{for } v/c = \tfrac{1}{2} \quad \gamma = 1.155 \quad \text{and} \quad 1/\gamma = 0.84$$

Then the field E and the force F are increased by a factor $\gamma = 1.155$. From the relation $F = ma$, the acceleration a is increased by that factor. Using the kinematic equation

$$\Delta z = \tfrac{1}{2} at^2 \quad \text{and} \quad t = \sqrt{2a\Delta z}$$

the time is decreased by a factor $\sqrt{1/\gamma} = 0.93$ to 0.93 seconds.

From the train system, both the charged plates and the ball that is moving in the platform system are at rest. Hence, the transit time is just 1.0 second, as in the station. But from the platform system, the train clock runs slowly by the time dilatation factor γ, and Oliver will see his platform clock tick off $\gamma = 1.155$ seconds.

Oliver calculates the same reality using his model of electric and magnetic fields. He agrees with Olivia that he will see a stronger electric field E than the field E_y they measured when the apparatus was at rest, and that $E = \gamma E_y = 1.155\, E_y$. Then his calculation of the transit time for the ball, which was initially at rest in his system, agrees with that of Olivia, $t = 0.93$ second.

For the other ball he finds a total force of

$$F = q(E - Bv/c) \quad \text{where } B = Ev/c$$

with

$$v/c = \tfrac{1}{2} \quad \text{and} \quad F = \sqrt{\tfrac{3}{4}}\, Eq = 0.84\, Eq.$$

But the mass of the ball has increased by a factor $\gamma = 1.155$ from the relativistic increase of mass with velocity. Hence, the entire acceleration is reduced by a factor of $\tfrac{3}{4}$ from the acceleration of the ball seen in the test in the yards. The elapsed transit time will then be $\sqrt{4/3}$ seconds or 1.155 seconds, which is the same result as from Olivia's calculation.

3. Multiplying both sides of the equation

$$B = \frac{1}{c} Ev$$

by dy we have

$$B \cdot dy = \frac{1}{c} Ev \cdot dy$$

where $v\,dy$ is the area of field E cut per second, and then $Ev\,dy$ is the number of lines of electric field cut per second.

4. Oliver calculates the magnetic field strength to be

$$B = Ev/c = 4\pi\sigma\gamma v/c$$

just as in the first experiment. With no electric field, the only force will be that due to the magnetic field:

$$F = qBv/c = qEv^2/c^2$$

The ball will be accelerated downward but with one-third the force observed in the first experiment. Hence, the time of 2 seconds Oliver measures is $\sqrt{3}$ times longer than the transit time he measured in the first experiment.

5. From her notebook describing the preparation in the station, Olivia knows that the electric field from the plates bolted onto the train will be

$$E = 4\pi\sigma$$

Having set up the second experiment with Oliver, she knows that they put a larger charge, generating a larger field, on the plates attached to the platform. The installed charge density of those plates was $\gamma\sigma$. Because the plates are now moving (backward, of course) with a velocity of v/c, the relativistic contraction leads to a higher charge density $\gamma^2\sigma$, and the field generated by that charge density is in the opposite direction—downward. The total field is then

$$E' = 4\pi\sigma(\gamma^2 - 1) = 4\pi\sigma/3 \qquad \text{for } v/c = \tfrac{1}{2}$$

The force on the charge is then one-third of the force when the experiment was conducted in the station, and the transit time is $\sqrt{3}$ times as long, or 1.732 seconds.

6. Oliver notes that the magnetic field in his platform system is equal to

$$B = Ev/2 = 4\pi\sigma\gamma v/2$$

But he expects that the field will be stronger on the train as the field is moving with a high velocity and will be compressed by a factor γ. Hence, on the train the magnetic field will be

$$B' = 4\pi\sigma\gamma^2 v/2$$

Then, using the relation $E = Bv/2$, the electric field on the train generated by the moving magnetic field will be

$$E' = 4\pi\sigma\gamma^2 v^2/4$$

which is just what Olivia measured—and calculated from her knowledge of the charge distributions using relativity.

7. The description in terms of "moving" **E** and **B** fields is unconventional (raising logical problems that need not concern us here) and is introduced for pedagogic simplicity.

8. Maxwell's laws are usually written in the more abstract and precise form of vector calculus. For interactions in a vacuum,

$$1. \quad \nabla \cdot \mathbf{E} = 4\pi\rho$$
$$2. \quad \nabla \cdot \mathbf{B} = 0$$
$$3. \quad \nabla \times \mathbf{E} = -\frac{1}{c}\frac{\partial \mathbf{B}}{\partial t}$$
$$4. \quad \nabla \times \mathbf{B} = \frac{4\pi\mathbf{J}}{c} + \frac{1}{c}\frac{\partial \mathbf{E}}{\partial t}$$

where ρ is the charge density and $\mathbf{J}$ is the current density.

9. The value of the electric field at the y-line can be taken as E, and thus there will be ELv lines of flux passing through the line in 1 second. From Ampere's law, the magnetic field multiplied by the length of the line L will equal the number of electric flux lines cutting the y-line divided by c; hence, $BL = EvL/c$ and $B = Ev/c$. Similarly, Faraday's law applied to the moving B field cutting the z-line tells us that $EL = BvL/c$ and $E = Bv/c$. Then

$$B = E\frac{v}{c} \quad \text{and} \quad E = B\frac{v}{c}$$

Finally,

$$E = E\frac{v^2}{c^2} \quad \text{and} \quad \frac{v^2}{c^2} = 1$$

9

The Problem
of Change—
The Second Law
of Thermodynamics

$$S = k \log w$$

The relation between entropy and probability carved on
Ludwig Boltzmann's gravestone in Vienna.

Time's Arrow—Order to Disorder

We have identified three fundamental conservation laws: the conservation of
energy, momentum, and angular momentum. When a ball falls off a picnic
table to the surface of the earth, momentum is conserved. In falling, the ball's
newly acquired momentum, directed toward the earth's center, is equal in
magnitude and opposite in direction to the earth's acquired momentum
directed toward the ball. The total change in momentum throughout the
course of the ball's path toward the earth, and during the succession of
bounces of diminishing amplitude until the ball finally lies at rest, is zero.
Likewise, the total energy is conserved. The potential energy of the ball with
respect to the earth is changed to kinetic energy; the kinetic energy of the ball
as a whole is changed, by a series of inelastic collisions (the bounces), to heat
in the ball and the earth—and also in the atmosphere, which absorbs small
amounts of energy through air resistance. The change in temperature of these
materials is the macroscopic measure of microscopic increases in the kinetic
and potential energies of a very large number (perhaps 10^{24}) of discrete atoms
and molecules. But the total energy is unchanged. The total angular momentum of the earth-ball system will also remain unchanged through the history
of the ball's fall from the table.

The total momentum is unchanged; total energy is unchanged; total angular momentum is unchanged; yet change has occurred! In our mechanics we
have considered only conservation and invariance. What have we lost in our

formulation of mechanics which differentiates between the ball on the table and the ball on the ground?

We might sharpen our identification of that property which differentiates between the initial and final position of the ball by examining the characteristics of the event in a motion picture. If the film is run normally through the projector, we observe that the filmed activity is in accordance with all the laws of mechanics we have formulated. Furthermore, the action looks normal. If the film is run through the projector backward, we see the ball, resting initially on the ground, begin to move in increasingly large hops until it jumps up onto the table. It is obvious to us that time is running backward—we cannot believe that the event really happened in the manner shown to us. Yet if we make measurements of the images (including detailed microscopic pictures of the matter involved), we will find no violation of the laws of mechanics. We *know* that the ball fell off the table and that it did not jump from the ground to the table. What have we neglected, somehow, in our formulation of mechanics which should tell us that the configuration of ball, table, and ground must change—and change in only one direction? How can we answer Aristotle's criticism that the model of Demokritos—and our model—does not encompass change? And what is it that tells us, intuitively and unerringly, the direction of time: using Sir Arthur Eddington's phrase, what is it that tells the direction of "time's arrow"?

First, could the ball arise again? What is required in order that the history of the ball's flight be *reversed?* It is clear that all the energy now distributed among the enormous number of molecules and atoms in the earth, the ball, and the atmosphere, would somehow have to act collectively in such a way as to eject the ball from the ground in increasing jumps until the ball was back on the table. An enormous number of microscopic interactions between molecule and molecule would have to be reversed according to a precise schedule. Is this impossible? We do not know that it is; in our picture of action, it is only improbable, but the level of improbability is such as to skirt the common idea of impossibility. Our measurements of the interactions of individual molecules and atoms show that all these individual interactions are reversible. But it is improbable that all the individual interactions of individual molecules would proceed in such a way as to put the ball on the table. If the universe were packed tightly with balls and picnic tables, and we had all time—the entire age of the universe—to observe the balls, the probability of seeing a ball leap onto the table would still be absolutely negligible. Time's arrow must have something to do with probabilities.

We can move from the extreme complexity of real life through the artificial simplicity and elegance of games toward models sufficiently simple to allow a quantitative discussion of the direction of change by considering some of the motions of billiard balls on a billiard table (with no pockets). Now, our motion picture might show a set of 15 colored balls arranged in the form of a triangle near one end of the table and a white cue ball moving rapidly toward the stationary triangular set. After the cue ball strikes the other balls, the energy of the cue ball is distributed among the colored balls in a

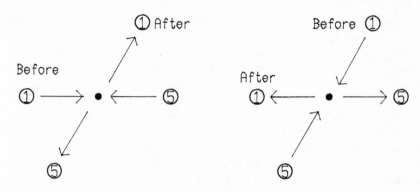

Figure 9.1 At the left, a collision of two billiard balls; at the right, the same collision with time reversed.

generally random or disordered manner. Many balls will be moving around the table, colliding with one another and roughly sharing the initial energy of the cue ball.

However, if the motion picture shows a randomly moving set of balls that collide with one another in such a way that finally the 15 colored balls come to rest in a triangle that fits a pool rack, and the white ball moves rapidly toward the far end of the table with all the kinetic energy, we know something is wrong: we know that the film is being run backward. We are seeing the action with time reversed. If the balls actually acted in the way we were shown, we would consider it most unusual—virtually a miracle! Nevertheless, it is definitely possible, but very very improbable: the scattering of the colored balls by the white ball is effectively irreversible.

All the individual interactions of elementary molecules and atoms seem to be reversible. Certainly, the individual collisions of ideal billiard balls are reversible in time if we neglect the friction of the balls moving on the table and neglect the small inelasticity (also friction) involved in collisions between real balls.

Figure 9.1 shows a collision between two moving balls. At the left, the 2-ball and the 5-ball are shown before the collision as they are moving toward the collision point (the arrows show the direction and magnitude of their velocities). Their positions and velocities are also shown at a time after the collision. At the right, the time-reversed collision is shown where the original initial and final positions of the balls are reversed and the balls have reversed velocities. The character of the collisions is independent of the direction of time: we could never tell the direction of the film (or the direction of time) by watching the collision of one ball with another.

The billiard table system lends itself to relatively simple quantitative analyses that serve to illustrate some of our considerations. Suppose we mark the table into three equal parts with chalk and take all but three balls off the table, leaving the yellow 1-ball, the black 8-ball, and the striped 14-ball. We set the three rolling about the table and exercise our poetic license to eliminate all friction so that the balls roll around indefinitely on the table. We then take a set of pictures of the table at random times and classify the pictures as to the configurations of the balls with respect to the three areas. These configurations are shown in Figure 9.2, together with the number W of different ways the configurations can be formed and then the probability that the balls will be in a given configuration: that is, the proportion of a very large set of the pictures that will show the balls in that configuration.

In the diagram numbered 10 in Figure 9.2, all three balls are in the third section of the table. There is only one such configuration, $W = 1$; labeling the balls, y, b, and s (for yellow, black, and striped), we write that pattern as [ ][ ][ybs], where the ordering of the balls in the section is irrelevant. Diagram 4 shows two balls in the first section and one in the second. There are three such configurations ($W = 3$): [yb][s][ ], [ys][b][ ], and [bs][y][ ]. If the balls are rolling about at random and each ball has an equal probability of being in each part of the table at any time, the probability of finding the balls in the fourth configuration is three times that of finding them in the tenth configuration, where all the balls are in the third section. Because the total number of equally probable combinations is $3 \cdot 3 \cdot 3 = 27$, the probability of finding the ball in the tenth configuration is 1/27, and the probability of finding the fourth configuration is 3/27. There are six different ways ($W = 6$) of putting the balls in the pattern shown in the first diagram, where there is a ball in each part of the table: [y][b][s], [y][s][b], [b][y][s], [b][s][y], [s][y][b], and [s][b][y], so that pattern has a probability of 6/27.

If the balls roll around on the table, continually changing positions, all combinations will occasionally occur. About 1/27 of the pictures will show them in the configuration [0, 0, 3], with all three balls in the third sector, whereas a much larger proportion, about 6/27, will show them in the pattern [1, 1, 1], with one ball in each sector. If we use the table and balls as a basis

Figure 9.2 Configurations of three balls on three equal parts of a billiard table together with the probabilities, P, for finding that configuration.

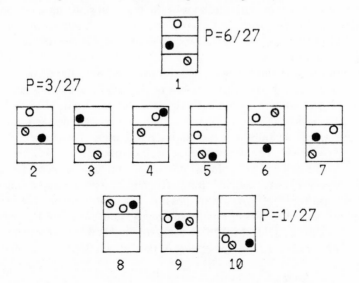

for a gambling game, we should pay odds of 27:1 to anyone who bets the balls will be in the pattern [0, 0, 3] at the time a random snapshot is taken, and only 27:6 or 4½:1 odds to those who bet on [1, 1, 1].

Now, if we begin with the balls in the highly ordered and unusual configuration [0, 0, 3], the balls will tend to go spontaneously to the more disordered pattern of [1, 1, 1], whereas if we start with the disordered configuration [1, 1, 1], there will be a much smaller chance that the system will spontaneously go to the ordered pattern [0, 0, 3]. Indeed, the probability of going from order to disorder—from [0, 0, 3] to [1, 1, 1]—is six times as great as the probability going from disorder to order—from [1, 1, 1] to [0, 0, 3]. The spontaneous reversal of the order-to-disorder change is improbable.

When we consider systems that are more complicated than the simple system of three balls rolling around on a table divided into three parts, the ratios of probabilities and improbabilities become much greater. If we had divided the table into 10 parts and used 10 balls, for example, the probability of going from an ordered pattern (such that all the balls were in one sector) to a disordered pattern (such that one ball were in each sector) is about $3.6 \cdot 10^6$ times as great as the reverse change from the disordered configuration to the ordered. But our illustrative model of 10 balls in 10 places on a pool table is still very, very simple compared to a world of enormous numbers of microscopic molecules and atoms moving in a three-dimensional manifold. Then the improbabilities we found for our simple systems become virtual impossibilities for analogous actions in the real, complex systems found in nature. If it is unlikely that the three balls will be found in one of the three divisions of the billiard table, it is essentially impossible that the randomly moving colored balls on the table will assume stationary positions in the triangle with the white ball racing away.

It is most important to notice that our construction of the relative probabilities of the various configurations on the billiard table is largely independent of any consideration of the detailed character of the interactions of the balls with each other or with the sides of the table. In general, the important statistical conclusions or statistical laws we develop are almost independent of the details of individual interactions. Of course, this is why the laws are so general and so powerful.

For expositional simplicity we have considered only the distribution of the positions of the balls; the momenta of the balls plays a similar role in analyses of probability. As the balls move about the frictionless table and occasionally collide, the momenta of the balls will interchange and there will be a probability distribution of such momenta. For any particular ball—say, the yellow 1-ball—we could measure the momentum of the ball at many random times. The ball would be almost stationary, and have almost no momentum, in very few measurements. Similarly, few measurements would show the 1-ball with nearly the maximum possible momentum (when the black 8-ball and striped 14-ball were almost stationary, and the 1-ball had most of the energy of the system). Most of the measurements would show the 1-ball with a momentum such that the energy of the ball would be about one-third of the total ball-energy. Moreover, a ball with very high or very low energy will tend to come to a state of average energy in time, whereas the reverse change will be improbable: a ball with an average energy or momentum is not likely to lose all its energy or take up all the energy of the balls on the table. The ball will tend to go from an unlikely momentum to a likely momentum and will be unlikely to proceed from probable to improbable momenta.

Even as an enlargement of the table would serve to increase the range of positions that the balls might have, thus increasing the number of ways the balls would be distributed, a change in the total energy of the balls, allowing a wider spread of momenta, would increase the number of possible momentum configurations and thus the range of probability.

Entropy and the Second Law of Thermodynamics

To make use of the principles presented here that lead to the concept that order goes spontaneously to disorder and not the reverse, we need to define what we mean by order and disorder in a quantitative way. Here again we proceed anhistorically, going from our discussion of probabilities to a consideration of the properties of macroscopic material rather than the historic path that was just the reverse.

We then define the quantity entropy in terms of the relative statistical probability of a system W by equating a change in the entropy with a proportional change in the number of ways the system can be formed

$$dS = k \cdot dW/W$$

where dS is the change in the entropy S, dW is the change in the probability W, and k is a constant of proportionality called Boltzmann's constant. (The value of k is not important to our immediate concerns.[1])

Because the forms dS and dW are properly used only for small changes (indeed, infinitesimal changes such that $dW \ll W$), our application to the billiard-ball model can only be suggestive. However, the change from the most ordered to the most disordered state will give a change in entropy on the order of

$$dS \approx k \cdot \frac{dW}{W} = k \cdot \frac{6-1}{(6+1)/2} = k \cdot \frac{5}{3.5} = k \cdot 1.43$$

where we approximate W by the average value of W for the two configurations.[2] The change in entropy for the reverse process—the change from the highly disordered state to the ordered state—is

$$dS \approx k \cdot \frac{1-6}{(6+1)/2} = -k \cdot \frac{5}{3.5} = -k \cdot 1.43$$

For the order-to-disorder change, the entropy increases; for the inverse disorder-to-order transition, the entropy decreases. This example illustrates our postulate, the *Second Law of Thermodynamics,* that isolated systems must change from ordered to less ordered configurations, expressed in the statement that *the entropy of isolated systems can never decrease.*

We emphasize the difference between ordered and disordered configurations because ordered systems have gross features that differ from those of disordered systems, and those features are quite important. Consider a box divided into two parts. There is a relatively ordered configuration of the system such that all the molecules of air in the box are in one part, and a relatively disordered configuration such that the air is equally divided between the two compartments. Although the total energy of the two systems are the same, we can derive useful energy or work from the first system by leading a tube from one compartment to the other and running a windmill or turbine by the impulse of the air passing from the high-pressure side to the low-pressure side. An experiment limited to the box could extract energy from the unbalanced configuration to run some mechanical device, but no energy could be extracted from the balanced configuration. The ordered configuration has available *free energy*. The relevance of statistical considerations to the relations between thermal processes and energy—thermodynamics—is clear.

We have noted that an increase of the mean energy of the balls leads to an increase in the number of ways the energy could be distributed, hence to an increase in entropy. This increase in entropy with the addition of energy or heat does not hold only for billiard balls but is quite general. For macroscopic systems, in terms of macroscopic measures of such systems, the definition of entropy we have stated can be written in the thermodynamic form

$$dS \geq dQ/T$$

where Q is the heat transfer to the system and T is the absolute temperature of the system. Upon the addition of heat energy dQ to a system, the increase in entropy dS must be equal to or greater than the additional heat divided by the absolute temperature (temperature measured from absolute zero).

With this definition of entropy in terms of heat transfer, we understand Clausius's expression in 1850 of the Second Law: *Heat cannot flow from lower to higher temperatures in a self-acting way.*

The thesis that any isolated system tends inevitably to move toward a state of maximum probability or maximum entropy does not mean that subsystems interacting with other subsystems of high order or low entropy may not increase their order or decrease their entropy. A refrigerator cools its contents (and heats the room in which it stands), thus reversing the flow of entropy and increasing the order within the refrigerator, but only at the expense of the increasing entropy of the power station producing the electricity that drives the refrigerator motor. The entropy of the entire system, refrigerator and power source, must not decrease—and, in practical matters, will increase.

Highly ordered crystals with very low entropy grow "spontaneously" in the evaporation of the liquid from saturated solutions. There, high order and low specific probability or entropy is achieved only at the expense of lower order and increased entropy of the surroundings. The total probability of entropy of the entire system (water, crystals, air, vapor, etc.) must increase although the entropy of a part, the crystals, decreases.

Life itself is a most striking example of high organization or low entropy derived from nominally less organized raw materials. It has been suggested that living organisms violate the Second Law and spontaneously bring order from chaos. According to this vitalistic hypothesis, such an attribute would differentiate between animate and inanimate things. But there is no evidence for such a violation of the Second Law, although it is not easy to analyze entropy changes in living processes because they are so complex. Certainly, there is no entropy change in crystal growth, an inanimate process similar to life in some ways.

The entire universe might possibly have the character of an isolated system. If it can be so considered, we would expect that after a sufficient time the total entropy of the universe would reach a maximum; a condition of maximum probability would occur. All energy would be evenly dispersed and the temperature would be the same everywhere. No change could occur. The universe would be dead. This eventuality has been called the "heat death of the universe." Actually, our knowledge of cosmology, the structure of the universe in space and time, is much too incomplete to regard such a hypothesis as more than an amusing fancy.

Phase Changes

Substances undergo certain structural changes as the temperature, or mean energy of the particles that make up the substances, changes. As the temperature drops, steam condenses to water, and water freezes to ice. In each of these *phase transitions* the symmetry of the water is reduced, the order is increased, and the entropy is reduced. For water in the gaseous form as steam, there is no preferred direction at all; the spatial symmetry is complete. In the solid form of ice, that symmetry is destroyed, and the water molecules are ordered into rows and columns making up the ice crystal planes. At each of these changes to phases of higher order and lower entropy, heat dQ is given off according to the relation

$$dQ \le T \cdot dS$$

where dS, the change in entropy, is negative, and a negative dQ means that heat is given off by the water. About 540 calories of heat are released upon condensation of steam to a gram of liquid; about 80 calories upon freezing the liquid to ice. Does the decrease in entropy accompanying the condensation and freezing violate the Second Law? No. Upon condensing or freezing, the released heat increases the entropy of the surroundings so that the overall entropy is not decreased.

Where did the heat energy emitted in condensation and freezing come from when the gas changed to water and then to ice? They came from the forces that constrain the water molecules and destroy the symmetry. Even as energy is required to tear the molecules making up a crystal from one another, energy is given off when free molecules are attached to the lattice. The Second Law tells us that energy is necessarily given off upon the increase in order accompanying a phase change that reduces symmetry.

Although we have discussed only water here, the general concepts are enormously broader. We have learned to describe processes occuring during the birth of the universe in terms of phase changes accompanying losses of symmetry.

Notes

1. Boltzmann's constant k is the ratio of a characteristic energy of a single particle to the temperature of the medium holding the particle. For example, the kinetic energy of a molecule of a gas at an absolute temperature T is $3/2\ kT$.

2. With better mathematics we can use the formula carved on Boltzmann's gravestone in Vienna, $S = k \log W$, and $\Delta S = k(\log 6 - \log 1) = 1.8k$.

10

Quantum Mechanics—
Determinism to Probability

> The underlying physical laws [of quantum mechanics] necessary for the mathematical theory of a large part of physics and the whole of chemistry are thus completely known.
>
> Paul Adrian Maurice Dirac

The Diffraction of "Particles"

All intellectual activity of either analytic or synthetic character takes place in the context of implicit frames of reference, and the character of those background structures strongly influences any results derived from that activity. Again, the fish that are caught depend upon the character of the net. The dramatic progress of physics in the nineteenth century followed from the successes of mechanical, causal, deterministic descriptions of natural phenomena. As a result, most physicists grounded their observations and analyses of nature on an intellectual set based on an implicit faith in the necessity of such descriptions of "reality." Celestial mechanics—indeed, the celestial mechanics of Newton—might be considered the exemplary paradigm of this "classical" physics. Given the position and momenta of a set of bodies at a given time, together with the rules concerning the forces among the bodies, one can predict their positions and momenta at any future time and deduce the positions and momenta at any past time. The future is determined and linked securely to the past; the actions of the particles are determined uniquely from initial conditions or causes.

This nineteenth-century world view, the very basis of thought for most physicists of that time, was shaken by Einstein's Special and General Theories of Relativity, published in 1905 and 1915, and upset completely by the development of quantum mechanics, beginning with Max Planck's analysis of black-body radiation in 1900 and culminating in the construction of the formal quantum mechanics by Bohr, Erwin Schroedinger, Werner Heisenberg, Max Born, Paul Dirac, and others circa 1925. The revolution in physics did not sweep away the minds of all physicists with it—not even the minds of

149

those who initiated it. Einstein, born in 1879, who received the Nobel Prize for his contributions to quantum mechanics (not relativity!), was not comfortable with quantum mechanics and has been called the "last of the classical physicists." (Planck and Schroedinger were also very much classical physicists.) In the same vein Bohr, born in 1885, might be considered the first of the "modern" physicists.

Although it is extremely interesting to trace the ideas of quantum mechanics, which finally mixed the classical pictures of the discrete (particles) and the continuum (waves) according to their historic evolution, an anhistoric approach better serves our purpose of describing quantum mechanics. Rather than recreate the historical experiments, we will instead look at a set of critical observations of nature, or experiments, and discuss the remapping of classical physics and the classical concepts of reality required by those observations. In particular, we will begin our consideration of quantum mechanics by studying two such critical observations of nature that suggest that the entities known classically as particles have properties similar to classical waves.

The results of these experiments force the conclusion that beams of electrons, protons, neutrons, and other "particles" behave in a manner previously considered characteristic of "waves." In particular, effects such as interference and diffraction are observed in particle beams. Conversely, other experiments demonstrate that entities heretofore thought of as waves behave somewhat as particles. Light, known to act as ripples on a surface of water since the time of Thomas Young two centuries ago, also sometimes acts like a stream of thrown pebbles. Indeed, the description of all the basic entities of nature, classically separated into sets of waves and particles, is now shown to have a form with aspects of each set. The observations wipe out the borders separating waves and particles. Quantum mechanics supplies the unified wave-particle description required by those observations. Moreover, this wave-particle aspect of nature codified in quantum mechanics requires a reexamination of the determinism implied by classical physics—a reexamination of concepts of reality, hence a radical revision of epistemology, the science of the nature of knowledge.

Wave motion is closely related to oscillatory motion, such as the motion of the bob of a pendulum or of a weight suspended from a spring. A wave such as a water wave can be considered a set of such oscillators coupled so that the oscillatory motion of each point on the surface of the water is correlated in space. A cork floating on the ripples passing through a pond will bob up and down as if the cork were a weight mounted on a spring. Each point on the surface of the water moves as a weight suspended from a spring or the bob of a pendulum. The square of the amplitude (or maximum excursion) of a water wave is a measure of the energy of the wave—just as the energy of motion of a pendulum is proportional to the square of the amplitude of motion—and the special name *intensity* is given to that quantity.

Sound waves traveling through a fluid such as air or water can be described in terms of variations of the pressure of the fluid. The deviations in the pressure over space and time form a pattern similar to the pattern of deviations in the height of the water surface in the presence of a water wave. Moreover, the intensity of the sound is proportional to the square of the amplitude of the pressure deviation, just as the intensity of the water wave is proportional to the square of the deviation of the surface of the water. Pressure in a fluid and height of a surface are scalar quantities defined by but one number at each place and time. The displacement of the position of a string or rope upon the passage of a wave along the string takes place in a specific direction; these are vector waves. Electromagnetic waves are also vector waves where at each point in space the vector electric (and magnetic) fields are changing in a pattern characteristic of a wave. The variation of each component of the vector fields varies in time and position very much as the surface of water when a set of ripples pass by. Also, the intensity of the electromagnetic wave (for instance, the brightness of light) depends on the square of the amplitude of the electric (or magnetic) field variation. There are also tensor waves transmitted in elastic media, where the disturbance in the media at any place and time depends on a set of numbers that transform as a tensor. Physicists now believe that there are waves in gravitational fields and that these are tensor waves.

When small-amplitude water waves from different sources cross, the amplitudes add. If there are two separate waves acting at a point, the amplitude at that point at any specific time will be just the sum of the displacements resulting from the two waves acting alone. Under some conditions the two displacements will have opposite signs and tend to cancel (destructive interference); under some conditions the two displacements will add (constructive interference).

At the top of Figure 10.1 is a wave, perhaps the displacement of a water surface along some line at a particular time. At any time the displacement of

Figure 10.1 (*a*) A typical wave form, perhaps the displacement of a water surface, along a line viewed at a definite time. (*b*) The addition of two waves of the same amplitude and wavelength that are in phase (the phase difference is zero). (*c*) The addition of two waves of the same amplitude that are out of phase.

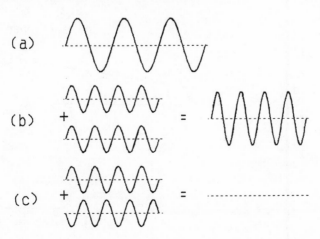

the water surface can be described in terms of the maximum displacement of the water affected by the wave, which is called the "amplitude," and the phase where the phase describes the position on the oscillation cycle.[1] We can use these concepts of amplitude and phase to consider the diffraction of water waves by gaps in a barrier. Figure 10.2 shows parallel water waves incident on a barrier with two equal sections of the barrier open to the waves. At the right is a graph that shows the relative heights of the waves that reach a distant line in the water parallel to the barrier. Maximum wave heights or amplitudes are shown for situations such that both apertures are open, and when only one (either one) is open. There are regions of the double-aperture (double-slit) pattern, such as that near point (*b*), where the wave height is double that which is measured when only one of the sections of the barrier is open. Here, the amplitudes of the waves passing through the two barriers are seen to add—when the wave from one aperture is at a maximum, so is the other; when the wave from one aperture is at a minimum, so is the other. At other places (*c*) along the measurement line, the water will be calm when both apertures are open, as the maximum from one aperture is met by a minimum from the other. When only one of the openings is open, the cancellation will not occur, and waves will be seen. Here the two waves interfere such that the resultant amplitude is the difference in the amplitude of the two waves, or zero in this case.[2]

If we were to conduct a set of experiments with different separation of the

Figure 10.2 Intensity patterns for the diffraction of water waves by apertures in a barrier. The amplitudes of the waves incident on the line to the right are shown by the graph based on that line. The curves at the right show the pattern with both slits open, and with but one slit open. The diagram is not drawn to scale; the barrier and apertures are magnified so that (*d*) is much smaller than the separation of the peaks at the measurement line.

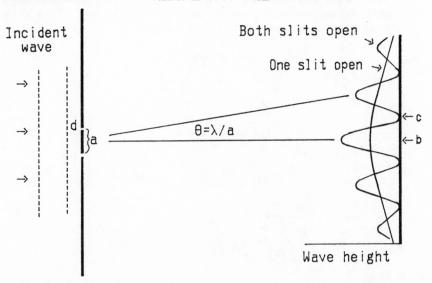

gaps in the barrier—and, perhaps, waves of different wavelength—we would find that the separation of the maxima at the measurement line (perhaps a beach or breakwater) can be summarized by the relation

$$\theta = \frac{\lambda}{a}$$

where γ is the wavelength, a is the distance between the apertures or slits, and θ is the angular separation of the maxima of the diffraction pattern as measured from the position of the slits. Briefly, if the wavelength is reduced by a factor of 2, the maxima in the diffraction pattern will be closer together by a factor of 2; if the distance between the slits is reduced by a factor of 2, the pattern will be broadened by a factor of 2.

At the left of the barrier, the water waves are traveling in a well-defined direction, precisely in the direction to the right. After passing through the slits (one or two), waves are "diffracted" in other directions. The original direction of motion of the segment of wave that passes through the slit cannot be reconstructed from a measurement of the direction of a segment of the wave to the right of the barrier in Figure 10.2; elements of the wave are traveling in different directions. Indeed, the mean angle ϕ that the diffracted waves make with the original direction is about $\phi = d/\lambda$, where d is the width of an angle slit.

Now this same experiment can be conducted, not with water waves incident on apertures separated by tens of meters, but with a beam of monoenergetic electrons incident on barriers (or scatterers) of crystal with spacings on the order of 10^{-8} cm. Instead of measuring wave intensities on a line distant from the barrier, electron densities are measured, perhaps by placing a special photographic emulsion in place so that an electron that strikes the emulsion leaves a dark spot on the developed film. The design of the experiment and the pattern of the density distribution on the emulsion, shown as a kind of reduced copy of the water-wave pattern of Figure 10.2, is illustrated in Figure 10.3. A beam of monoenergetic electrons, from a source far to the left, passes through a double slit onto emulsion. The similarities between the water-wave pattern and the electron pattern are striking; the main difference is that the electron density pattern appears to correspond not to the pattern of the water-wave amplitude but to the water-wave intensity, the square of the amplitude. The energy carried by a wave is proportional to the square of the amplitude. We can then equate the density pattern of the electrons to the intensity distributions of electron waves.

Because electrons and water waves (and other waves) act similarly, we can use the equations relating wavelength and aperture separation to the separation of the intensity peaks, to find some kind of effective wavelength for the electrons. Using the relation described previously for other waves, $\theta = (\lambda/a)$, we find for electron waves

$$\theta = \frac{h/mv}{a}$$

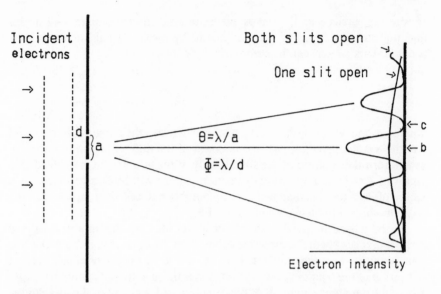

Figure 10.3   The scattering of a beam of monoenergetic electrons incident on a double slit. The pattern of the intensity of electrons incident on a film to the right is shown by the graph at the right.

hence

$$\lambda = \frac{h}{mv}$$

where h is found to be a constant of nature, called Planck's constant, and mv is the momentum of the electrons in the beam; m is the electron mass and v the electron velocity. In short, the electrons *act like waves* with a wavelength inversely proportional to their momentum.

The wavelike character of the electron beam is illustrated further by an additional experiment. We might close one of the slits and remeasure the electron intensity. This single slit intensity is also shown in Figure 10.3. We see that closing a slit increases the intensity in some regions. The number of electrons striking the film near point (b), with one slit closed, is greater than with both slits open. This cannot be understood if electrons were to behave in the way we usually attribute to particles; it is easily understood if electrons act, in some circumstances, as waves.

Moreover, we note that at the left of the slit, the electrons are traveling in a precisely defined direction—to the right. After the diffraction of the electrons by the slit, the electron directions are dispersed such that the mean angle ϕ of deflection with respect to the original direction is about equal to

$$\phi = \frac{\lambda}{d}$$

where d is the width of the slit.

We note that the constraint by the paths of the electrons by the slit modified the direction of the electrons even though the slit material *did not interact with the electrons in any direct way;* the electrons did not hit the slit edges but were deflected by the act of constraint. Such an effect can be easily understood for classical waves but is quite inconsistent with a picture of classical particles.

Experiments precisely like the diffraction experiments discussed here are quite difficult, but results that are similar in principle were obtained in about 1926 by C. Davisson and L. W. Germer in the United States and by G. P. Thomson in England, who scattered electrons from the ordered arrays of atoms in crystals. The crystals acted like a grating of many apertures. Very nearly the same kind of measurements have been made using neutrons, and related experiments show that other particles act as waves in the same way.

The Time "Diffraction" of Particles

The second model experiment exhibits and defines the wave character of particles by considering, again, a parallel experiment conducted with ordinary waves, this time with sound waves. Figure 10.4 shows sound of a definite frequency passing through a very large aperture containing a shutter that can serve to block the sound. We may choose, as a detector that will determine the frequency of the sound in an operationally valid manner, a set of tuning forks, each set to a slightly different frequency. Suppose we design the apparatus so that the oscillator frequency is set at 440 cycles per second and with tuning forks set for integral frequencies near 440; that is, with forks set for 435, 436, 437, ... , etc. When subjected to sound waves that are of the same frequency as the natural frequency of a fork, the fork will begin to oscillate sympathetically in resonance with the sound wave, and the amplitude of

Figure 10.4 Sound waves from an oscillator passing through an aperture onto a set of tuning forks used as a frequency analyzer.

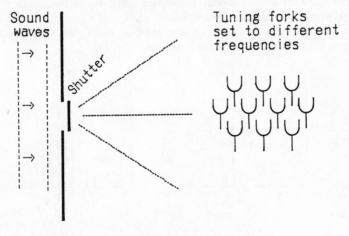

Sound
waves

Shutter

Tuning forks
set to different
frequencies

oscillation of the fork will be a measure of the intensity of the wave at the natural frequency of the fork. Hence, to determine the intensities of the sound at different frequencies, we can measure the amplitudes of vibration of the different tuning forks.

For our first measurement, we leave the shutter open for a long time (many seconds) and measure the intensity of vibration of the different tuning forks that oscillate in resonance with the sound from the oscillator. A survey of the tuning forks will show a large intensity for the tuning fork that has a natural frequency of 440 cps and almost no intensity for the others that have slightly different natural frequencies. The intensity distribution is shown on the left graph of Figure 10.5.

Then we repeat the measurement, but this time we leave the shutter open for only a very short time dt, a small fraction of a second. Now we find a smaller amplitude for the 440 cps fork but considerable amplitudes for the forks with natural frequencies near 440 cps. The graph at the right of Figure 10.5 shows the distribution of amplitudes of the tuning forks excited under this condition. This distribution will have a width df, and we find

$$df \approx \frac{1}{dt}$$

The shorter the time the shutter is open, the broader the frequency spread of the transmitted sound. Even though the initial frequency of the sound is sharp, exactly 440 cps, the transmitted sound, cut off in time, contains a spread of frequencies.

Now we conduct a similar experiment where monoenergetic particles with an energy E are directed through a shutter onto an apparatus designed to measure the energy of the particles. When the shutter is open for a long time, a measurement of the energy distribution of the particles passing the aperture shows, as illustrated on the left of Figure 10.6, that the particles all have nearly the same energy E as they had originally. However, if the shutter is open only for a very short time dt, there will be a spread of final energies as

Figure 10.5 The distribution of frequencies of the sound passing through an aperture: at the left, when the aperture is open continuously; at the right, when the shutter is open only for a short time dt. The solid lines indicate the amplitudes of vibration of the different tuning forks. Note the different amplitude scales.

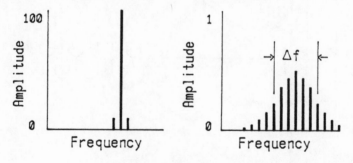

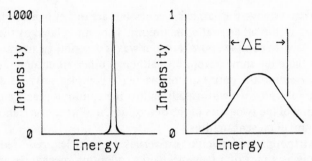

Figure 10.6 The spread in energy of monoenergetic particles of energy E passing through a continuously open shutter at the left and, at the right, for a shutter open only for a time dt. Note the different intensity scales.

suggested by the right side of Figure 10.6. The spread of the energies can be expressed by the relation

$$dE = \frac{h}{dt}$$

where, again, h is found to be equal to Planck's constant.

Comparison of the intensities of sound waves constrained to a short time and the distribution energies of particles constrained in time indicates that the particles of energy E act as waves, where the frequency f of the particle waves is

$$E = hf$$

Although the particles incident upon the aperture governed by the shutter all have the same energy E, those admitted for a short time by the opening of the shutter are found to have a spread of energies *even though the shutter did not interact with those particles directly.* Somehow the act of constraining the particles to the time dt spread the energy without an overt interaction. Again, such an effect on waves is understandable but quite inconsistent with any classical view of the properties of a particle.

The time diffraction experiment, as described above, is impractical. However, analogous measurements have been made on the energy spread of particles emitted from sources with naturally short lifetimes, and these observations are in accord with the wave description and the equation of E with hf. The model experiments discussed here should be thought of as condensed and simplified versions of actual experiments that have been conducted.

The results of the two experiments we have discussed can be expressed concisely by describing any beam of particles in a manner almost equivalent to the description of a simple, classical wave.[3] Although the particle wave (also called the de Broglie wave or Schroedinger wave) is like a classical wave, there are important differences: the properties of these matter-waves require a description in the form of not one wave but two separate yet related waves.

These two waves have the same frequency but are out of phase; when one is at a minimum, the other is at a maximum. They are related as the real and imaginary parts of complex numbers. For waves describing a free particle, the two waves have the same amplitudes although other amplitude ratios occur in general. Because the entire double wave (or complex wave) is difficult to illustrate graphically, we will usually follow the common practice of considering just one of the two parts of the description in illustrations, and will take care not to misrepresent matters.

This attribution of wavelike properties to particles, required by such observations, is a part of a new mechanics, quantum mechanics, defined by the large set of microscopic observations of nature made during the last half of the nineteenth and the first quarter of the twentieth centuries. These experiments show that the entities usually called particles and regarded as the fundamental pieces of matter exhibit properties that are similar to those considered specific to waves. We must be careful to view all statements concerning the character of such particles as statements concerning experimental results. The names we assign to the constructions that evolve from the collection of experimental results, as well as to the general relations we use to correlate these results, are important—we need to use common names such as waves or particles in order to communicate. But we must not make the error of confusing the specific results of experiments and the specific character of the theoretical models used to correlate these results (be they right or wrong) with semantic properties of the names themselves. Certain properties of the entity we call an electron are similar to the properties we attribute to primitive particles such as stones or balls. Other properties of electrons are like those of waves. It is not necessary, however, to make a decision as to whether an electron is a wave or particle. It is an entity to which we give the name electron, with properties that are well understood and lead to no contradictions, and that are somewhat like those evident in a ripple on a surface of water and somewhat like those of a pebble that might be thrown into the water.

Of course, the discovery that classical mechanics does not correctly describe the small-scale events of nature cannot mean that classical mechanics is inadequate to describe the macroscopic observations of nature—after all, classical mechanics was invented to describe large-scale observations. Must we then have two unrelated systems of mechanics, one for the macroscopic world and one for the microscopic world? Certainly that is an unaesthetic hypothesis—and probably quite unworkable. The dichotomy must be eliminated, and the obvious reduction must be found in the character of quantum mechanics. The quantum mechanical description of nature must correspond to the classical description when the value of h, Planck's constant, the characteristic measure of quantum mechanics, is negligible compared to the magnitudes of relevant dynamical quantities. There is only one system of mechanics, quantum mechanics. Classical mechanics is a limiting case of quantum mechanics valid when h can be neglected. This is a statement of Bohr's Correspondence Principle.

The Probability Interpretation of Wave Intensity

The results of the experiments discussed in the last section could be under-stood in terms of a model of a beam of particles that describes the beam very much as a plane wave, where the intensity of the wave is interpreted as the density of particles. Further observations of the experimental results will necessitate our adopting an extraordinarily interesting modification of this interpretation, a modification that will require us to readjust our concept of reality.

The large difference between the density of particles at points (b) and (c) in Figure 10.3 represents the large difference between the intensity of the wave at points (b) and (c). It is possible to reduce the intensity in experiments such as this so that one might change the recording film so rapidly that most of the film has no track at all on it. If the negative film is quite transparent, we might hold all the separate films together in a stack and look through the many layers, as in Figure 10.7. We would then see the intensity variation across the film described in Figure 10.3. However, if we analyze the distri-bution of electrons on the separate films and count how many have no track, one track, two tracks, and so on, we will find that the distribution is precisely that governed by the laws of probability for random events.

For example, if one spins a roulette wheel 37 times (a Monte Carlo wheel with 36 numbers and one zero), it is probable that 7 will come up once. If one conducts this set of 37 spins 1000 times, 7 will come up nearly 1000 times. Not one number 7 will come up during about 370 such sequences; one 7 will occur in about 370 other sequences; two 7s will appear in about 185 of the sets of spins; 7 will come up 3 times in about 60 sequences; there will be 4 in about 37 sets; and so on. If we select a region of the transparency such that there will be one electron per 37 films on the average, the distributions

Figure 10.7 The distribution of hits of single electrons that pass through a double slit onto emulsions: at the left, an individual transparency with one electron spot; in the center, stacked transparencies show the intensity pattern; at the right, a plot of the density across the stack.

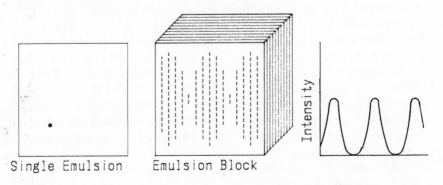

Single Emulsion   Emulsion Block

of 0, 1, 2, 3, and so forth electrons in sets of 37 will be the same as for the roulette wheel.

We must therefore interpret the intensity of the de Broglie wave, $|\psi(x, t)|^2$, as a measure of the probability that a particle is at the position x at the time t with a numerical value that is equal to the probable number of particles in a unit volume at the point x and the time t. This definition extends, rather than contradicts an earlier definition of ψ^2 as the density of particles per cubic centimeter. Indeed, when very large numbers of particles are involved, the two definitions are nearly identical. If the average number of particles per cubic centimeter in a certain region is $\psi^2 = 1,000,000$, the actual number is not going to be very different if the distribution follows the laws of probability (the uncertainty would be about 0.1%).

Statistical models of natural behavior were not unknown when Born proposed this statistical interpretation of the intensity of the de Broglie wave function in 1925. Before this application to quantum mechanics, the statistical treatments used commonly in physics were designed to consider the effects of very large numbers of actions where it was impractical to follow each individual action. Einstein's explanation (see Chapter 4) of the Brownian movement of pollen grains in terms of statistical fluctuations of a rather large number of collisions of molecules with the grains is characteristic. According to classical physics, the behavior of each molecule can be determined, but that behavior is complicated, and it is a reasonable approximation that the behavior of a large number of molecules should follow statistical laws. This use of statistics is rather like the use of statistics in considering probabilities in such card games as bridge or poker. The arrangement of the cards in the deck is not fundamentally random, but—in the absence of a card sharp—is unknown to all players and is practically random. But the quantum probability is of a previously unknown fundamental nature. The fluctuations in the number of electrons striking the film of Figure 10.7 are fundamental, a characteristic of nature, and uncontrollable.

According to our present understanding of quantum mechanics, there is absolutely no measurement that can be made in principle on the electrons before they enter the beam-defining slit present in the apparatus shown in Figure 10.3, which will decrease the spread of the electrons across the emulsion.[4] In particular, no measurement performed on an electron just after (or before) the electron passes through the slit can be used to determine the section of emulsion that the electron will strike. This dispersion of the electrons and loss of certainty is not an assessment of ignorance, but it is an assessment of the character of nature. One might cheat and look through the deck of cards before they were dealt. But one could not cheat by making measurements on an electron just after it passes through the slits so as to allow one to deduce where it will strike the emulsion. The measurement would provide a last cosmic shuffle generating a new pattern determined not by the slits but by one's measurements.

Conversely, we might consider the past history of that electron which struck the emulsion at the right in Figure 10.7. Can we determine, or even ask, through which slit the electron passed before it struck the emulsion? Operationally, such a question has no answer except as a consequence of a possible measurement. Because any such measurement would destroy the pattern of which the position of the electron in the emulsion is a part, and would almost certainly send the electron to a different sector of the emulsion, the measurement will not and cannot answer our question. If we cannot know the slit through which the electron passed, or the position of the electron at the slits, does that position even have a meaning? To a physicist, the answer is no.

Again, this description of events concerning particles in terms of probability, a description we have presented in a partial, simplified form, is determined by experiment—that is by controlled observations of nature—and is sufficient to fit all the experiments that have been conducted. To the classical physicists of the first part of this century, such as Einstein, this interpretation of quantum mechanics was the most radical physical hypothesis of their time—and perhaps all time. The statistical interpretation of quantum mechanics required a deep reassessment of the very meaning of reality.

We should add here that it is not certain that this statistical description is complete or correct, in the sense that a new and different description might also fit equally well every observation that has been made. Any such alternative description of nature can be really new, and not just a restatement of the conventional description, only if it predicts a different result for some kind of possible observation than is predicted by the conventional description. Attempts to construct such an alternative description of reality have not been successful. Most physicists do not believe that any such alternative will ever be found; indeed, most physicists believe that the statistical features of quantum mechanics describe reality completely.

The Heisenberg Uncertainty Principle

The probabilistic description of nature we have adopted, required by the necessity of including the results of observations of the type considered in the first sections of this chapter, leads immediately to a most interesting restriction on the character of the information we can obtain concerning the physical universe. If quantum mechanics represents a correct and complete description of mechanics, pairs of *complementary quantities* exist such that it is not possible to know, or measure, the values of both quantities to an unlimited precision. Classical mechanics has always admitted the possibility of measuring the quantities of mechanics, such as position and momentum or energy and time, to any degree of accuracy; it is therefore permissible to ascribe an exact momentum and an exact position to a particle, as it is implicitly assumed that exact measurements could be made of those quantities. If the description of nature provided by quantum mechanics is correct, such measurements cannot be conducted *in principle* with unlimited accuracy. Hence, from an operational view such as we adopt in physics, a particle cannot have an exact position and momentum. The resultant uncertainty in the values of complementary quantities, such as momentum and position, was particularly evident in Werner Heisenberg's matrix formulation of quantum mechanics—hence the *Heisenberg Uncertainty Principle*.

It is useful to examine the constraints nature imposed upon the position and momentum of a particle by analyzing the results of an attempt to identify the position and momentum in some direction. In particular, consider particles moving in the x-direction as shown in Figure 10.8. Suppose that we "measure" the position of a particle in the y-direction by selecting a particle that passes through the slit that defines the y-position. Qualitatively, we note that the passage through the slit will result in a diffraction or spread of the particles, so that after the slit the particles will have a spread of velocities or momenta in the y-direction. The determination of the y-position, by requiring passage through the slit, will impose a spread or uncertainty in the y-component of momentum. Moreover, the narrower the slit, the more precise the determination of the position, the broader the diffraction pattern, and the greater the uncertainty in the y-component of momentum.

Thus, the uncertainty in the momentum dp_y in the y-direction is greater if the uncertainty dy in the position of the wave (the slit width) is smaller. Quantitatively,[5] we have

$$dp_y \cdot dy \approx \frac{h}{2\pi} = \hbar$$

The combination, $h/2\pi$, Planck's constant divided by 2π, is found so often that it is conveniently abbreviated as $\hbar$ (in speech, "aitch-bar").

Although we have described only one specific method of determining the position and momentum of a particle, the general results obtained by analyzing that procedure also hold for any other technique. It is impossible to determine both the position and momentum of a particle at any particular time with unlimited accuracy: the product of the uncertainty of position of a par-

Figure 10.8   De Broglie wave, normally incident on a plane, passing through a slit in the plane.

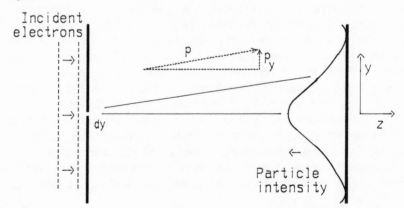

ticle dy at some particular time, multiplied by the uncertainty in the component of momentum in that direction, dp_y, cannot be less than $\hbar$. We say that a pair of quantities that are related by the Uncertainty Principle, such as dp_y and y, are complementary quantities. One can learn the value of y only at the expense of knowledge of p_y; as the slit is closed down, the determination of the position y of the particles that pass through is improved, but the angle of diffraction, and therefore the spread in the values of p_y, is increased. Conversely, the diffraction and spread in p_y can be reduced by increasing the size of the slit, but this entails a loss in the accuracy of knowledge of y.

Energy and time constitute another set of complementary quantities; we cannot determine the exact energy of a particle or any other system at a precisely determined time. Speaking imprecisely, as an illustration, a very quick measurement of the energy will disturb the energy in an unpredictable manner even as a very short opening of the shutter discussed earlier results in a spread (or time diffraction) of the energy of the particles passing through the aperture. There will be a minimum uncertainty dE in any measurement of the energy of a system conducted over a time interval dt such that

$$dE \cdot dt = \hbar$$

Although this uncertainty is extremely small on the macroscopic scale, it is quite important on the microscopic scales relevant at atoms, nuclei, and elementary particles.

The uncertainty in energy that results from a limitation of the time of observation does not depend on whether the time limitation is imposed externally by some obvious action of an observer (who might activate a shutter) or whether the time limitations are intrinsic, because the system may have a definite lifetime for a transition into another state. Therefore, particles that are unstable and decay within some well-defined lifetime do not have a definite energy. Because mass and energy are related through the relativistic mass-energy relation $E = mc^2$, such systems also do not have a definite mass. If the lifetime of the state is τ, from the uncertainty principle

$$dE \cdot \tau \approx \hbar \qquad \text{or} \qquad d(mc^2) \cdot \tau \approx \hbar \qquad \text{or} \qquad dm \approx \frac{\hbar}{\tau c^2}$$

where dm is the uncertainty of the mass of the particle. For example, the neutron decays into a proton, an electron, and a neutrino after a mean time of about 1000 seconds. This leads to an uncertainty in the mass of the neutron of about $\hbar/\tau c^2$, where $\tau = 1000$ seconds. This uncertainty is about one part in 10^{27}, and that is not detectable by any means now available to us.

However, some of the more exotic particles have very short lives, and the spread in mass is considerable. For example, a particle now labeled the Δ, discovered about 1951 by Enrico Fermi and his associates, decays into a proton and another elementary particle called a pion in an average time of about $5 \cdot 10^{-24}$ seconds. The mass spread or uncertainty that results from this short lifetime is about 10% of the average mass of the particle (which is about 30% greater than that of the proton).

In summary, the difference between the classical description of nature, which is so near to our intuition but is incorrect, and the quantum description can be expressed in a number of ways. In the classical model of the universe, particles are discrete, and all dynamic quantities such as time, velocity, momentum, and so on are continuous. There is a granularity of particles, but no granularity of dynamic variables. These distinctions are blurred in quantum mechanics. The dynamic quantity "action," equal to momentum multiplied by distance or energy multiplied by time, is quantized into integral multiples of $\hbar$, whereas particles have continuum properties like those of a water wave or sound wave.

The Interaction of the Observer and the Definition of Reality

Although there are surely uncertainties in any measurement, before the discovery of the Uncertainty Principle or the wave nature of particles, it was assumed that there was no limitation *in principle* to the accuracy of measurements of dynamical quantities such as position and momentum, or energy and time. But the quantum description of nature leads to definite limits on the accuracy with which sets of certain complementary quantities can be known. The act of measurement itself of a quantity (such as the y-coordinate of the position of a particle) to a specified accuracy (dy) can be considered to disturb the complementary quantity (the y-component of momentum, p_y) in an indeterminate manner; thus, the value of that quantity must be fundamentally uncertain (dp_y) to a degree such that the product of the uncertainties ($dy \cdot dp_y$) cannot be less than h, Planck's constant. Hence, if quantum mechanics is the correct description of nature, the internal consistency of the description can be maintained only by restricting the minimal interaction of the observer to the amount expressed by the Uncertainty Principle. The disturbance that results from the act of observation or measurement by the experimenter cannot be reduced indefinitely by the use of more and more subtle techniques.

According to the classical description of nature, if one knows the positions and the momenta of a set of objects and has an understanding of the forces between them, in the absence of any external interaction one can in principle predict or calculate the configuration of positions and velocities at any later time. Because this set could be the complete set of all the particles in the universe, the classical world picture suggests that the future could be predicted in principle and, more important, that a specific and unique future exists. Then, the entire course of the universe, including the entire course of our lives, would be predetermined or predestined. If this argument were valid, the concept of free will would have to be questioned.

Now physicists believe that we are limited in principle in our measurements of complementary quantities by the Uncertainty Principle, and we cannot know the exact positions and momenta of all particles. According to the operational view of reality, used in practice by most physicists, a particle does not then have a definite position and momentum, as no meaning can be attributed to a property that cannot be measured in principle. If a particle does not have a precise initial position and momentum, it is not possible to predict with certainty where the particle will be at a later time. However, the quantum mechanical world view does not preclude, in principle, a set of measurements that determine the exact probability distribution of particles in the future. One could then predict, exactly, the probability of the particles being in any particular configuration in the future. And this probability depends only on the initial situation as determined by our instruments. Of course we have no more (or less) free will this way than in the classical argument. Instead of having the future exactly determined, independent of our volition, now we are at the mercy of chance—of the throw of the dice—again independent of our volition.

Physicists acting as physicists are not much interested in such arguments. In the realm of physics the arguments are empty until it can be shown that it is possible in principle to gather all the information required. Because that looks difficult, physicists generally leave predestination with philosophers and theologians.

We should not leave the subject of the Uncertainty Principle without pointing out that a small but respectable group of physicists do not accept the probability interpretation of quantum mechanics as representing the ultimate reality. This was Einstein's position—he has been quoted as saying that he could not believe that "God plays dice with the world."

Standing Waves and Stationary States

We have amplified the classical description of beams of particles by describing those beams in terms of waves. But how do we describe the particles themselves made up, as they are for the most part, of other particles? How do we describe a static system such as a hydrogen atom made up of a proton and an electron, or the proton made up largely of three quarks? Running waves—such as a train of waves moving across a surface of water—constitute an appropriate description of beams of particles. We find a parallel to bound or stationary states of particles in the concept of standing waves, such as the waves or vibrations of a guitar string.

In general, waves undergo some reflection at any interface where the physical situation produces a change in the velocity of the wave. The reflection of light waves off a glass surface and the reflection of sound waves from a wooden panel are examples of such reflections. Typically, the velocity of light in glass is about two-thirds of the velocity of light in air; as a consequence, a portion of light incident on a pane of glass is reflected and a portion is transmitted. Similarly, sound travels at a different (higher) velocity in wood than in air, and therefore, sound incident on a wood panel is reflected as well as transmitted. Under circumstances such that there will be no transmission of the wave but also no substantial absorption, there will be nearly full reflection of the waves. Light waves are almost fully reflected by a good mirror; sound waves can be almost fully reflected by appropriate structures.

The (wave) velocity of a de Broglie plane wave describing a beam of particles is one-half of the classical particle velocity.[6] When forces act on the particles in a manner that would change the classical velocity, the wave velocity will also be changed, and the de Broglie wave will be reflected—generally partially reflected. Because nuclear matter exerts a force on a free neutron, thereby changing the neutron velocity, a neutron will be elastically scattered by an atomic nucleus under some conditions. This scattering can be described in terms of the scattering of the de Broglie waves, describing a beam of neutrons, at the surface of the nucleus. A quantum mechanical treatment of the scattering of neutrons by spherical nuclei is thus very much like the calculation of sound scattering from a spherical object or of light scattering from a very small spherical droplet of water. Although the scale of dimensions of wavelength and object varies greatly, the mathematical or logical analyses are almost identical.

If there is total reflection at a boundary, "standing waves" may be set up, where the term refers to wave patterns that do not change their position in time but "stand" still. Waves in a taut string are totally reflected at a secured end, and then standing waves can be set up in a string tied at both ends. The vibrations of a piano or guitar string represent familiar forms of one-dimensional standing waves (the line of the undisturbed string defines but one dimension). Two-dimensional standing waves can be set up on the surface of a drum and by the ripples on a small, enclosed surface of water such as the surface of the water in a glass. Organ pipes, as well as other wind instruments, set up three-dimensional standing waves of sound—with some leakage of sound from the instrument, of course. Quantum standing waves or standing de Broglie waves describe a large part of nature. An electron bound to an atomic nucleus can be considered a three-dimensional standing wave reflected at a distance from the nucleus as a consequence of the electrostatic force between the positively charged nucleus and the negative electric charge held by the electron. Similarly, a neutron or proton held in a nucleus through the nuclear forces exerted by other neutrons and protons can be described in terms of a standing wave. And the three quarks that make up a neutron or proton held together by the strong forces of quantum chromodynamics can be considered to exist in the nucleon as standing waves. Perhaps it is better said that molecules, atoms, nuclei, and the nucleons—neutrons and protons—*are* standing waves of their immediate constituents. Stationary waves, it should be noted, do not describe stationary particles but stationary trajectories; the electron held to the proton to make up the hydrogen atom moves about the proton somewhat as the earth moves in its orbit about the sun, but the pattern of movement of the electron does not change just as the earth's orbit does not change.

Although we will touch upon the standing wave structure of atoms, nuclei, and especially elementary particles, these structures are subtle and complex. It is useful to consider the simple set of standing waves describing states of a particle in a one-dimensional box (a system not found in nature) to illustrate the general qualitative features of static systems. (A three-dimensional box with two dimensions that are very large compared to the third—the dimension of interest—constitutes a good approximation to a one-dimensional box for those who develop claustrophobia in one dimension.)

The possible particle standing waves in such a system are similar to the set of standing waves for a taut string tied at both ends. For particle wave or guitar string, there must be an integral number of half-waves along the extent of the system, along the length of the guitar string or along the length of the box. For the guitar string the different modes of vibration shown in the upper diagrams of Figure 10.9 correspond to different harmonics of the string—the fundamental vibration with a wavelength twice the length of the string and the first harmonic, where the wavelength is equal to the length of the string, one-half that of the fundamental, and the frequency is double that of the fundamental or one octave higher. For the de Broglie waves, the diagrams can be considered to show the lowest energy "ground" state with the longest wavelength—twice the length of the box—and lowest momentum and the first "excited state," where the wavelength is one-half the wavelength of the ground state and the momentum is twice as great. For the vibration of the strings, the diagrams can be considered pictures of the string at the moment of greatest vibrational extension. For the depiction of the de Broglie waves, the diagrams show the maximum amplitudes along the length of the box.

For de Broglie waves or guitar strings, the wavelength λ_1 of the fundamental is twice the length of box or string; the wavelength λ_2 of the first harmonic is equal to the length of the box or string, or twice that length divided by two; the wavelength λ_3 corresponding to the second harmonic is equal to twice the length of the box or string divided by three; and in general,

$$\lambda_n = 2L/n$$

where n is an integer—$n = 1, 2, 3, \ldots$—and L is the length of the box or the string. The momentum of a particle is inversely proportional to the de Broglie wavelength of the particle. Hence, the momentum of the particle in the state corresponding to the first harmonic is twice that for the particle in the fundamental[7] or "ground" state, the momentum of the particle in the second harmonic state is three times that in the ground state, and so on. But if the

Figure 10.9 De Broglie standing waves for a particle in a one-dimensional box (or amplitudes of a guitar string) are shown in the upper diagrams. A schematic representation of a ball in the one-dimensional box is shown at the lower left, and a realistic representation of a three-dimensional box, which will have the lowest states, will vary much as would the one-dimensional box shown at the lower right.

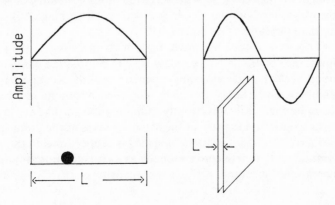

wave is stationary, what is meant by the particle momentum? If we were to reach in abruptly and measure the momentum of a particle described by a particular standing wave, we would find the momentum calculated here. But sometimes the particle would be moving from left to right and, with equal probability, sometimes from right to left. Like a ball bouncing from cushion to cushion on a frictionless billiard table, the average motion of particles described by a standing wave is zero—but the particle is moving.

The kinetic energy of a particle is proportional to the square of the momentum; therefore, the kinetic energy of the particle in the first harmonic state is four times that of the ground state corresponding to the fundamental. Similarly, the energy of the state corresponding to the second harmonic is nine times the ground state energy. Quantitatively, the energies of the different states can be expressed as

$$E_n = n^2 E_1 \quad \text{where } E_1 = \frac{h^2}{8L^2 m}$$

and where m is the mass of the particle. This set of possible energies constitutes the energy-level *spectrum* of the system of a particle in a one-dimensional box.[8] The system can have no other energies; in particular, the energy cannot be zero!

The existence of such a minimum "zero-point" kinetic energy is characteristic of systems of particles held together by forces between them. The value of this minimum energy can be estimated by considering the size of the system. Because the wavelength of a particle contained in the system cannot be much greater than the size of the system, there will be a lower limit on the momentum of the particle that will be inversely proportional to the size of the system and a commensurate least kinetic energy that will vary as the inverse square of the system dimension.[9] The smaller the box, the larger the energy.

The increase of the energy of the system with the number of wavelengths n is characteristic of a wide variety of systems. If there are more wavelengths packed into the system, the wavelength must be shorter, the momentum greater, and the kinetic energy greater.

For other standing wave systems, the detailed energy spectrum will be different than for the box, just as the wave functions describing those stationary states will differ from the simple sine-waves[10] describing de Broglie box-states (or the vibrations of guitar strings).

If the energy, and thus the wavelength or momentum, of the particle is well known, the position must be indeterminate because the momentum and the position of the particle are complementary quantities connected through the Heisenberg Uncertainty Principle. For such a system of well-defined momentum or energy, the probability of finding the particle at a position x is proportional to the intensity of the standing wave at the point x, $I(x) = |\psi(x)|^2$, where, as for all waves, the intensity is proportional to the square of the amplitude written here conventionally as ψ. It is the intensity that has a direct operational meaning, because the intensity is directly observable. If there are very many particles in the box occupying the same state, the average intensity of the wave over a small region will be proportional to the number of particles in the region; if there is just one particle in the box and that particle has an energy corresponding to the state n, the intensity $I_n(x)$ multiplied by the length of the region about the point x represents the probability that the particle will be found there.

We can give operational meanings to the concept of a state of definite energy and the intensity of the de Broglie standing wave associated with that state by considering a set of rather fanciful measurements in a "quantum land" where h is very large. Figure 10.10 shows a gunner who is attempting to hit a tank stationed in a canyon with a flat bottom and high walls on each side. His intelligence service has recognized that the tank in the canyon is very much like a ball in a box and that a measurement of the kinetic energy of the tank in its motion from wall to wall will provide information concerning the position of the tank. (We ignore any motion of the tank up or down the canyon, toward or away from the gunner.) This energy is then determined by scouts who measure the force or pressure on the wall of the canyon (which is proportional to the kinetic energy of the tank) and find that the tank has the second lowest energy possible. Incidently, the gathering of that information affects the tank so that previous intelligence concerning its position has to be thrown out.

The shell from the gun is known to be heavy and moving fast; it has a sufficiently high momentum that the wavelength of the shell is quite short compared to the width of the canyon. Hence, for the measured tank energy, the gunner can neglect the uncertainty in position of the shell with respect to

Figure 10.10 The upper graph shows the amplitude for a particle in a stationary state of a one-dimensional square well. The particle can be a ball in a box or a tank in a high-walled canyon. The lower graph shows the variation of intensity over the span of the box or canyon and a gun prepared to shoot a high-momentum particle at the tank in the canyon. The graph at the right shows the probability of finding the tank with a certain energy—or in a certain state—after it is hit at point a.

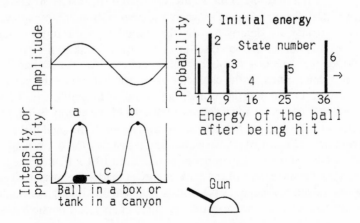

the uncertainty in position of the tank. When the gun is fired, the shell may strike the tank as it passes through the canyon; but however expert the gunner, he cannot be sure of a hit since the position of the tank is not well defined. If the position of the tank is uncertain, where should the gunner aim? Is there a best strategy? Yes, there is. The plot of the intensity in Figure 10.10 represents a plot of the probability of the tank being hit as a function of x, the distance along the canyon from wall to wall; the plot also shows the probability of the tank with the measured energy being at a position x. The shell will have a relatively high probability of hitting the tank near points a and b, and almost no possibility of hitting the tank near point c. Firing along the canyon, the gunner should aim at point a or b to maximize the probability of a hit. If he aims accurately at the exact center of the canyon, he could shoot all day and never hit anything. Although the tank may be found at either side of the center, it will never be found exactly in the center of the canyon. Of course, if the tank occupies other states corresponding to different measured energies, the gunner should adopt different strategies.

The stationary states of the particle in the box (or tank in the canyon) are states for which the energy can be known precisely because the energy measurement can be conducted over a long time. Such states correspond to the characteristic standing waves of a vibrating string, where the frequencies are precisely defined because the vibration persists over a very long time. However, if we measure the position of the particle at some particular time, rather than the energy, we will have a different, complementary kind of information. What can we know about the momentum (and energy) of the tank after the gunner locates it by hitting it with a shell from the gun (which does not destroy the heavily armored tank)? We will know the position of the tank in the canyon when it is struck by the shell—at that time, the tank must be where the shell hits it and no other place—but we will have lost our information concerning the complementary momentum and energy of the tank. Momentum from the shell may be transferred to the tank. There is a correspondence between the state of the tank with a known position and the state of a string where the initial configuration of the string is in some particular shape—perhaps as a result of the application of a bow. Just as the string then vibrates with different frequencies where the distribution of intensities of the various overtones are determined by the character of the initial state (the point of application of the bow), so the energy of the tank in the canyon will be described in terms of a set of probable energies where the probability distribution is determined by the character of an original state—the position of the tank when it is hit and the momentum of the shell that hits it. If a string is plucked near one end, we might find that 70% of the energy is fed into the fundamental, 20% into the first harmonic, and so on. If the shell hits the tank near point (a) in Figure 10.10, we would find a distribution of probable energies such as that shown at the upper right of Figure 10.10. If, after the tank is hit, a second measurement of the tank energy is made, the measured energy will be one of the box-energies previously calculated. Although it would not be possible to predict the value of the measured energy, it is possible to assign relative probabilities to the different possible energies. It is twice as likely that the tank will have four times the minimum energy than the minimum; it is almost impossible that the tank will have 16 times the minimum energy, although such an energy is available to quantum tanks in general.

To continue the story of the tank and gunner, after the tank is hit and the energy remeasured (by measuring the force on the canyon wall) the gunner loses the position information gained by the first hit on the tank, and he must again adopt a statistical strategy as at the beginning of the battle. After the momentum (hence the energy) of the tank is measured, the position is indeterminate—that is, defined only statistically; after the position of the tank is measured, the momentum (and energy) are known only statistically.

If Planck's constant is very small compared to any possible momenta of the tank, as it is in the world we live in, the position of the tank would not be very diffuse, and any sensible intelligence service will attempt to ascertain the (almost) exact position of the tank. In terms of the wave picture, the tank will be described by a very large set of standing waves with slightly different numbers of nodes (which will be very large numbers). The resultant tank will appear as a constructive interference of the waves at the point the scouts see the tank, and the waves will cancel out quite precisely over the rest of the canyon.

Barrier Penetration by Waves

As mentioned, the velocity of a de Broglie wave describing the motion of a particle is one-half the velocity of the particle, and all waves are reflected at interfaces where the wave velocity changes. Hence, when the forces on a particle are such as to change its velocity, we can expect some reflection of the de Broglie wave and some reflection of particles. If the intensity of the reflected wave is 10% of the incident intensity, the probability of the particle being reflected at the discontinuity will be just 10%. The nonclassical character of this reflection is suggested by the scenario of Figure 10.11, where a cart moving from left to right on the level passes up a hill to another level region. If the cart is moving fast enough to coast up the hill, classically we can expect it to continue on the upper level, albeit with reduced velocity. The cart will never be reflected—that is, it will never go partway up the hill and roll back down, going to the left with its original speed.

Figure 10.11 At the left, a cart rolling on frictionless wheels approaches a hill that is low enough so that the cart can coast up the hill and continue. At the right, the cart approaches a downhill region from which it can be reflected.

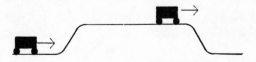

However, if we procure a very small quantum-cart with a very small momentum such that the de Broglie wavelength of the cart is not much smaller than the length of the hill, the cart will sometimes be reflected[11] at the hill even though it has more than enough energy to reach the top. The same kind of reflection will affect the quantum-cart at a downhill interval where the velocity of the cart must increase. Again, there will be some probability of reflection of the cart by the discontinuity in velocity (even as light is reflected when it leaves the window glass for the air—from a region of low velocity to high velocity).

The reflection of particles at surfaces separating regions of different particle velocities is manifest in quite different natural regions. Electrons moving in conductors are usually reflected to some extent at an interface of two different conductors, just as the energies of the electrons differ in the different materials rather as the cart moves at different elevations in Figure 10.11. At energies a million times larger, neutrons and protons scatter at the surface of nuclei. These nucleons are reflected at nuclear surfaces, hence scattered from nuclei, as a consequence of the nuclear forces (generally attractive) that act on nucleons at the surfaces. Here, the nucleons have much larger energies and velocities inside the nucleus than outside.

Even as the quantum-cart will be reflected where a classical cart will not, the quantum-cart will "tunnel" through barriers insurmountable by the classical cart. Figure 10.12 shows the cart approaching a hill too high to roll over; the cart has only enough kinetic energy to roll halfway up. But it does have enough energy to roll on the other side of the hill if a tunnel is constructed. The cart will occasionally pass through—the de Broglie wave will provide some penetration of the barrier. The penetration is much reduced[12] if the barrier is increased in height or width or if the cart energy is reduced. But no tunnel is needed for the quantum-cart to occasionally "tunnel" under the hill.

The right side of Figure 10.12 suggests the character of an alpha particle in a nucleus where the barrier is too high for the particle to escape classically. The barrier is a consequence of the very strong repulsive force between the

Figure 10.12 At the left a cart approaches a hill with too little energy to coast up to the top, but the cart will be found—occasionally—on the other side. At the right, an alpha particle has only enough energy to "slide" up the potential hill to the arrow so that (classically) it can never get out of the uranium nucleus. But it will—occasionally—be found outside.

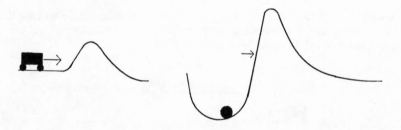

positively charged nucleus ($+92$ e for uranium) and the positively charged α-particle (with a charge of $+2$ e) that dominates outside the range of the very strongly attractive nuclear forces. The height of the barrier is about 30 *MeV*, whereas the energy of the alpha particle in the nucleus is only about 8 *MeV*. Classically, an α-particle with an energy of 8 *MeV* could never climb the hill. But in the real quantum-world, it occasionally leaks through. The barrier works both ways. Classically, an α-particle with an energy of only 8 *MeV* moving toward a uranium nucleus would be stopped by the electrostatic repulsion and never strike the uranium nucleus. But, as a consequence of quantum mechanical barrier penetration, the alpha particle will occasionally reach the nucleus.

The penetration probability can be very small and still be significant. We can describe the alpha-particle decay of heavy nuclei adequately in terms of a model of a free alpha particle moving inside the nucleus. An alpha particle inside a uranium nucleus hits the surface wall perhaps 10^{21} times a second and penetrates the wall about once in a billion years. Yet this leads to most natural radioactivity.

Waves to Particles—Photons

According to classical electromagnetic theory, a beam of light that carries an energy E will carry a momentum p such that $p = E/c$, where c is the velocity of light. The Special Theory of Relativity defines a relation between the mass m of a particle and the energy E and momentum p of the particle such that this ratio $E = p/c$ holds for a particle of zero mass[13] that must travel at the velocity of light. Hence, the momentum-energy relationship obtained classically for electromagnetic waves is the same as might be expected for massless particles traveling at the speed of light.

Moreover, the relation between momentum and wavelength of a particle required by quantum mechanics, together with the relation between frequency, wavelength, and wave velocity which hold for any wave lead to a de Broglie wave frequency that is the same as the electromagnetic frequency.[14] Therefore, the electromagnetic wave has the form of a de Broglie wave for a massless particle; that particle is called a *photon*. Such a photon would carry an energy equal to hf and a momentum hf/c, where h is Planck's constant, f is the frequency of the electromagnetic wave, and c is the velocity of that wave—that is, the velocity of light.

If the electromagnetic wave is indeed the de Broglie wave for a photon, the probability of finding a photon in a small volume must be proportional to the intensity of the wave at the position of the volume. With each hypothetical photon carrying an energy equal to hf, the intensity of the de Broglie wave will then be proportional to the energy density carried in the wave. But the energy density carried by an electromagnetic wave is proportional to the intensity of the wave. Again, quantum mechanics and special relativity suggest that the electromagnetic wave is a de Broglie wave for a particle without mass that does, however, carry momentum and energy. It can also be shown that the particle—the photon—must also carry angular momentum; a photon has a spin or intrinsic angular momentum of $1\hbar$.

It was Einstein who first concluded that this hypothetical electromagnetic particle, the photon, which fitted special relativity and quantum mechanics so well must be real, and electromagnetic waves must have particle-like properties.

The particle aspect of these packets of electromagnetic energy or photons is particularly evident when collision-type processes between radiation and free electric charges, such as electrons, are considered. Figure 10.13 illustrates some aspects of the scattering of an electromagnetic wave by a free electron. The classical view of such a process is as follows: the oscillating electric field **E** generates an oscillating force on the charged electron, which causes the electron to move in oscillatory motion; the kinetic energy of the moving electron is supplied by the energy in the field. In turn, the oscillating electron sets up an oscillating field **E'** and radiates energy. Moreover, the electron is pushed away gradually by the interplay of forces involving the two fields **E** and **E'** and the electron charge.

According to the (correct) quantum mechanical description of the process, the absorption of the energy by the electron does not take place gradually and continuously; the electron absorbs energy and momentum in impulses just as if the photons collided elastically with the electrons. When the frequency of the electromagnetic radiation is very large and the energy of the photon is large, single collisions can be observed. The constraints of the conservation of momentum and energy then result in definite relations between the momenta and energy of the scattered particles.[15] Such a collision is shown in Figure 10.13.

We see that the scattered photon loses energy just as the electron recoil carries off some of the original energy. Because the frequency of a photon is proportional to the photon energy through the relation $E = hf$, the energy loss manifests itself in the reduced frequency of the scattered photon. These discrete collisions of photons with electrons were first observed by Arthur

Figure 10.13   To the right, a classical view of the scattering of electromagnetic waves by an electron. To the left, the quantum mechanical picture.

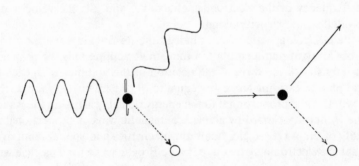

Compton in 1923, and this particle-like scattering of the photon by an electron is called *Compton scattering*. With the observation of Compton scattering, the particle nature of electromagnetic waves, considered a continuum classically, could no longer be doubted.

When low-frequency electromagnetic radiation such as radio waves is incident on an electron, we find that the subsequent motion of the electron is very much what we might expect from the classical viewpoint. If the frequency of radiation is very low, the energy hf of a single photon, which is proportional to that frequency, will be extremely small. The energy taken up by the electron, which can be relatively large, must then be transferred through the collisions of very many photons, and these photons act together, correlated by the form of the electromagnetic wave, so that the results of the very many collisions follow the classical prescriptions as we might expect from the correspondence principle.

The inverse of the Compton effect, the scattering of high-energy charged particles by photons, also occurs in nature. High-energy photons (X-rays) are found as part of the cosmic radiation that fills the interstellar regions of the galaxy, and some part of these high-frequency electromagnetic radiations appears to be derived from collisions between the charged particles and low-energy photons of starlight, where the photons from the starlight are propelled to high energies by the collisions.

Although Compton's demonstration of single photon-electron scattering served to complete the proof that electromagnetic waves had particle-like properties, an earlier analysis of the photoelectric effect had convinced Einstein and others of the necessity of such a particle description of light. When electromagnetic waves of sufficiently high frequency are incident upon metal, electrons are immediately ejected from the metal. The maximum kinetic energy of the electrons, E_m, is equal to the photon particle energy hf minus a small energy ϕ, characteristic of the material required to extract the electron. This *photoelectric equation* is written as

$$E_m = \tfrac{1}{2}mv^2 = hf - \phi$$

No matter how intense the radiation, no electron will be ejected if $hf < \phi$. In contrast, even if the intensity of radiation is very small, if $hf > \phi$, electrons will begin to be ejected from the metal as soon as the radiation strikes the metal: the emission of electrons begins instantaneously after the radiation begins.

According to a classical explanation, the free conduction electrons in the metal would gradually absorb energy from the field until the electrons had enough thermal energy to escape. Such heating does occur, and electrons are emitted from metal heated by incident electromagnetic waves according to the classical prescriptions; however, the metal reaches a sufficiently high temperature to emit electrons only after a long irradiation. The immediate emission of electrons, with their energy related to frequency, can be understood only if discrete packages or quanta of energy are delivered to individual electrons. This *photoelectric effect* is thus, again, a quantum mechanical effect not explicable in classical terms. Einstein received the Nobel Prize (in 1921) for his explanation in 1905 of the photoelectric effect in terms of such quantum effects. (Einstein was not satisfied with his discussion of the photoelectric effect and did not push to the conclusion that the quanta acted like particles. He seems to have felt that his explanation was ad hoc and regretted his inability to place it in a broader context.)

An inverse to the photoelectric effect is also known. High-energy electrons striking metal or any other material generate the high-energy photons called X-rays. Classically, we can understand the generation of electromagnetic radiation as a consequence of the acceleration or deacceleration of the charge carried by the electrons as they stop. However, there is a well-defined maximum frequency of the radiation that cannot be explained classically but follows naturally from the particle or photon description of electromagnetic radiation required for consistency with the quantum mechanics of particles. That maximum frequency f_m is obtained from a rewriting of the photoelectric equation

$$hf_m = E_m + \phi = \tfrac{1}{2}mv^2 + \phi$$

where f_m is the maximum frequency of the photons produced by electrons of mass m and velocity v. (For simplicity, we consider nonrelativistic electron energies such that $v \ll c$.) The maximum energy of the photon is thus just equal to the energy of the electron plus the small additional energy of absorption of an electron by the metal described by the work function ϕ.

The establishment of the particulate properties of the electromagnetic waves and the concept of the photon as a particle, indeed an elementary particle, was revolutionary. This development, extending from 1905 with Einstein's paper on the photoelectric effect through 1923, when the thoroughly particle character of the photon was established by Compton, took place before the work by de Broglie, published later in 1923, even suggested that particles have wavelike properties. (The term "photon" appears to have been first used as late as 1926.) Now, with the quantization of fields, and with the wave description of particles, fields and particles are joined inextricably, and the primitive concepts of waves and particles, the continua and the discrete, are forever blurred.

Angular Momentum and the Spin of Particles

The classical electromagnetic wave carries polarization information; for example, a light wave traveling in a z-direction may be polarized such that the electric field is directed in the x-direction. The interpretation of the electromagnetic wave as a de Broglie wave, where the intensity of the wave is proportional to the probability of finding a photon, does not ostensibly carry such polarization information. That directional capability must be provided by some property of the photon. We find that photons and also other particles carry an intrinsic angular momentum or *spin*. One can think of the particles as spinning about an axis that defines a direction by its alignment.

The character of the elementary particles are determined by only a few properties. Each particle has a specific mass, holds specific charges (electric charge and other varieties of charge), and has a specific value or rate of spin. This spin, determined by the fundamental structure of the particle, is a subtle property, and the general quantum constraints on angular momentum, which apply to spin, are best introduced by analyses of classical orbital angular momentum.

Quantum mechanics places certain interesting restrictions on the properties of particles moving in orbits with respect to another particle—or with respect to any special point in space. Here we use the word "orbit" to emphasize a special interest in particles that move in a manner that corresponds to closed classical paths. Such a particle can be described by a localized de Broglie wave constrained to the region about the classical orbit. The wave description will not change in time, just as a planetary orbit does not change in time.

For pedagogical definiteness, let us consider the wave description of an electron in an orbit about an atomic nucleus, perhaps a proton. If we follow any closed path about the proton, or any other point, at some specific time and map the amplitude of the de Broglie wave, we expect that the amplitude will go through zero an even number of times (to get back to the original value) as we complete the circuit. There must be an integral number of wavelengths about the path. As an especially simple example we might take a circular path of radius r, as shown in Figure 10.14. There must be n wavelengths about the path of length $2\pi r$, where $n = 1, 2, 3, \ldots$; hence, the wavelength must be such that $\lambda = 2\pi r/n$. From the relation between wavelength and particle momentum, the momentum of the electron in such a path will be $p = \hbar/\lambda$. Because the angular momentum j of a particle moving in a circle is equal to its momentum multiplied by the radius of the circle, $j = pr$, and the angular momentum of an electron about that path is $n\hbar$. Properly, this is the component of angular momentum in the direction of the axis of the circle (for electrons moving counterclockwise about the path). In general, the component of angular momentum about any axis must be an integral number of wavelengths and then an integral number of units of $\hbar$. Angular momentum is quantized.

For any atomic state of the electron, there will be a maximum number of wavelengths j for any complete path and then a maximum component of angular momentum $j\hbar$ in the direction of the axis of that path. A classical orbit can be constructed such that the axis of rotation lies in any direction, and the component of angular momentum in that direction can have any value less than or equal to a maximum that is the angular momentum of the particle. (The maximum value would be such that the particle in the orbit were moving counterclockwise in a plane normal to the direction chosen.) The quantization conditions require that the component in any direction be equal to an integer number of $\hbar$ units. If the maximum is j (corresponding

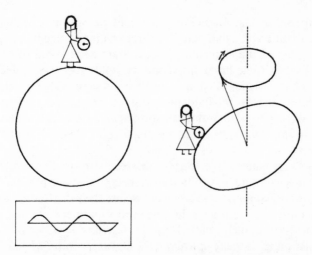

Figure 10.14   The solid line at the left describes a circular path about the proton where the de Broglie amplitude of an electron is defined. Below left is a map of the (real part of the) amplitude of the de Broglie wave about the circle constructed by the small observer who has walked quickly about the path measuring that amplitude. Because there are two waves about the circumference, the angular momentum in the direction of the axis of rotation is $2\hbar$. The diagram at the right presents a classical description of the uncertainty of the direction of the axis of the angular momentum with respect to a chosen z-direction. The vector can be considered to lie at random positions about the small circle.

to a counterclockwise rotation in the plane normal to the chosen direction), the minimum will be $-j$ (corresponding to a clockwise rotation in the normal plane). Summing all the different possibilities $(j, j - 1, j - 2, \ldots, -j)$, there will be $2j + 1$ possible values.

We will follow common usage and, and for a state such that the maximum component of angular momentum in any direction is $j\hbar$, call j the angular momentum of the state. Thus, we will label states and particles as having angular momenta of 0, 1, 2, and so on units (of $\hbar$).

It is useful to consider that a state of angular momentum $j\hbar$ can assume only $2j + 1$ different directions. This is a consequence of the Uncertainty Principle and implies that the direction of the angular momentum can be known only to an accuracy of $1/(2j + 1)$ of a full circle.[16] If the angular momentum $j\hbar$ is such that $j = 2$, the angle corresponding to one-fifth of a circle $1/(2j + 1)$ is $72° = 360°/5$, and the uncertainty in direction is roughly $\pm 36°$.

It is difficult to discuss spin in an accessible way even by metaphor, as the intrinsic spin of elementary particles is a quintessentially nonclassical concept. However, if quantum mechanics is to be consistent, the same rules established for orbital angular momentum must hold for intrinsic spin. For a particle of spin $s\hbar$, there must be just $2s + 1$ differentiable orientations of the spin (axis). But the requirement that $2s + 1$ be an integer allows s to take on half-integral values, and this niche is filled by nature. While all particles associated with forces—such as the photon—have integral spins, those fundamental particles associated with matter all have spins of $1/2$ ($\hbar$). These conserved matter-particles, such as the electron, with spins of $\hbar/2$ (that is, $s = 1/2$ and $2s + 1 = 2$) can be oriented in only two different ways with respect to *any* direction.

It is clear that a closed course, such as shown in Figure 10.14, which contains a half-integral number of wavelengths will not close in one circuit. Indeed, if the description is correct, one circuit will leave the sign of the de Broglie amplitude reversed, and one must go around twice to return to the starting value. If the particle is rotated 360°, the de Broglie amplitude describing the particle must change sign. The state will be back to where it was originally only if the electron undergoes two complete rotations!

This most unclassical behavior of these half-integral (spinor) particles can be observed. Consider the double-slit diffraction of polarized electrons shown in Figure 10.15. Here we place a solenoid magnet, made by winding a current-carrying wire on a hollow cylinder, in front of one slit so the spin of the electron will rotate in the magnetic field. If the electron is rotated once, the sign of the amplitude of the wave from that slit will be reversed, and the diffraction minima will change to maxima and the maxima will go to minima, as

Figure 10.15 Polarized electrons diffracted by a double slit, where a solenoid magnet in front of one slit rotates the electrons that pass through that slit. The solid line shows the characteristic double-slit diffraction pattern on the screen that is seen if the solenoid is off or if the field is such as to rotate the electrons about twice. The dotted line shows the intensity if the electrons are rotated once.

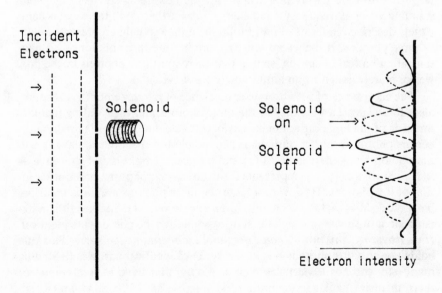

shown by the dotted lines in the figure. However, if the magnetic field is doubled, rotating the electrons two full revolutions, the diffraction pattern will return to the original.

Identical Particles and the Exclusion Principle

The intensity of a standing de Broglie wave at any point is proportional to the probability of a particle being in a small volume about that point. The sum of the intensity of the wave over all small volumes will be equal to the total number of particles represented by the standing wave. We might consider a specific standing wave such that the total intensity over the entire volume occupied by the wave corresponds to one particle in the state: this might be the state of a particle in a box. Then, if the amplitude of the wave is increased by $\sqrt{3}$, the intensity, which is proportional to the square of the amplitude, will be increased by a factor of 3. There will be three particles in the box, all in the same state.

Assume that the box, with perfect mirror walls, contains an electromagnetic standing wave in a mode such that the frequency of the wave is f and the energy stored in the box is $3hf$; that is, three photons of the same frequency (or color) are stored in the box. The energy may have been introduced into the box through three distinguishable, discrete operations. The box may have been opened to starlight to trap an amount of radiation sufficient to raise the energy from zero to hf; then the box may have been opened to sunlight to trap a second photon and raise the energy to $2hf$; and finally the box may have been opened to the light of a candle to catch the third photon and raise the energy to $3hf$. (In all this we allow a certain poetic license; the experimental procedures are hardly practical.) Now the box holds an electromagnetic standing wave of frequency f and energy $3hf$. Furthermore, the wave is completely described in terms of the amplitude of the wave that represents a state of three photons. It makes no sense to number the three photons—all photons are identical[17]—and by stating that there are three photons in the box, we are merely assigning an amplitude to the wave.

The acceptance of the absolute equivalence of the three photons is equivalent to assigning an identity to the three photons and then stating that any ordering of the three cannot make any observable difference. We might number the photon from the starlight as (1), the photon from the sun as (2), and the photon from the candle as (3) and write the amplitude of the wave as $A(1, 2, 3)$ where (1, 2, 3) represents some order—perhaps the order of reception of the photons. However, because the three photons are indistinguishable, $A(1, 2, 3) = A(1, 3, 2) = A(2, 3, 1)$, and so on. The final result is independent of the order in which the state was filled, as the photons are identical. *If we exchange any two photons, the amplitude remains the same.* This rule holds not only for photons but for every set of identical particles (including manifestly composite particles such as atoms) that have integral values of their intrinsic angular momentum or spin—$0\hbar$, $1\hbar$ $2\hbar$, and so on. (Recall that the photon has a spin of $1\hbar$.) These particles are called *bosons* after the Indian physicist S. N. Bose, who wrote on these properties of photons in 1924.

We can gain some understanding of the observable consequences of this absolute identity of particles by considering certain characteristics of simple systems. Assume that we have a large set of systems such that each system has two states—for example, two standing waves in a box or two different orbits of the hydrogen atom—and these systems hold two particles distributed randomly among the states. In general, some of the systems will have two particles in one state and none in the other, and some systems will have each state occupied by one particle. If the particles are not identical, what is the probability that each state is occupied by one particle? The right side of Figure 10.16 shows the various equivalent possibilities: it is clear that half of the systems will have a particle in each of the two states.

If the particles are two identical bosons, the two configurations with one particle in each state will not be distinguishable and will represent but one possibility instead of two. The probability of two states being occupied will then be 1/3 instead of 1/2. The numbers or labels attached to individual bosons are irrelevant.

Although the *statistics* of the nonidentical particles (called Boltzmann statistics) are what we should expect classically, the *Bose-Einstein* statistics of the identical bosons have no clear classical counterpart. If we throw two balls, colored red and green, into a box divided into two equal parts by a partition, the probability of finding the two balls in different parts of the box will be 1/2 if the throws are made randomly. If we make the balls more nearly identical by painting both red, the probability will not change. Even if the balls

Figure 10.16 The different possible distributions of two particles among two different states or standing waves. Particles that are not identical are labeled (*a*) and (*b*).

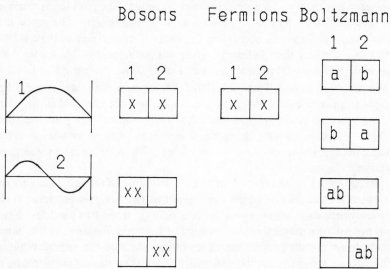

are manufactured to be so nearly identical that we could not differentiate between them by any measurement we might make in practice, the probability of finding the balls in different sectors would remain at the value of 1/2.

The balls are not really identical; if we made more subtle tests, we would surely find some difference. And the two sides of the box represent not two different states but two different sets of states, where each set represents an extremely large number of differentiable states: there are very many different ways in which one ball can rest in one side of the box. When we talk of states in the microscopic realm, where we must use quantum mechanics, we mean only one unique wave or orbit; and when we say that the particles are identical, we use the word "identical" with a rigor that has effectively no classical counterpart. The Bose statistics of the two elementary particles in the two states at the left of Figure 10.16 is a quantum mechanical concept.

Although some particles, such as the photon, have integral spins, other particles, such as the electron, are observed to have half-integral values of $\hbar$ as their intrinsic angular momentum: $1/2\ \hbar$, $3/2\ \hbar$, $5/2\ \hbar$, and so forth. If $s\hbar$ is the value of the angular momentum, there are $2s + 1$ different orientations, as was the case for particles with integral angular momentum. The electron has a spin of $1/2\ \hbar$, and there are then two differentiable orientations of the electron spin direction. Particles with half-integral spin, such as the electron and proton, are called *fermions* after Enrico Fermi.

Aside from their half-integral spins, fermions have another unusual property: two identical fermions are never observed in the same state. If we perform the model experiment in Figure 10.16 with fermions, we will find that all the sytems have the two particles in different states. We are able to include this property of fermions in a quantum mechanical description of nature by postulating that the amplitude of a state of two (or more) identical fermions changes sign upon the interchange of the two (any two) fermions. If two identical fermions occupy the same state, interchanging the two cannot change anything—but the wave function must change sign! The paradox is removed only if two identical fermions cannot occupy the same state. The probability of two identical fermions occupying the two different states of Figure 10.16 is therefore 1 rather than 1/2 for the different particles, and 1/3 for the identical bosons. We say that fermions obey *Fermi-Dirac statistics*.

We note one caveat. Particles with different spin orientations are differentiable in principle and in practice, and are then not identical in the sense that concerns us. When we speak of identical fermions or bosons, we speak implicitly of particles with the same component of spin in some direction defined by the observer—that is, particles with their spins in the same direction.

The nearest classical equivalent to an occupation of the same state by two particles is the occupation of the same point in space by two particles. If two electrons occupy the same region, an interchange of the two particles cannot result in any change; yet because the sign of the wave function must change, two electrons cannot occupy the same region. As one can see from Figure 10.16, just as fermions avoid each other more than nonidentical particles, bosons are more likely to be found together.

This "exclusive" property of fermions is known as the *Pauli Exclusion Principle* after Wolfgang Pauli, who first clearly expressed this property of fermions. We cannot expect to relate these properties of fermions, the Exclusion Principle and half-integral spins, to classical concepts and our macroscopic experience. Pauli was able to show that the two properties are related to each other and to the requirement that no signals travel faster than the speed of light even in the microscopic world of quantum mechanics. Physicists admit these ideas as experimental facts or constructs that directly account for experimental observations.

Objective Reality and the EPR Paradox

Einstein and others felt that quantum mechanics, although an accurate description of nature, must be an approximation to some more fundamental concept. Although Einstein's early disaffection with quantum mechanics followed from his reluctance to accept the statistical interpretation of quantum mechanics and the idea that "God played dice with the world," he was especially concerned later with his conclusion that the results of quantum mechanical predictions of certain correlations could not be made consistent with views he held strongly concerning the nature of reality. The conflict between these quantum mechanical predictions and plausible definitions of reality is genuine. Perhaps more important, because those predictions appear to be in accord with observations, it seems that there are properties of nature that conflict with such definitions of reality, independent of quantum mechanics!

We will illustrate the conflict between quantum mechanics (and nature) and a plausible view of reality by discussing a problem with correlated observations first pointed out by Einstein, Boris Podolski, and Nathan Rosen in 1935. Here we look at a variant of the EPR "paradox" similar to that proposed by David Bohm.) The observation that the correlations predicted by quantum mechanics—correlations that have now been observed—are inconsistent with simple definitions of reality, independent of the validity of quantum mechanics, was demonstrated by J. S. Bell in 1964. (Our analysis of the conflict between observations and "reality" is taken, with minor variations, from an ingenious pedagogical presentation of N. D. Mermin.)

We proceed by conducting a thought experiment, as imaginary experiments are easier than real experiments. However, this is almost a real experiment, and the results are quite similar in principle to the results of measurements from real experiments. Figure 10.17 shows a schematic view of the apparatus designed to make measurements of the correlations between the planes of polarization of the photons emitted in the two-photon decay of positronium. Positronium is the name given to a kind of atom made up of an electron and positron bound together by their Coulomb attraction. Such a state will decay, quite quickly, through the annihilation of the positron and electron, particle and antiparticle, to two photons. If one measures the planes of polarization of the two photons, one finds that the planes are perpendicu-

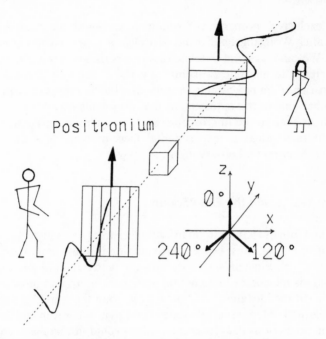

Figure 10.17 Photons from the decay of positronium passing through polaroid-like polarimeter filters. The arrows attached to the polarimeters define a reference direction that is marked 0° in the diagram.

lar. The system described in Figure 10.17 consists simply of a source of decaying positronium and two polarization filters that transmit light plane-polarized in one direction and absorb light plane-polarized in a direction perpendicular to the direction of transmission. (The polarimeters can be presumed to act in the same manner as polaroid does for much lower energy photons.) Figure 10.17 shows a particular situation such that both photons are transmitted; the photons are shown (fancifully) as classical waves so as to illustrate the relative position of the planes of polarization.

The apparatus of Figure 10.17 is further arranged so that each of the polarimeters can be rotated so that the orientation arrow shown on each will be pointed at an angle of 120°, or at 240°, or at 0° as shown.[18] We might also add appropriate instrumentation so that a photon absorbed by the polarimeter at the left registers a "no," and one that passes through the polarimeter—to strike a detector that is not shown—would register a "yes"; indeed, red and green lights could be flashed indicating no and yes. And the right-hand arm of the "coincidence" apparatus could be arranged in the same manner. Moreover, we can arrange our detectors and associated electronics so that results are recorded only if there is a time coincidence between the signals so that we are certain that the two photons are emitted in the same process.

Now we can calculate, using quantum mechanics, the results *we will find experimentally* for the correlations under different conditions of orientation of the polarimeters. If the polarimeters are aligned—if both are set at the same angle—half of the positronium decays that emit photons directed toward the polarimeter detectors will result in a "yes" (or green light) in both arms, and half will produce a "no" (or red light) in both arms. The correlation will be 100%. If the polarimeter elements are not aligned, only one of eight events will produce a yes-yes response in the two arms, and one in eight will give a no-no result. However, three in eight will give a yes-no and three a no-yes signal, so that 25% of the signals from the two arms will be positively correlated and 75% negatively correlated. We can restate these observations as Table 10.1.

Now we add the results of one more observation: half of all photons directed toward either polarimeter are transmitted, and half are absorbed by that polarimeter.

The stage is now set for the philosophers. For the purpose of illustration, consider the proceedings from the view of simple idealism and simple realism. The idealist believes that nothing exists without observation; if, for instance, no one is watching a tree in the Oxford Quad, the tree cannot be said to exist. Ronald Arbuthnott Knox (a distinguished cleric) wrote this poem:

> There once was a man who said "God
> Must think it exceedingly odd
>   If he finds that this tree
>   Continues to be
> When there's no one about in the Quad"

The realist considers that an objective reality exists quite independent of any intercession of a conscious observer and insists that the tree is always in the Quad whether or not it is observed by Oxford dons. That position is stated in an anonymous answer to Knox's limerick:

> Dear Sir, Your astonishment's odd:
> I am always about in the Quad.
>   And that's why the tree
>   Will continue to be,
> Since observed by yours faithfully, God.

The question raised in the spirit of Einstein concerns not the tree in the Quad but the polarization of the photons. Does that polarization exist without observation? Accepting at this time the uncritical use of "exist" and "observation," our answer must be "no." Rather than engage in philosophical debate, we define "exist" and "observe" operationally. If each photon can be considered to carry with it the information required to define the response of the polarimeter detector to the photons' subsequent passage, independent

Table 10.1

Polarimeter settings	*yy*	*yn*	*ny*	*nn*
Same angle	50%	0%	0%	50%
Different angle	12.5%	37.5%	37.5%	12.5%

of the orientation of those detectors, we can say that the polarization of the photons exists independent of the observation by the detectors. If not, we say that such a polarization does not exist—without observation.

Now we leave quantum mechanics, and we even leave physics to consider not the question of the validity or interpretation of quantum mechanics, but the constraints that observations of real phenomena place on definitions of objective reality. If the polarization of the light quanta are real, according to our hypothetical definition, each photon must carry information that defines its behavior upon striking the polarimeter, independent of the orientation or even the existence of the polarimeter (which measures or observes that polarization). A use of the pathetic fallacy (giving the photons a fanciful consciousness) may clarify matters. If the photon polarization has a *real* polarization, the photon must "know" what it will do—be absorbed or transmitted—upon meeting any polarimeter, whatever the orientation of the polarimeter. Because there are three possible settings of the polarimeter in our experimental design, and for each setting the photon may be transmitted (a "yes" or green-light response) or absorbed (a "no" or red-light response), we can designate the information a photon must carry by three indices set at either yes (*y*) or no (*n*). For example, if the information package is *yny*, the photon will be transmitted (*y*) if the polarimeter is set at 0°; absorbed (*n*) if the polarimeter is aligned at 120°; and transmitted (*y*) by a polarimeter at 240°.

Of course, the results of the photons passing through the two-arm polarimeter must be in accordance with the correlations actually observed, which were listed above. The 100% correlation between the absorptions or transmissions of the two photons when the polarimeters are set at the same angle immediately requires that the package of information carried by the left photon must be identical with that carried by the right photon: if the left photon carries a pack of information that reads *yny* according to our code, the right photon must carry the same packet, *yny*.

With that immediate simplification imposed by the correlation results observed when the polarimeters are aligned, we can fill in Table 10.2, which lists all possible sets of photon information packets.

If we accept the realism of the photon polarization, we must assume that the positron-electron system, upon decay, emits up to eight different kinds of photons with eight different programs of response to the three polarimeter settings we are to use. Nature must assign probabilities to each of these possibilities. We can show that *no* set of probabilities so assigned can account for the correlation results that are calculated and measured; hence, nature is not in accord with this kind of realism. We proceed most simply by considering the symmetry of the polarimeter positions and the symmetry of the transmission (*y*) and absorption (*n*) patterns. The three orientations must be logically equivalent, and the observations demonstrate a symmetry between *y* and *n*. It then follows that

$$P(yyy) = P(nnn) = \alpha$$

where $P(abc)$ is the probability of emission of a photon that responds to the 0° polarimeter with a response $a = y$ or $a = n$; b is the response to the 240°

Table 10.2 Possible information carried by the two photons emitted in the positronium decay. The first index defines transmission (y) or absorption (n) if the polarimeter is set at 0°; the second, for a polarimeter at 120°; and the third, for a setting of 240°.

Left Photon	Right Photon
yyy	yyy
yny	yny
yyn	yyn
ynn	ynn
nnn	nnn
nyn	nyn
nny	nny
nyy	nyy

polarimeter; and c is the response to the polarimeter aligned at 240°. Again from the symmetries,

$$P(yny) = P(yyn) = P(ynn) = P(nyn) = P(nny) = P(nyy) = \beta$$

Setting the left polarimeter at 0° and the right polarimeter at 120°, the following probabilities are defined by the observed correlations discussed previously:

$$P(yyy) + P(yyn) = 12.5\% = 1/8 = \alpha + \beta$$
$$P(nnn) + P(nny) = 12.5\% = 1/8 = \alpha + \beta$$
$$P(yny) + P(ynn) = 37.5\% = 3/8 = 2\beta$$
$$P(nyy) + P(nyn) = 37.5\% = 3/8 = 2\beta$$

From the last two entries giving the two negatively correlated responses, we must conclude that $\beta = 3/16$. Then, from the first two entries giving the positively correlated responses, we find $\alpha = -1/16$, an absurd result as no such probability can be negative. *Before observation, the photons do not have a definite polarization.* Using our (and Einstein's) definition of reality, the polarization of such a photon is then not *real*.

In summary, we have shown that the description of realism we have adopted leads to a contradiction with observations. Although the results of the observations are calculated correctly from quantum mechanics, because the conflict is between observations of nature and constraints from our definition of reality, the conflict is independent of the validity of quantum mechanics. Quantum mechanics does not force a peculiar view of nature on us; nature is peculiar, and quantum mechanics is only odd insofar as it correctly predicts the strangeness of nature.

In reviewing the conflict between the observations of the correlations and the definition of reality we have used, we see that the problem lies squarely in the *correlation* results. No matter what the orientation of either polarimeter, one-half of the emitted photons traveling to the left are absorbed by the left polarimeter and one-half of the photons moving to the right are absorbed by that polarimeter. Observations of only one polarimeter reveal no peculiar behavior.

To further illuminate the problems associated with views of reality and the photon polarization correlations, we present a statement concerning the polarization correlations that is made from an implicit philosophical position concerning reality. If the unobserved photons cannot be assigned a polarization, the measurement of the polarization of either photon fixes (a semantic trap!) the polarization of the other! Even though the two photons are far apart, the determination of the polarization of one photon immediately defines the polarization of the other. Moreover, this information (another semantic trap!) is transmitted instantaneously.

Before discussing the statement in detail, let us look at the thought experiment in more detail. We have our physicist friend Oliver far off at the left measuring the polarization of the photons that come his way. In the course of his measurements, he rotates the polarimeter at different angles, and he also varies the distance of the polarimeter from the source. He records the results of the passage of the photon—absorption or transmission, y or n, the flash of a green light or red—and also takes down the time. Then, to the right, his colleague Olivia makes similar measurements on the photons that come her way, and she writes those results down in her notebook. Later, perhaps over the weekend, the two observers get together over dinner and compare the results they have accumulated. Although both observers find that their notebooks show approximately equal numbers of yes and no responses, red or green lights, they find that these responses are distributed randomly in each notebook. However, they do find correlations between the entries in the two notebooks corresponding to the correlations we have discussed.

Returning to the statement that described the measurement of one photon "fixing" the polarization of the other photon, we see that the use of fix is invalid if "fix" is meant to imply action. Did the measurements of Oliver fix the results observed by Olivia? Or was it the reverse? Neither did any fixing; nature arranged correlations that they observed. Did information travel instantaneously? No information traveled at all. Nothing in Olivia's notebook tells her that Oliver was even making any measurements. Nothing Oliver did or can do transmits any information to Olivia. However, perhaps a signal did pass from photon to photon that determined the correlations, but in a manner such that no information could pass from Oliver to Olivia or the reverse. Upon inspecting their notebooks, Oliver and Olivia would find that sometimes the apparatus to the left was closest to the source and the left photon was measured first, so perhaps a signal could pass from the left photon to the right, saying "I just got measured and I found that I am polarized in the x-direction. You had better quickly arrange yourself to give the proper result when you are measured." Sometimes the right polarimeter was closer, and that measurement took place first; presumably, the message had to go from right to left. And sometimes the two polarimeters were about the same distance apart and no useful signal could be translated either way except if its velocity exceeded the velocity of light. In this way the measurement of one photon would physically, through the signal, "fix" the result of the measurement of the second photon, although the measurements were far apart in space and perhaps in time.

Now our idealistic philosopher friend, who doubts the existence of the unobserved tree in the Quad might conclude that idealism has conquered. The polarizations of the photons from the decays have no meaning and do not exist unless observed by someone. If the electron-positron system is set up in the Oxford Quad but Oliver and Olivia take the weekend off to attend a play in London and do not attend the polarimeters, the unobserved polarization of the photons does not exist. But our philosopher's realistic colleague, having studied Niels Bohr and the Copenhagen school on quantum mechanics, says no. He says that one must and can consider the whole of a system in any discussion of reality; there may be no significance to a part. If the entire apparatus—positron-electron source, polarimeters, and all—were set up in the Quad, the red and green lights would flash on merrily whether observed or not by anyone in the Quad. Human or sentient observation is irrelevant. The world is real.

Part of the problem of observer intercession lies in the very use of the word "observe," which is usually reserved for the interaction of systems with human consciousness. But the term "observed" is used here and elsewhere in the language of quantum mechanics to refer to the interaction of nominally isolated phenomena with artificial or natural structures. Bohr on occasion expressed his dissatisfaction with the loose use of the term. In the Copenhagen or realistic view, for the system of photons and polarimeters, the polarimeters perform the "observation" or definition of the polarization quite independently of the existence of any person who might record the results. Polarimeters, not people, observe; human intercession is irrelevant, and the universe exists independently of sentient observers.[19]

Both Bell's theoretical work and the experimental work confirming the strangeness of nature were conducted after Einstein's death in 1955. What would he have said, had he known of these results?

Notes

1. At any given time, we can describe the wave along a line in an x-direction by the equation

$$y = A \sin (\phi x + \alpha)$$

where y is the displacement of the surface from the equilibrium position, and x is the distance along the surface from a particular point. Here, A is the amplitude and $\phi x + \alpha$, the argument of the sine function, is the phase of the wave at the point x.

2. In general, at any point on the line, the two-aperture amplitudes are the sum of the two single-aperture amplitudes when the relative phases of the waves are taken into account. These relative phases of the waves at a point depend on the differences in distance between that point and the origin of the waves as measured along the paths to the two different apertures. If the difference between the distance from a point to the two apertures is an integral number of wavelengths, the two waves will be in phase and the amplitudes will add; if the difference is an integral number of wavelengths plus one-half a wavelength, the resultant amplitude will be the difference between the two single-aperture amplitudes.

In general, the rules for the addition of waves, taking into account the phases of the waves, are identical to the rules for the addition of complex numbers (and for two-dimensional vectors). We then write the amplitude of a wave as

$$\mathbf{A} = Ae^{i\phi} = \operatorname{Re} A + i \operatorname{Im} A = A \cos \phi + iA \cos \phi \qquad \text{where } i = \sqrt{-1}$$

With this description of a wave amplitude, we have a rule (essentially from Christian Huygens, a contemporary of Newton) that the amplitude of a wave $\mathbf{A}_q$ at a point q is just the sum of all the contributions, $\mathbf{A}_i$ from waves arriving at q over different paths i added according to the rules for the addition of complex numbers (or vectors):

$$\mathbf{A}_q = \sum_i \mathbf{A}_i$$

3. For exposition we write the quantum mechanical wave equation for a beam of particles traveling in an x-direction where the energy of the particles is E and the momentum is mv as

$$\psi(x, t) = \psi_0 e^{i(kx - \omega t)} \qquad \text{where } \omega = \frac{2\pi E}{h}$$

and

$$k = \frac{2\pi}{\lambda} = \frac{2\pi mv}{h} \qquad \text{where } \psi(x, t)$$

is the instantaneous amplitude of the wave and $|\psi(x, t)|^2$ is the intensity at time t and position x. Usually, we call ψ_0 the amplitude and $|\psi_0|^2$ the wave intensity. The energy (density) of the wave is proportional to this intensity.

4. The probability spread can be increased, however—for example, by placing a narrower slit just before the slit shown in the figure.

5. Consider the passage of particles with momenta p_x through a slit of width $2 \cdot dy$ that tells us the position of the particle with an uncertainty of $\pm dy$. From the description of the diffraction of particles by a slit, we know that the mean angle of deflection will be about equal to $\theta/2$, where

$$\theta = \frac{\lambda}{2 \cdot dy} = \frac{h/p_x}{2 \cdot \lambda}$$

where θ is the angle from the slit to the first minimum of the central diffraction pattern. The mean angle of diffraction, about equal to $\theta/2$, will be about equal to the mean momentum in the y-direction, dp_y, divided by the momentum in the x-direction, p_x. Or

$$\frac{dp_y}{p_x} = \frac{\theta}{2} = \frac{h/p_x}{2 \cdot dy} \qquad \text{or} \qquad dp_y \cdot dy = \frac{h}{4}$$

If we calculate the root-mean-square values of both the uncertainty in y and the uncertainty in p_y, we can write the uncertainty relation in a more conventional form

$$dp_y \cdot dy = \frac{h}{2\pi} = \hbar$$

6. The wave velocity is $\omega/k = f \cdot \lambda = E/p = v/2$, one-half of the classical velocity.

7. The momentum of the wave is $p = mv = h/\lambda$. Therefore we have

$$p_1 = \frac{h}{2L} \quad \text{and} \quad p_n = np_1$$

where n is the number of nodes or zeros of the wave function (counting one end of the box as a node). Perhaps it is best to think of this *quantum number n* as the number of (half) waves contained in the box.

8. Taking the kinetic energy of a particle as

$$E = \frac{1}{2} mv^2 = \frac{p^2}{2m}$$

and using the de Broglie relation between momentum and wavelength, $\lambda = p/h$, we have for the spectrum of possible energies of a particle of mass m in a one-dimensional box of length L

$$E_1 = \frac{h^2}{8L^2m} \quad \text{and in general} \quad E_n = \frac{n^2h^2}{8L^2m} = n^2E_1$$

9. Given a system of the size (such as a radius) of about a centimeters, the lowest momentum will be about $p = h/a$ and the lowest kinetic energy T about

$$T = \frac{p^2}{2m} \approx \frac{h^2}{2ma^2}$$

10. We write the analytic form of the wave functions for the different states of the particle in the box; if we take one end of the box as $x = 0$ and the other end as $x = L$, the standing wave amplitudes have simple sine-wave forms

$$\psi_n(x) = \sqrt{\frac{2}{L}} \sin(n\pi x/L)$$

The frequency of vibration is given as

$$f = \frac{E}{h} = f_n = \frac{E_n}{h}$$

11. If the interval Δx over which the velocity changes is small compared to the wavelength, the reflection probability for waves incident normally on an interface will be

$$R = \frac{(v - v')^2}{(v + v')^2}$$

where v and v' are the velocities at different sides of the interface. If Δx is larger than the wavelength, the reflection probability will be reduced. If the hill is longer and less steep, the cart will be less often reflected.

12. The penetration probability is about equal to

$$P = 10^{-pw/h} \quad \text{where } p = \sqrt{2m(V - E)}$$

and w is the width of the barrier. Here V is the height of the barrier in energy units and E

the energy of the particle. Because E is less than V, the particle could never get up the "hill" classically.

13. From the Special Theory of Relativity,

$$E^2 = (mc^2)^2 + (pc)^2$$

It follows that in the limit of a massless particle $(m = 0)$, $E = pc$.

14. The momentum and wavelength of a de Broglie particle are related according to the relation

$$\lambda = \frac{h}{p}$$

which, for a massless particle, is

$$\lambda = \frac{h}{E/c}$$

Again, from quantum mechanics, writing $E = hf$,

$$\lambda = \frac{h}{hf/c} = \frac{c}{f}$$

which is just the relation between frequency, wavelength, and velocity for any wave and, in particular, for electromagnetic waves.

15. Using the notation of Figure 10.13, from the conservation of energy,

$$hf = hf' + \tfrac{1}{2} mv^2$$

where m is the mass of the electron and v is the electron velocity. The energy hf' of the scattered photon, plus the kinetic energy of the recoiling electron, is equal to hf, the energy of the incoming photon. The angular relations between the scattered photon and the recoiling electron are determined from the conservation of transverse and longitudinal momenta:

$$mv \sin \theta = \frac{hf'}{c} \sin \theta' \quad \text{and} \quad \frac{hf}{c} = \frac{hf'}{c} \cos \theta' + mv \cos \theta$$

16. Correctly, it is the square of the angular momentum J that is exactly defined or quantized, and $J^2 = j(j + 1)\hbar^2$.

17. Actually, two photons may have different polarizations and then be different. Similarly, bosons and fermions with their spins in different directions can be differentiated experimentally, hence are different particles in the sense used here. Our discussion then considers identical particles as particles otherwise identical with their spins in the same direction.

18. The intensity of plane-polarized light transmitted by polaroid is proportional to $\cos^2 \theta$, where θ is the angle between the direction of the polarization and the direction of (maximum) transmission of the polaroid. If the incident intensity is I, the transmission for $\phi = 240°$, equal to the transmission for $\phi = 120°$, is $I \cdot \cos^2 \phi = I/4$. For individual photons the probability of transmission of the photon is $\cos^2 \phi$. The values of the table can be calculated with this information. For example, assume that the polarimeters are at different angles. Then the probability of transmission of a photon traveling to the left is 50%. If the photon is transmitted, the right photon must be polarized the same way and will be transmitted by the off-angle polaroid with a probability of 25%. Hence the probability of yy is $(0.5) \cdot (0.25) = 0.125$ and $ny = yn = (0.5) \cdot (0.75) = 0.375$.

19. A technical note is in order. As described by quantum mechanics, the two-photon system generated by the decay of the positronium state is a state of definite (odd) parity and definite (zero) angular momentum. This (complete) description does not define the separate polarizations of the two photons. The two-photon description cannot be factored into two one-photon descriptions. After the polarization measurement, there is no longer a two-photon system with definite parity and angular momentum, but rather two one-photon states.

11
The Atom—
A Quantum Laboratory

The study of atoms . . . not only has deepened our insight into a new domain of experience, but has thrown new light on general problems of knowledge.

Niels Bohr

The Nuclear Atom

The study of the atom, interesting in itself, is particularly important because the atom constitutes an especially simple system of particles that behave in a manner that cannot be described classically. The atom can then be considered a kind of laboratory for the understanding of quantum mechanics and composite systems. Nuclei, composed of neutrons and protons, and neutrons and protons, composed of quarks, have qualitative features similar to those of atoms made up of electrons and a nucleus.

Throughout the first years of the twentieth century, the atom was generally considered to be an indivisible particle, although analyses of the spectra of light emitted by atoms in magnetic fields demonstrated that the atom held electrically charged particles that had properties similar in some ways to free electrons. However, the entire atom was electrically neutral. Although the distribution of positive and negative electricity in atoms was unknown, it was a popular conjecture that the negative electrons were spread through some massive positive matrix, rather as raisins in a pudding.

All such ideas were swept away by the experiments of a remarkable New Zealand-born Englishman, Ernest Rutherford, who, while working at Cambridge in 1911, showed that most of the mass of the atom is contained in a small, positively charged nucleus and that the electrons must then form some kind of light outer shield.

Rutherford and his collaborators bombarded thin gold foils with alpha particles from naturally radioactive sources. For his purposes the alpha particles, which are the nuclei of helium atoms, could be considered simply as heavy, small, positively charged particles traveling at a considerable veloc-

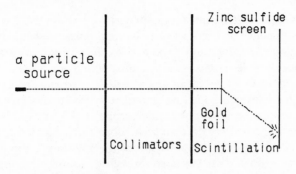

Figure 11.1 A schematic showing the character of the Rutherford scattering experiment.

ity—about 3% of the velocity of light. Rutherford believed that only electrical forces would act on an alpha particle as it shot through the atoms in the foil, and in this he was correct, because the nuclear forces play no important role. If the charges were spread out over the entire atom, even if in small, discrete pieces, he expected that the alpha particles would never be strongly deflected by the atoms—the atoms would be "soft" and semitransparent. But if the mass and charge were concentrated somewhere in the atom in a "hard" core, the alpha particles would sometimes be scattered at large angles.

Figure 11.1 illustrates Rutherford's experimental design. Alpha particles

Figure 11.2 A sketch suggesting the character of the Rutherford nuclear atom. The density of points in the outer atom is meant to show qualitatively the density of electrons in that region. The diagram is not to scale; the nucleus drawn to scale would be about 200 times smaller than shown and visible only with a microscope.

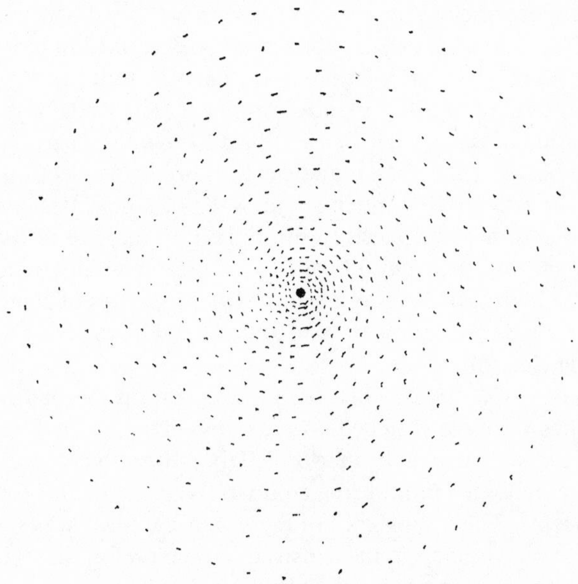

from a natural radioactive source (radium derivatives) were used to bombard a thin gold foil; gold was used because the total positive charge of the gold atom was known to be large, and it was easy to make very thin gold foils. The scattered alpha particles struck screens coated with a zinc sulfide compound that scintillated when struck by an alpha particle. The large-angle scattering of the alpha particles could then be seen directly by the observers watching the scintillations through low-power microscopes.

Such large-angle scatterings were seen, albeit rarely. Knowing the thickness of the foil and the approximate size of the gold atom, Rutherford calculated that each alpha particle passed through thousands of atoms as it traversed the foil. However, only about 1 alpha particle in 10,000 was scattered strongly; hence, the nucleus was surely very small. Rutherford's detailed analysis of the distribution of the angles of scattering of the alpha particles demonstrated that the scattering was just that which would be expected if the force on the alpha particle was the electrostatic force that would be exerted by a small atomic nucleus, 10,000 times smaller than the atom, which carried all the positive charge and almost all the mass of the atom. Figure 11.2 suggests the character of the atom made up of a small, heavy nucleus surrounded by a light cloud of electrons.

The Bohr Model of the Hydrogen Atom

A single proton forms the nucleus of the hydrogen atom, the lightest and simplest atom, and a single electron comprises the outer region. The hydrogen atom is thus a compound of a proton, holding a positive charge, and an electron, carrying an equal negative charge, held together by electrostatic forces. Most of the mass of the atom is held by the proton, which has a mass about 1835 times that of the electron.

We have seen that a particle in a box must be described in terms of standing de Broglie waves if the momentum of the particle multiplied by the length of the box is on the order of h, Planck's constant. The walls of the box constrain the position of the particle by exerting a force on the particle. The electrostatic force between the proton and electron holds the electron in a small region about the proton very much as the walls of a box. Because the characteristic momentum of the electron multiplied by the size of the "box" or atom is small and of the magnitude of h, we must eventually understand the structure of the hydrogen atom in terms of the properties of standing waves of the electron inside the region allowed and defined by the force between the proton and the electron.

However, in 1913, a decade before formal quantum mechanics—and the wave mechanics we have sketched—were constructed, Niels Bohr, a young Danish physicist working in Rutherford's laboratory, developed an understanding of the hydrogen atom using comparatively elementary, if heuristic, quantum concepts. The importance of these seminal ideas was so great, and the approximations implicit in the constructions so useful, that we must now review the simple basic structure of Bohr's developments.

Following Bohr, we presume that the atom can be described in terms of an electron rotating about a proton in a circular orbit. Because the proton is so much heavier than the electron, we can consider that the proton is at the center of this orbit of radius a. The force F_e between the electron and proton is expressed simply by Coulomb's law

$$F_e = \frac{e^2}{a^2}$$

where e is the unit charge, the electron charge is $-e$, and the proton charge is $+e$. From Isaac Newton, the centrifugal force F_c, which must be equal to the electrostatic force if the orbit is to be stable, is

$$F_c = \frac{mv^2}{a} \quad \text{and then} \quad F_e = F_c \quad \text{and} \quad \frac{e^2}{a^2} = \frac{mv^2}{a}$$

where m is the mass and v the velocity of the electron. With simple algebra[1] we can then determine the total energy of the system (the sum of the potential energy and kinetic energy) as

$$E = -\frac{1}{2}\frac{e^2}{a}$$

This energy is called the *binding energy*. The negative sign means that energy must be added to free the particles from one another; hence, the calculated energy is the energy binding together the particles (in this case, the electron and proton).

All this is classical physics similar to that employed by Newton in his analysis of the motion of the moon around the earth. But here Bohr introduced a bold conjecture. He assumed that only specific, discrete, angular momenta were allowed; angular momentum must be quantized and equal to integral multiples of Planck's constant divided by 2π. In the language of concepts developed later, the circumference of the orbit must be equal to an integral number of de Broglie wavelengths. Then

$$n\lambda = 2\pi a \quad \text{and since } \lambda = \frac{h}{mv}, \quad mva = n\hbar$$

where n is an integer, $n = 1, 2, 3, \ldots$.

Using the classical expression for the energy of the system and the relation expressing the quantization of the angular momentum, we can follow Bohr and develop further consequences of the *Bohr model of the hydrogen atom* with a little algebra.[2] We find that the energy is *quantized;* that is, only discrete energies are allowed. The lowest energy (for the most strongly bound state) is 13.6 electron volts[3] (eV). The next state is bound one-fourth as strongly, the next one-ninth, and the next one-sixteenth. Generally, the binding energy of the nth state is $(13.6 \text{ eV})/n^2$.

The radii of the orbits are also discrete. The smallest orbit, corresponding to the most strongly bound state, has a radius of $0.53.10^{-8}$ centimeters. The next most tightly bound state has a radius four times larger, the next is nine times larger, and so on. In general, the radius of the orbit for the nth most strongly bound state is $n^2 \cdot 0.53 \cdot 10^{-8}$ centimeters.

The Hydrogen Atom According to Wave Mechanics

In the decade following Bohr's introduction of his model of the hydrogen atom, this description of the atom was extended to more complex systems through procedures of "classical quantum mechanics"—a set of ingenious recipes for extending the use of quantum concepts—although a consistent unifying theory did not yet exist. In particular, elliptical orbits were introduced such that a set of orbits could have the same energy, proportional to the square of the quantum number n, but with angular momenta less than $n\hbar$. However, after the development of formal quantum mechanics, including the wave mechanics we have touched upon, an almost completely satisfactory description of the hydrogen atom in terms of standing de Broglie waves of the electron was quickly constructed.

In considering the standing-wave atom, we first show that quantum considerations preclude the possibility that an electron can be bound tightly to a proton by electrical forces to form a heavy neutral object, such as a neutron. An electron whirling around in an orbit with constant energy should escape falling into the proton in very much the same way that the earth escapes falling into the sun. A particle moving in a circle undergoes a centripetal acceleration toward the center—in Newton's words, the particle in orbit is always falling (accelerating) toward the center (but gets no closer as the tangential motion carries it away from center). But an accelerating, charged particle radiates electromagnetic energy. Classically, one must then expect that the electron would not stay in a stable orbit as Bohr required, but would spiral into the center as it loses energy through the radiation of light. (Similarly, the earth would spiral into the sun if it were to lose its kinetic energy by the radiation of gravitational waves. But gravitational forces are so weak that the radiated power from the earth of about 200 watts of gravitational wave energy is negligible compared to other effects.)

However, quantum mechanical considerations preclude such a fate for the electron. As we have noted, the potential energy of an electron in the electric field of the proton is inversely proportional to the distance between the proton and electron. Yet, if the electron is held very close to the proton (as in a very small box), the wavelength of the electron must be correspondingly small; the electron momentum, inversely proportional to the wavelength, must be large; and the kinetic energy of the electron, proportional to the square of the momentum, will also be large. If the box is reduced in size by a factor of 2, the energy of electrostatic attraction doubles but the kinetic energy increases fourfold.[4] If the kinetic energy is larger than the potential energy, the electron cannot be held in by the box. The smallest box that will hold the electron by electrostatic forces corresponds to a mean distance between electron and proton of about $a_0 = 0.53 \cdot 10^{-8}$ centimeters, the Bohr radius. The electrostatic forces between the proton and electron are not strong enough to hold the electron any closer.

Although detailed analyses of the standing-wave description of the hydrogen atom require more mathematical apparatus, the understanding of three-dimensional standing waves in "boxes" with different kinds of walls was well advanced even in nineteenth-century mathematics and physics. Hence, the analysis of standing de Broglie waves describing electrons held to a proton by electrostatic forces was no great challenge to quantum physicists of the 1920s. Remarkably (and fortunately), classical calculations involving forces that vary inversely with the square of the distance, such as electrostatic forces, lead to almost exactly the same results as the correct quantum calculations. Both Rutherford's calculations of the scattering of alpha particles by the electrostatic field of the nucleus and Bohr's quasi-classical calculations concerning the hydrogen atom are then hardly modified by the considerations of a complete quantum mechanics. In particular, the energy formula for the energies of the different states (orbits) of the electrons is unchanged even though the physical description of the states is considerably modified.

It is difficult to make a useful drawing of a three-dimensional standing wave (requiring four dimensions, including the amplitude) on a two-dimensional surface. We therefore enlist a small mapmaker—a quantum Olivia—to map the amplitudes of particular electron standing waves about a proton. Figure 11.3 suggests the pattern of procedure. Olivia walks along three different paths recording the de Broglie amplitude with her ψ-meter and writes down her readings in her notebook, which she consults later to construct maps of the amplitude. At the right in Figure 11.3 are her maps of a particular standing wave. Upon inspection of this set of maps, Olivia finds that there are three nodal surfaces: the amplitude goes to zero on a sphere at infinity (where the radius r is infinite) and is zero on a smaller sphere with a radius

Figure 11.3 At the left are the paths along which Olivia, the quantum mapmaker, will construct maps of the de Broglie wave for an $n = 3, j = 2, m = \pm 1$ state of the hydrogen atom. The maps she makes are shown at the right.

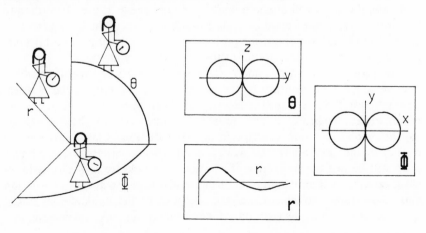

equal to that shown as a zero of amplitude on her map of the amplitude versus the radius. The x-z plane is also a null surface.

Later, after mapping a number of atoms and consulting a description of the Bohr model, Olivia realizes that the total number n of nodal surfaces corresponds to the *principle quantum number n* in the Bohr model, and the number of nodal surfaces in her angular coordinate maps corresponds to the angular momentum of the electron in its Bohr orbit. With three null surfaces, the principle quantum number is 3; with one null surface in angular coordinates, the angular momentum of the electron, j, is equal to 1 (unit of $\hbar$).[5]

Broad features of the map of the atom follow from general properties of waves in a spherically symmetric system. If the real Olivia had used a pressure meter to map the standing sound waves inside a hollow metal sphere several meters in diameter resonating with a certain frequency, she could have found a wave such that her maps for θ and ϕ were identical to those drawn by her quantum sister. And her map of the amplitude dependence on r would have been only slightly different. The resonant frequencies of this spherical drum would increase with the number of nodal surfaces very much as the energies of the hydrogen atom states increase with the number of such surfaces. However, for the spherical drum (unlike the atom), the frequency will also depend to some extent on the distribution of nodal surfaces between the radial and angular coordinates.

Electromagnetic Radiation from the Atom

At ordinary temperatures (such as 20° Celsius or 68° Fahrenheit), the average kinetic energy of atoms and molecules is about $\frac{1}{40}$ of an eV. Because the energy difference between the lowest energy state of the hydrogen atom (the ground state) and the first excited state (the state with the next lowest energy) is about 10 eV, 400 times larger, the atom will almost always be found in the ground state. However, if the atom is excited in some manner, such as through bombardment by high-energy electrons passed through hydrogen gas as an arc, the atom will often be found in higher energy states. According to the Correspondence Principle, the classical description of the hydrogen atom as an electron moving about the proton in a circular or elliptical orbit must provide an adequate description of those high-energy atomic states defined by large quantum numbers n and j, where n is the principle quantum number and the angular momentum of the electron is $j\hbar$. Moreover, the classical description of the electromagnetic energy radiated by the electron as it accelerates[6] toward the proton in orbit must correspond to (that is, be very similar to) the correct quantum description of that radiation.[7] Classically, the frequency of the electromagnetic waves radiated by an electron in a circular orbit will be just the frequency of rotation of the electron. Electrons in an elliptical orbit will emit radiation at the fundamental frequency of rotation and at overtone frequencies that are multiples of the fundamental. From quantum constraints, the electron can have only certain discrete energies and then can lose energy through the emission of radiation only in discrete amounts dE as it "jumps" from a higher to a lower energy state. From the Correspondence Principle, the frequency of that radiation f must correspond to the classical frequency of rotation of the electron in large orbits. By equating the energy difference $dE = E_n - E_{n-1}$ between adjacent states to the orbital frequencies, f_n and f_{n-1}, of the electron, we have[8]

$$dE = E_n - E_{n-1} = hf_{n,n-1}$$

where h is Planck's constant and $f_{n,n-1}$, the frequency of the electromagnetic wave or photon, is intermediate between the electron orbital frequencies f_n and f_{n-1}—which differ very little for large values of n. This is the relation between the energy and frequency of a photon, suggesting (correctly) that the electron emits photons as it changes from higher energy states to lower energy states (or spirals in from large orbits to smaller orbits).

If either (or both) of the classical orbits are elliptical—hence generating harmonics of the fundamental rotation frequency—the principle quantum numbers may change by more than one unit, and the foregoing relation can be expanded to the form

$$E_j - E_k = hf_{jk} \quad \text{where } j - k = 1, 2, 3, \ldots$$

Again, the photon frequency f_{jk} will be intermediate between the Bohr orbital frequencies of the electron f_j and f_k. The classical description of electromagnetic radiation is also a guide to other properties of the quantum emission of photons. The time[9] required for the transition from one orbit to another can be estimated from the classically calculated rate of energy loss; the distribution in direction of the photons corresponds closely to the distribution of radiated energy calculated classically; and the orientation of the spin of the photons corresponds, in a less transparent manner, to the polarization of the classical radiation.

From a classical or nonquantum view, an electron losing energy through radiation as it passes from an orbit with one energy to another with a smaller energy must have held every intermediate energy at some time during the transition, just as the possible energies of such an electron form a continuum. A value about equal to the average of the initial and final energies would be expected from a measurement of the electron energy at a time halfway through the transition. Although the energies available to the electron, when considered properly according to quantum mechanics, are discrete, the transitions are still continuous, although in a somewhat different sense. A measurement of the electron energy during the transition will give either the higher or lower energy, but the probability of finding the lower energy will increase continuously with time; at a time halfway through the transition, the probability of finding the lower energy will be about one-half.

This continuous change with time in the probability of finding the system in different states is reflected in the properties of the photon emitted during the transition. The wave nature of the photon—and the concomitant Uncertainty Principle—preclude a view of a photon emitted gradually during a transition; at no time will one-half of a photon have emerged. But if we proceed in a complementary fashion, considering wave properties rather than particle properties, by measuring interference between photons (as in the double slit interference pattern shown in Figure 10.2), we find that photons have a finite length and take a finite time to pass a point—a time corresponding to their emergence over a finite transition time! If mirrors are placed such as to lengthen the path from one slit so that it is much more than 1 meter longer than the path from the second slit, there will be no interference effects from incident light generated by typical atomic transitions. For most visible light, that photon "length" is about a meter, which corresponds to a typical transition time of about $3 \cdot 10^{-9}$ seconds.

A change of orbits in the Bohr model of circular orbits demands not only a change of energy but also a change of one unit $\hbar$ of angular momentum. To retain the conservation of angular momentum, we must assume that the radiation carries off that angular momentum. The emission and absorption of radiation is thus associated with the emission and absorption of angular momentum as well as energy. Just as a quanta of radiation contains an energy equal to hf, it carries an angular momentum equal to $\hbar$: the photon has an intrinsic spin of $\hbar$.

Energy Levels and Spectra

The set of energy levels of the hydrogen atom and the important radiative transitions from level to level are shown conveniently in Figure 11.4, which is an energy-level diagram. Energies of the different states (corresponding to classical orbits) are labeled by the principle quantum number n and the angular momentum quantum number j. Each line corresponding to a state of angular momentum j actually represents $2j + 1$ lines designating states with different values of the component of angular momentum in a specified direction. In the absence of a magnetic field, these states all have the same energy; if the atoms are subject to a magnetic field, the energies of the different members of the multiplet will have different energies.

The energy scale of Figure 11.4 is given in eV. From the relation $dE = hf$, a transition such that dE is large represents the emission of high-frequency light; a transition such that dE is small corresponds to low-frequency light. For orientation, if $dE = 1.8$ eV, the light will be red; the light will be violet for $dE = 3$ eV. The set[10] of frequencies (or wavelengths or photon energies) of radiated light from the hydrogen atom comprises the spectrum of light emitted by the atom.

Energy-level diagrams are convenient expositions of the different possible energies of systems. At the left in Figure 11.4, our researcher Olivia is climbing about on a set of gravitational energy levels (perhaps part of an Inca temple complex) that extends below the surface. If she disturbs a rock from the second level on which she is standing so that the rock falls to the first (ground) level, she might decide (as a physicist) that the rock had made a transition

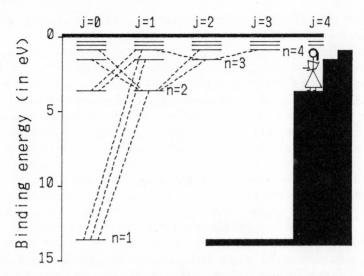

Figure 11.4 Lower energy levels of the hydrogen atom, together with the transitions between the levels. The solid lines represent allowed transitions where the angular momentum of the atom can change by one unit of $\hbar$. The principle quantum number is denoted by n, whereas the orbital angular momentum number is j. At the right is a similar macroscopic set of gravitational energy levels.

from the first, excited level to the ground level. As she stands on the second level, she need not exert too much energy to reach the surface—her binding energy is small. If she were on the "ground" level, much further below the surface, she would have to exert considerable energy to reach the surface—her binding energy would be large.

Because the photon carries off one unit of angular momentum, only transitions between states that differ in angular momentum by one unit are allowed; the selection rule is $\Delta j = \pm 1$, where $\hbar \Delta j$ is the change in the angular momentum $j\hbar$. The other transitions do occur, but the probability of these "forbidden" transitions is much reduced.

Heavy atoms with a large nuclear charge hold many electrons in "orbits." Because the force on an electron in any particular orbit derives from the nucleus and the other electrons, it is most difficult to describe heavy atoms accurately. For many purposes, and for atoms that are not too heavy—and composed of too many electrons—it is useful to consider that the electrons fill the Bohr orbits sequentially. Because the electrons are fermions and two fermions with their spins in the same direction cannot occupy the same state, only two electrons—with opposite spin directions—occupy each orbital state. Hence, as many as $2n$ electrons can occupy the nth quantum level, after which the next electron must go to a higher level. From this view there are closed "shells" after $2 \cdot 1 = 2, 2 + 2 \cdot 3 = 8$, and $8 + 2 \cdot 5 = 18$, electrons are added to a nucleus.

This is a simplistic model of the atom. When the interactions between the electrons are considered in more detail, the description becomes greatly complicated. Although the forces between electrons and between electrons and the nucleus are known quite precisely, and although the (quantum) mechanics is precisely defined, the many-body problem calculations required for the understanding of atoms, molecules, and chemistry are quite intractable. Exact solutions are far beyond any mathematics we have in hand, and the complexity of exact numerical calculations of the simplest properties of molecules would test any conceivable computer even if it were to run for all the time left for the universe. Hence, although chemistry is grounded in the physics of electromagnetism and quantum mechanics, it remains an art in itself aside from physics.

Even as an excited atom emits light of specific frequencies in the course of transitions that return the atom to the ground state, atoms in the ground state—and excited states—absorb light of specific frequencies in the course of transitions from the ground state to excited states. Even in the outer photosphere of stars, such as the sun, the temperatures are not such as to propel many atoms into higher states. Therefore, the dominant absorption occurs through transitions from the ground state to higher energy states. For hydrogen, the frequencies associated with such transitions are in the ultraviolet. But for many heavier atoms, such transitions are in the visible, marking the continuous spectrum from the stars with dark Fraunhofer absorption lines named after the early nineteenth-century optician who first noticed them.

Analyses of the spectral lines that mark the light from far-off stars—and from far-off galaxies of stars—lead to determinations of the constitution of stars and galaxies and conditions (temperatures and pressures) of the emission of the light. In particular, the observation of the shift toward the red of certain well-defined absorption lines (especially a pair of lines from absorption by calcium atoms) is especially important inasmuch as the magnitude of the shift is an indication of the velocity of recession of the objects—galaxies and quasars—that emit the light.

Notes

1. The potential energy of the electron at a distance a from the proton is

$$V = -\frac{e^2}{a}$$

where e is the charge of the electron. From the equality of the electrostatic force and centrifugal force,

$$\frac{e^2}{a^2} = \frac{mv^2}{a}$$

And then, multiplying both sides of this equation by $a/2$, we have

$$\frac{1}{2}\frac{e^2}{a} = -\frac{V}{2} = \frac{1}{2}mv^2 = T$$

where T is a common notation for the kinetic energy. The total energy of the system, E, is equal to the sum of the kinetic and potential energies:

$$E = T + V = -\frac{e^2}{a} + \frac{1}{2}mv^2 = -\frac{1}{2}mv^2 = -\frac{1}{2}\frac{e^2}{a}$$

2. Squaring the angular momentum,

$$m^2v^2a^2 = n^2\hbar^2 = 2ma^2\left(\frac{1}{2}mv^2\right) = 2ma^2\left(\frac{1}{2}\frac{e^2}{a}\right) = e^2ma$$

Dividing the first and last member of the chain of equalities by m^2v^2a, we find for the radius of the orbit

$$a = \frac{n^2\hbar^2}{me^2}$$

Using the relation for the radius a, the energy is found to be

$$E = -\frac{1}{2}\frac{e^2}{a} = -\frac{1}{2}\frac{e^4m}{n^2\hbar^2}$$

3. For layman or scientist, it is difficult to relate atomic units to human experience. The electron volt, defined as the energy required to move an electron through a potential difference of 1 volt, is equal to $1.6 \cdot 10^{-19}$ joules. A small amount of energy on a human scale (although under proper conditions, the human eye can detect a flash of light carrying an energy as small as 10 eV), 1 eV is substantial on an atomic scale. The mean kinetic energy of atoms at the surface of the sun is about 1 eV; the energy required to turn hot water to steam is about ½ eV per atom of water.

The atomic unit of length of 10^{-8} centimeters (one Angstrom) is even more elusive. The smallest distance that can be resolved under a high-power microscope is as much larger than an atom as a football field is larger than an ant crossing the field.

4. If we estimate the wavelength as $\lambda = r$, the momentum will be $p = \hbar/\lambda$ and the kinetic energy

$$T = \frac{p^2}{2m} = \frac{\hbar^2}{2m\lambda^2} = \frac{\hbar^2}{2mr^2}$$

Requiring the total energy $T + V$ to be zero (or less) so that the electron is bound, $T = -V$ where $V = -e^2/r$, and

$$\frac{\hbar^2}{2mr^2} = \frac{e^2}{r} \quad \text{and} \quad r = \frac{\hbar^2}{2me^2} = a_0$$

This rough estimate of the size of the smallest possible compound of proton and electron gives the Bohr radius.

5. Although the energy of a state or orbit depends only on n, the number of nodal surfaces and the form and placement of the surfaces depend on the angular momentum of the electron and then the angular momentum quantum numbers. Listed below are relations among quantum numbers.

(a) For a given value of the principle quantum number n, there will be n states with values of angular momentum j running from $0\hbar$ to $(n - 1)\hbar$.

(b) For a given value of the angular momentum j, there will be $2j + 1$ states with values of the component of angular momentum j_z in a chosen direction of quantization (taken here as the z-direction, for definiteness) ranging from $-j\hbar$ to $+j\hbar$.

For the sake of completeness, we add a third rule, which concerns the spin and statistics of the electrons.

(c) Each such state, defined by quantum numbers n, j, and j_z, will be divided further into two states, one such that the electron spin is in the z-direction of quantization, the other with the electron spin in the opposite direction.

6. The acceleration a is equal to F/m, and $F = e^2/r^2$ from Coulomb's law.

7. The energy radiated per second by a charge e moving with acceleration a is equal to

$$\frac{dW}{dt} = \frac{2}{3}\frac{e^2a^2}{c^3}$$

where c is the speed of light.

8. The energy difference is

$$dE = E_n - E_{n-1} \qquad \text{where } E_n = -\frac{1}{2}\frac{e^4m}{n^2\hbar^2} \qquad \text{and} \qquad E_{n-1} = -\frac{1}{2}\frac{e^4m}{(n-1)^2\hbar^2}$$

Then

$$dE = \frac{1}{2}\frac{e^4m}{\hbar^2}\left(\frac{1}{(n-1)^2} - \frac{1}{n^2}\right)$$

Rewriting the quantity in the parentheses,

$$\frac{1}{(n-1)^2} - \frac{1}{n^2} = \frac{2n-1}{n^4 - 2n^3 + n^2} \rightarrow \frac{2n}{n^4} = \frac{2}{n^3} \qquad \text{for large } n$$

and in the classical limit, when n is large,

$$dE = \frac{e^4m}{n^3\hbar^2}$$

From the quantization of the angular momentum

$$mva_n = n\hbar \qquad \text{or} \qquad v = \frac{n\hbar}{ma_n} \qquad \text{where } a_n = \frac{n^2\hbar^2}{me^2}$$

The frequency f is equal to the velocity of the electron divided by the distance traveled in one revolution, which is the circumference of the orbit. Then

$$f = \frac{v}{2\pi a_n} = \frac{n\hbar}{2\pi ma_n^2} = \frac{e^4m}{2\pi n^3\hbar^3}$$

And, using the expression written previously for dE,

$$dE = 2\pi\hbar f \qquad \text{or} \qquad dE = hf$$

From a rather simple calculation we then find that the energy of the atom is lost in discrete amounts or *quanta* equal to hf, where f is the frequency of revolution of the electron, hence the frequency of oscillation of the electromagnetic field.

9. The transition time can be estimated using the Correspondence Principle and the results of expressions in the previous footnotes. The lifetime of a state—and the photon—will be

$$\frac{E_n - E_{n-1}}{dE/dt} \qquad \text{where} \qquad E_n - E_{n-1} = \frac{e^4m}{n^3\hbar^3} \qquad \text{and} \qquad \frac{dE}{dt} = \frac{2e^2A^2}{3c^3}$$

where A is the centripetal acceleration equal to e^2/ma^2. Putting everything together and noting that the orbit radius is $a = (n^2\hbar^2)/me^2$,

$$\tau = \frac{3c^3 n^5 \hbar^6}{2e^{10}m} \approx n^5 \cdot 10^{-10} \text{ seconds}$$

This pedagogical discussion has been limited to the analysis of radiation from electrons moving in perfectly circular orbits in the classical limit of the Correspondence Principle. Even for classical orbits, some radiation will be emitted at higher frequencies, harmonics of the fundamental, if the orbits are elliptical. This radiation would correspond to changes of several units of the principle quantum number n in the Bohr model, just as the frequency would be several times the frequency of the fundamental transition where n changes by 1.

10. In detail,

$$dE_{ik} = E_1 \cdot \left(\frac{1}{n_i^2} - \frac{1}{n_k^2} \right) = hf_{ik} \qquad \text{where } E_1 = \frac{e^4 m}{2\hbar^2}$$

Small excited atoms can lose energy through collisions as well as through radiation. If the frequency of collisions is high—as it is for a gas at high pressure—the characteristic time for radiation will be sharply reduced, and as a consequence of the Uncertainty Principle, $\Delta E \cdot \Delta t \approx \hbar$, the photon energy will be very much broadened. The characteristic discrete spectrum will be transformed to a continuous spectrum described better by thermodynamics than by atomic physics.

12

Fundamental Particles and Forces— An Introduction

> We want to describe how we stand today in the age-old attempt to explain all of nature in terms of the workings of a few elements in infinite variety of combinations; in particular, what are the elements?
>
> Richard Feynman

The Fundamental Forces and the Elementary Particles

The study of the electromagnetic field has provided some insight into the relationships between forces and particles. In particular, it appears that the mass of particles is partially or wholly derived from the energies of the fields for which the particles serve as a source; the mass of the electron appears to be largely the mass corresponding to the energy of the electromagnetic field generated by its charge. From this view, a particle is essentially a source of force, or a source or singular point in a field; implicitly, the field is the primary concept. Conversely, a field of force can be considered to consist of particles, just as the electromagnetic field can be viewed as a spectrum of photons emitted and absorbed by charged sources. Thus, the concept of the field might be considered a secondary device for describing the behavior of a set of particles.

Furthermore, it is evident that no operational meaning can be given to a particle except in terms of its interaction with other particles and thus in terms of the forces that act on the particle. If a particle were acted on by no force whatsoever, the particle would be undetectable in principle and could have no physical meaning. A force or field that acted on no particle would be equally meaningless. Clearly, then, inquiries into the nature of particles and the nature of the forces that act on them, or of the fields that act as a description of those forces, are joined. We cannot consider particles and forces or fields separately in any sensible manner.

If we analyze the force between two protons, we find that we can separate the total force into four different parts in a meaningful way. There is a gravitational force, an electromagnetic force (once considered two forces, electrostatic and magnetic), a weak interaction force, and a strong interaction force—where "weak" and "strong" are names, not adjectives. Of course, we might hope that there is but one true force between the two protons and it is Man who has separated God's one force into separate bits because men do not yet understand God's connection between pieces. But we are learning, and most physicists believe that we are at a threshold of a unified theory of fundamental forces. Maxwell and Faraday unified the electric and magnetic forces; recently Steve Weinberg, Abdus Salam, Sheldon Glashow, and others have largely unified the weak and electromagnetic forces. There is now reason to believe that we are at the threshold of a unification of the electro-weak forces, the strong interaction forces, and gravity into a Grand Unified Theory (actually a collection of theories denoted by the unlovely acronym GUTS). There has been less progress in bringing gravity securely into the fold. Unified theories or no, we will begin here in a more primitive light by treating separately the forces we have named while recognizing their important similarities.

Consideration of the properties of fields such as the electromagnetic field requires a theoretical structure consistent with the tenets of quantum mechanics and relativity. As a consequence of efforts extending over more than half a century, a powerful quantum field theory has evolved that largely fulfills these requirements. In spite of considerable efforts, however, not all the problems associated with this description of nature have been solved. Given the degree of self-consistency of quantum field theory and its success in describing electromagnetism, physicists are confident that present field theory provides an accurate description of a large, albeit limited, region of nature. However, we are forced to use the theory far beyond the region of experience over which our observations support the model, since it is the only tool we have.

Although there is longstanding interest in providing an axiomatic basis for quantum field theory (as Euclidean geometry is axiomatic), that program is far from complete, and field theory is customarily approached from a variety of directions. In our discussion of the different forces of nature, we will very often use language and pictures related to the heuristic field theory methods developed by Richard Feynman. Later, when we consider efforts designed to create the ultimate Grand Unified Theory, God's One Equation—we will place more emphasis upon the *Lagrangian* formulation of that Equation. In these discussions, it is important to keep in mind that the pictures and language we use are at best but a rough description of complex calculations based on well-defined precepts.

Electromagnetism as a Paradigm

The electromagnetic forces between two charged particles can be described in terms of an exchange of photons: one charge emits the photon and the other absorbs the photon, or vice versa. The character of the electromagnetic field is implicit in the properties of the photon: the photon carries no electric charge, has a spin of $1\hbar$, and has a rest mass of zero. The diagram at the left of Figure 12.1 shows this exchange of a photon between two charged particles (taken here, for definiteness, as two electrons) in a schematic way; it is called a world diagram or, more commonly, a Feynman diagram. It illustrates the path in time of particle configurations where the sole coordinate, time, is usually used only to express the order of events and has no quantitative meaning. At the bottom of the diagram, at the earliest time, there are two electrons. Later, moving forward in time and upward on the diagram, one electron emits a photon. For a brief time there are two electrons and one photon in existence. At a slightly later time, the second electron absorbs the photon and there are again—at the top of the diagram—just the two electrons, although with somewhat different momenta as a consequence of the exchange.

The force between the two electrons can be considered to derive from the momentum transfer arising from the exchange of the photon, and the charge of the electron is a measure of the probability of the emission or absorption of the photon. Hence, the force between two electrons, which is proportional to the rate of momentum transfer between the two particles, is proportional to the square of the charge of the electron.

As a homely analogy, not to be pushed too far, consider two skaters on ice throwing a heavy snowball back and forth between them. As a consequence of their actions, they will be propelled apart—by the recoil when they throw, by the impact of the catch when they are on the receiving end. There will be an effective force F between the skaters derived from the exchange of the snowballs.[1] Moreover, that repulsive force will be proportional to the rate of exchange of the snowballs (in snowballs thrown and caught per second).

At the right in Figure 12.1, a Feynman diagram describes the exchange of a snowball between the two skaters. At the bottom of the diagram, at an early time, only the two skaters are shown. At a later time, the skater at the left

Figure 12.1 At the left, a Feynman diagram shows the exchange of a photon, shown as γ (gamma), between two charged particles (here, electrons). The ordinate is the time coordinate. At the right, a comparable diagram describes a macroscopic system of two ice skaters exchanging a snowball.

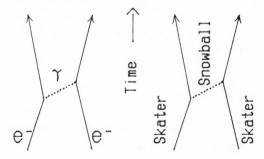

"emits" a snowball. For a brief time there are two skaters and a snowball in existence. Then the skater at the left "absorbs" the snowball, after which, at the top of the figure, there are again only the two skaters—however, with somewhat different momenta. No meaning is to be attributed to the distances between the lines representing the skaters or the precise times indicated; only the time order of the configurations has meaning.

The strength of the field is defined by the magnitude of the electric charge, that is, the strength of the coupling between the photon and charged source. This coupling strength is conveniently defined in terms of a dimensionless charge strength

$$\frac{e^2}{\hbar c} \approx \frac{1}{137} = \alpha$$

where e is the ordinary electric charge and α is a standard notation or abbreviation for the quantity $e^2/\hbar c$. This ratio has the form of a pure number and is thus independent of choices of units. A civilization residing in a far-off part of the galaxy will find the same value for it as we do. We are not certain why charge is found only in units of the electron charge, although if single magnetic poles (magnetic monopoles) exist, quantum mechanics requires such a quantization of charge—and of pole strength. (Monopoles are discussed in Chapter 13.) We do know that there are two kinds of charge, and we are able to fit that fact neatly into a description of electromagnetic forces in terms of vector fields. There can be sources emitting electric field lines and sinks absorbing electric field lines corresponding to positive and negative charges.

The value of the charge is a measure of the probability of emission or creation of a field quanta or photon by a charge; equally, it is a measure of the probability of absorption or annihilation of such a photon by a charge. Photons, and field particles in general, are not conserved. However, charge appears to be conserved. In any closed system, the total positive charge minus the total negative charge seems to be an invariant.

Let us review the process in terms of our imperfect metaphor of skaters and snowballs. The skaters are source particles (such as electrons) and the snowballs are field particles (such as photons). Source particles have properties similar to those we normally associate with matter, and we will sometimes call these "matter" particles. Similarly, we will sometimes call the field particles that result from the quantization of the fields of force "force" particles. The rate of throwing—and catching—snowballs by each skater is proportional to the square of the effective (electric) charge. The snowballs are not conserved; the skaters scoop up snow to make them and discard the fragments they catch. But the two skaters are conserved; they never leave the ice and never stop throwing and catching. In general, source (or matter) particles are conserved, and field (or force) particles are not.

The fundamental description of the electromagnetic force is now largely complete. Maxwell's equations are contained implicitly in that description. If we know quantum field theory—that is, quantum electrodynamics (QED)—we should be able to calculate the properties of all electromagnetic processes if we know how to do the calculations. Indeed, most processes can be calculated with great accuracy, but this is largely the case because the charge is small ($\alpha = \frac{1}{137}$, very much less than 1), thereby allowing certain mathematical approximations. Nevertheless, the accuracy of electromagnetic calculations is impressive.

Consider, for example, the calculation of the gyromagnetic ratio for the electron. This is the ratio of the magnetic moment of the electron and the angular momentum. Electrons are charged and they spin, hence they can be expected to generate a circumferential current that in turn generates a magnetic field; the electron is then like a small bar magnet. The magnetic moment of such a magnet, defined as the product of the pole strength multiplied by the distance between the poles, can be written as $g(e/2mc)$, where g, a dimensionless number, is calculated to be

$$g/2 = 1.001\ 159\ 652\ 460$$

to be compared to the measured result of

$$g/2 = 1.001\ 159\ 652\ 200$$

where the difference between the results of about 2 parts in 10^{10} (equivalent to an error of about a millimeter or $\frac{1}{32}$ of an inch in a determination of the distance from New York to Paris) follows largely from experimental uncertainties.

Physicists believe that all the force of nature can be described in much the same manner as electromagnetic forces. For the strong nuclear interactions, the weak nuclear interactions, and even for gravity, the forces can be considered to follow from the exchange of nonconserved field particles (counterparts to the photon) by conserved force particles (counterparts to the electron), where the rate of exchange is defined by a charge (such as the electric charge).

The Strong Interaction

The major part of the force between the two protons results from the strong interaction. It is not the protons per se that are sources of the strong interaction; the source is found in pointlike *quarks,* which can be considered the fundamental constituents of the proton. The quarks are especially curious because they hold discrete electric charges that are smaller than the smallest free charge we have seen, the charge of the electron or proton. In particular, quarks hold charges of $\pm e/3$ and $+2e/3$, where e is the charge of the (anti) electron; antiquarks hold opposite charges. We find that quarks also hold strong interaction charges and generate strong interaction forces very much as electric charge generates electromagnetic forces. Quarks exert forces on other quarks through the exchange of *gluons.*

Quarks are source or matter particles, and gluons are field or force particles that play a role in the strong interactions similar to that of charged particles and photons in electromagnetic forces. Figure 12.2 shows such an exchange generating a force between two quarks. Moreover, like photons, gluons seem to have zero rest mass and a spin of 1 (unit of $\hbar$). However, there are important differences between photons and gluons. Whereas photons are neutral electrically—they do not hold an electric charge—gluons, although electrically neutral, do carry strong interaction charges and produce strong interaction fields. And there are important differences between the strong interaction charges and electrical charges, as well as important similarities. Although there is but one electric charge (and its negative), there appear to be three different, but related, strong interaction charges, together with their negatives. And although there are no electrical forces between photons that do not carry an electrical charge, there are forces between gluons; gluons carry strong interaction charges.

The strength of the strong interaction force between the two quarks is probably not much different than the electromagnetic force between two electrical charges at very small distances; however, as a consequence of effects connected to the charge of the gluons, the force is much greater than electrical forces at larger distances (of the size of the nucleons—that is, 10^{-13} centimeters). The effective strength of the strong charge g can be expressed as

$$\frac{g^2}{\hbar c} \approx 0.2$$

Hence, the effective strong interaction charge strength is about 5 times larger than the electromagnetic force, and the square of the charge is 25 times greater.

Like electrical charge, the strong-interaction color-charge is conserved by the strong interactions (but not by weak interaction effects). Again, quarks are conserved just as electrons are; indeed, each of the six varieties of quarks is conserved by the strong interactions. And, like photons (the field quanta of the electromagnetic field), gluons (the field quanta of the strong fields) are created and annihilated in emission and absorption processes. Moreover, like photons, gluons have a spin of 1 (unit of $\hbar$). However, because of the strong forces between these particles, it seems that they cannot escape each other;

Figure 12.2 Feynman diagram showing the exchange of a gluon between two quarks.

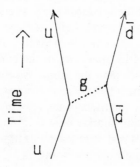

quarks and gluons appear to be *confined,* and we do not find free quarks. (A persuasive experiment suggests that free quarks may exist but are very rare. If they exist at all, they must make up less than 1 quark in 10^{19} nucleons.) As a consequence of containment, although an accelerating electric charge radiates photons, an accelerating quark cannot radiate gluons.

Because of the strength of the quark-quark forces, the strong interaction field is more complicated than the electromagnetic field and less well understood. Moreover, as confinement largely precludes the transfer of individual quarks and gluons between the composite particles found in nature, called elementary particles, the forces between these particles derive largely from the exchange of sets of quark-antiquark combinations called *mesons.* It is then customary, and very useful, to discuss the strong interactions in terms of a field theory of composite particles. The sources of the fields are then particles like the proton and neutron and are called, generically, *baryons;* they are made up of sets of three quarks. In this picture the field quanta are bosons made up of sets of one quark and an anitquark (mesons). The sets of bosons and mesons that take part in the strong interactions are labled *hadrons.*

Figure 12.3 shows the exchange of a *pion,* or π-meson, between two protons contributing to the force between the protons. In this description of nuclear forces, the strong interaction force between the two protons follows from the exchange of mesons, primarily pions, between the protons. The mesons have various integral spins and various rest masses. The pion, which is the lightest meson, has a rest mass (of 139 MeV/c^2) about 270 times that of the electron (0.511 MeV/c^2) and about ⅐ the mass of the proton (940 MeV/c^2). (Here, MeV is the conventional symbol for 1,000,000 eV; a GeV is 1000 MeV. From the relation $E = mc^2$, MeV/c^2 is a unit of mass.)

Because the pion has zero spin rather than 1 unit, the character of the force is somewhat different than for electromagnetism; however, the effective strength is rather large and can be described by an effective charge g such that

$$\frac{g^2}{\hbar c} \approx 1$$

Some of the limitations of this kind of field description of hadrons are seen when the "size" of the hadrons is considered. Like other composite particles such as atoms or nuclei, hadrons have a definite spatial extent. Measurements of the scattering of electrons off nucleons and mesons show that the charge distribution of these hadrons can be characterized by a radius of about 1 fermi (or 10^{-13} centimeters). This is a reasonable measure of the mean spacing of the quarks that make up the hadrons. Then, just as atoms cannot be considered simple pointlike particles in the consideration of phenomena that probe distances smaller than the atomic size of about 10^{-8} centimeters, the structure of hadrons must be considered if distances smaller than a fermi are important. The quarks themselves are very much smaller: the scattering of high-energy electrons off quarks shows no spatial extension at all; the charge is concentrated in a region at least 100 times smaller than the size of a hadron, so the quark is effectively a point.

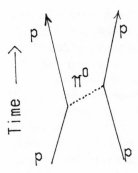

Figure 12.3 A diagram shows the exchange of π^0 between two protons contributing to the nuclear force between the protons.

The Weak Interaction

Like other forces, the fundamental *weak interaction* force can be described in terms of the emission and absorption of field quanta by source particles. The weak interaction source or matter particles are quarks and *leptons*.[2] Quarks and leptons are spin $\hbar/2$ fermions that differ inasmuch as the leptons have integral values of electric charge (historically defining what is meant by an integral electrical charge!) and no strong interaction charge, whereas the quarks have electrical charges of $\frac{1}{3}$ and $\frac{2}{3}$ of the unit electronic charge and carry strong interaction charges. However, leptons and quarks hold the same weak interaction charge. Leptons come in families (as do quarks); the lightest set is composed of an electron (e^-) and an electron-neutrino (ν_e) together with their antiparticles, the positron (e^+) and the antielectron-neutrino ($\bar{\nu}_e$). The positive and negative muons (μ^+, μ^-) and their associated neutrinos (ν_μ, $\bar{\nu}_\mu$) form another set. Relatively recently, we have learned of a third set, the positive and negative tau leptons (τ^+, τ^-) with their associated neutrinos ($\bar{\nu}_\tau$, ν_τ). The mass of the muon is about $\frac{1}{9}$ the mass of the proton (or 105 MeV/c^2), and the mass of the tau is about 1.5 times the mass of the proton (or about 1.5 GeV/c^2). The neutrino masses are quite small, perhaps (probably?) zero, and the neutrinos are electrically neutral, whereas the other leptons have integral charges.

The spatial distribution of the electric charge held by electrons, muons, and quarks has been measured with resolutions better than 0.01 fermi (10^{-15} centimeters), and there is no indication of any spread. The electron and the other leptons—and the quarks—are then taken as pointlike with no discernible structure.

The field quanta or force particles of the weak interaction are massive. There are three different field quanta (discovered in 1983 at CERN in Geneva by Carlo Rubia, Simon van der Meer, and others), which we call W^+, W^-, and Z^0, where the superscript represents their charge (in units of the electron charge). The W-mass is about 80 proton masses (or 75 GeV/c^2), whereas the Z-mass is about 100 proton masses (or about 95 GeV/c^2). These particles then have rest masses greater than that of a copper atom, which has a mass equal to about 64 protons (or about 60 GeV/c^2)! From the character of the processes observed (such as nuclear β-decay), we know that these *intermediate vector bosons* have an intrinsic spin of $1\hbar$ and then, like photons, are bosons and (again like photons) are associated with vector fields. The force between the two leptons or quarks derived from the exchange of a weak interaction field quanta will be described adequately by Figure 12.4 *if the two source particles are very close together.* Because of the great mass of the field quanta, the importance of the weak interactions at lower energies is found in processes described by more complicated diagrams that we will consider later.

Although the weak interaction is quite weak except at very high energies such that the energy of collision is about equal to the rest energy of the W-particle, the weak charge f is not extremely small, and the coupling strength is then not especially small:

$$\frac{f^2}{\hbar c} \approx \frac{e^2}{\hbar c} = \frac{1}{137}$$

The similarity between the magnitude of the weak charge and the electrical charge is not accidental and follows from the deep relation or equivalence between the weak and the electromagnetic forces. The striking difference between these forces stems from the difference between the great mass of the weak interaction field quanta and the zero rest mass of the photon. Although an accelerating electric charge emits photons, an accelerating weak interaction charge can emit W or Z particles only if very large energies are available to create the heavy particles. Also, the large mass of W and Z limits the weak force to an extremely small range.

The Gravitational Interaction

We have described the gravitational interaction in geometric terms as a consequence of the distortion of space by the presence of mass. This description

Figure 12.4 The exchange of W and Z bosons between quarks and leptons. Note the exchange of charge and identity accompanying the exchange of the boson.

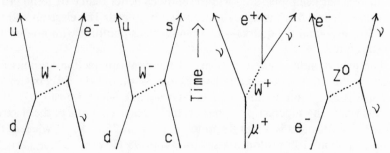

is a "classical" description inasmuch as quantum effects are not included. Because gravitational effects are on a scale that is very large compared to the size of Planck's constant h, quantum effects are unimportant for all applications of gravity that are now known and measurable. However, physicists do not really believe that gravity can be deeply and fundamentally separate from the other forces; therefore, it is important to be able to cast gravity into the same quantum mechanical form used to discuss the other forces if we are to proceed successfully toward the development of a unified description of all forces. In such a program, we must describe the gravitational interaction between two masses as a consequence of the exchange of gravitational field quanta between the masses as suggested by the Feynman diagram of Figure 12.5.

We know something of the character of the quanta that are required if the quantum picture of gravity is to reproduce the results of the geometric description. The electrostatic field E generated by an electric charge can be described at each point by three numbers defining the three components of the vector field. Hence, the quanta of the field, the photon, has a spin of $1\hbar$ where the photon amplitude is described by the $2s + 1 = 3$ numbers required to define the state of a particle with spin quantum number $s = 1$. But the gravitational tensor distortion of space at a point requires more than three numbers for a complete description, and thus a quantum description of gravity requires a field quanta with spin quantum number $s = 2$, with $2s + 1 = 5$ states. Therefore, the quanta of the gravitational field, which we name the *graviton*, must have a spin of $2\hbar$.

This difference in character between gravity and the other forces—the difference between a vector and a tensor field or between spin $1\hbar$ and spin $2\hbar$ field quanta—leads also to the difference that, unlike the vector fields, there are no repulsive gravitational fields. All gravity is attractive. It is this property of attraction only, combined with gravity's long range, which makes gravity by far the weakest of the fundamental forces microscopically, by far the most important on the larger scales of planets, solar systems, galaxies, and the universe as a whole. The other interactions generate both attractive and repulsive forces, which for electrical forces cancel quite precisely, and the weak and strong interaction forces also have a very short range.

Figure 12.5 The exchange of a graviton between two mass points, which generates the gravitational force between the two masses.

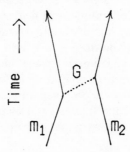

Using the quantum exchange description of gravity, we can define the strength of the gravitational force between two protons in a manner analogous to that used to describe the other forces:

$$\frac{GM_p^2}{\hbar c} \approx 10^{-38}$$

where G is a constant of nature called the *gravitational constant*, and M_p is the proton mass. Gravitational forces are extremely weak at microscopic distances compared to other forces and will not play an important part in any elementary particle interactions at energy scales that are now accessible. For the most part, it is not necessary to consider gravitational forces in any analyses of particle behavior and phenomenology.

In order to set the scale of magnitudes of time and space where the quantum character of gravity will become important, we define a *Planck mass M_g* such that the gravitational force generated by such a mass is as great as the forces from other charges:

$$\frac{GM_g^2}{\hbar c} = 1$$

The Planck mass is about 15 micrograms or about the same mass as 10^{19} protons. Although this is a small mass on a macroscopic scale, it is an enormous mass for a pointlike elementary particle. Using the Heisenberg Uncertainty Principle, $dt \cdot dE = \hbar$ at times as small as dt_g and at distances as small as dx_g, where

$$dt_g = \frac{\hbar}{M_g c^2} \quad \text{and} \quad dx_g = c \cdot dt_g = \frac{\hbar}{M_g c}$$

gravitational effects must be as important as the effects of electromagnetic or nuclear forces. From these relations, $dt_g \approx 10^{-43}$ seconds and $dx_g \approx 10^{-33}$ centimeters. Although these are extremely small distances and extremely small times, fundamental truths may lodge in such regions.

A comparison of the quantum description of gravity with similar descriptions of other forces provides some similarities and many differences. Although the graviton carries no strong, electromagnetic, or weak charges, and the graviton, like the photon, has zero rest mass, the graviton does carry energy just as the gravitational field, like the electromagnetic field, holds energy. Because energy is equivalent to mass according to the relation $E = mc^2$, that energy is a source of gravitational field itself. In the quantum picture, the graviton carries a gravitational charge (mass).

A more striking difference between gravity and the other forces is found in the lack of quantization of the gravitational charge or mass. The other charges, strong, weak, and electromagnetic, seem to come in parcels of definite size. Although we have never observed an electric charge that is other than an integral multiple of the fundamental charge (of $1.6 \cdot 10^{-19}$ Coulombs), we find all magnitudes of mass and no evidence of mass quantization at all.

As a consequence of the differences between gravity and the other fields, the spin $2\hbar$ quanta, and the lack of quantization of charge, physicists are now less sanguine about the possibility of soon understanding the deep connections between gravity and the other forces we believe (as a matter of faith) exist, than we are about understanding the unification of the other three forces.

Feynman Diagrams and Fundamental Processes

The description of forces by world diagrams or Feynman diagrams in terms of the emission and absorption of nonconserved field particles (such as photons) by conserved sources (such as electrons carrying charge) is nontrivial because the whole of a complicated interaction of interest can be broken down into parts describable by the conjunction of three line segments representing trajectories in time. The study of the fundamental interactions of particles is reducible to the study of those interactions which can be described graphically by three-line Feynman diagrams. A representative set of such fundamental pieces is shown in Figure 12.6. The first four diagrams of the figure show the entire set of different time-ordered emissions and absorptions of a

Figure 12.6 Feynman diagrams showing the different time-ordered emissions of field (force) particles by source (matter) particles. (*a*) The emission of a field particle *f* by a source particle *s*, which changes to *s'*. (*b*) The absorption of *f* by *s* which changes to *s'*. (*c*) The production of the source-antisource particles *s* and *s̄* by *f*. (*d*) The annihilation of *s* and *s̄* to form *f*. (*e*) The emission of a field particle *f'* (or *f''*) by *f* which then changes to *f''* (or *f'*). (*f*) The conjunction of two field particles *f* and *f'* to form a third field particle *f''*.

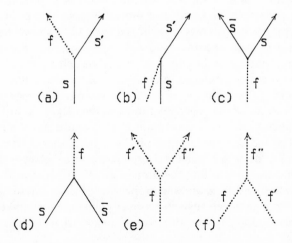

field particle by a source particle. These four diagrams are closely related—an understanding of one brings about an understanding of all. Note, especially, the general result that a particle moving forward in time is largely equivalent to an antiparticle moving backward in time. If the source particles s and s' are electrons and f is a photon, we are discussing electromagnetism. Figure 12.6a shows the emission of a photon by an electron; Figure 12.6b represents the absorption of a photon by an electron; and Figures 12.6c and 12.6d present, respectively, the production of an electron-positron pair by the annihilation of a photon, and the emission of a photon upon the annihilation of an electron-positron pair. The positron is the antiparticle corresponding to the electron with a positive charge rather than a negative particle. As noted earlier, the charge (in the case of the electron or positron) is a measure of the probability of emission or absorption of the photon.

For many processes just described, the masses of the particles are such that energy and momentum cannot be conserved. For example, an electron cannot emit a photon, as in Figure 12.6a, and conserve energy.[3] However, such a process can take place as a part of a total transition, such as the exchange of a photon by two electrons shown in Figure 12.1, which does obey the conservation laws. Because the photon is massless, energy and momentum cannot both be conserved for any electromagnetic processes described wholly by one of these diagrams. However, these diagrams are important as a part of more complicated diagrams defining electromagnetic reactions that can obey the conservation laws.

The last two diagrams (Figures 12.6e and f), show the emissions and absorptions of a field particle by another field particle. The electromagnetic field particle, the photon, carries no charge; therefore, such diagrams do not contribute to the description of the Abelian electromagnetic forces. The non-Abelian forces—the weak force, strong force, and gravity—have, as field quanta, the W and Z, the graviton, and the gluon carrying electric, gravitational, and strong charges, respectively. For such fields the quanta themselves are sources of other quanta, as shown by the Feynman diagrams.

We may note some of the heuristic properties of Feynman diagrams by considering certain characteristics of processes described by more complicated diagrams made up of sets of vertices. Figure 12.7a shows the interaction of one electron with another in elastic scattering; Figure 12.7b shows one electron emitting a photon in the course of scattering; and Figure 12.7c shows an electron-positron pair produced in the course of the scattering. We can now *estimate* some of the probabilities of occurrence of such processes using the qualitative features of the processes shown in the diagrams. In Figure 12.7a the force producing the scattering results from the emission of a photon from one charge and the absorption of the photon by the other charge. In this way a momentum transfer is created that produces a force between the particles. We see from the diagram that the probability of emission of the photon is proportional to $e^2/\hbar c = \alpha$, and the probability of absorption is also proportional to $e^2/\hbar c = \alpha$, which suggests (correctly) that the probability of scattering is proportional to the fourth power of the electric charge carried by the particles, that is , e^4.

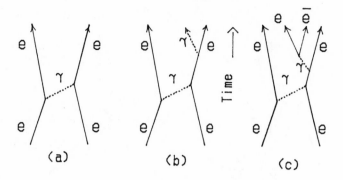

Figure 12.7   Diagram (*a*) shows the scattering of one electron by another through the exchange of a photon labeled γ (gamma); (*b*) shows the emission of a photon upon scattering; (*c*) shows the production of an electron-positron pair in the course of the scattering.

Now we can estimate the probability of the emission of a photon during the scattering (Figure 12.7*b*) as $\alpha = \frac{1}{137}$. That is, about 1 extra photon will be generated for each 137 scatterings. The probability of producing an electron-positron pair (Figure 12.7*c*) will be about α^2 times as probable as the scattering; in other words, for each $137 \cdot 137 = 20{,}000$ scatterings, an electron-positron pair will be produced. We estimate the probabilities by taking the product of the probabilities of each vertex. This is an estimate, and under some circumstances it can lead to quite incorrect results—as when, for example, symmetry rules forbid the process. Nevertheless, this kind of counting is a good guide for the estimation of the probability of processes if the coupling constant is sufficiently small.

Range of Forces

Aside from the strength of the coupling $g^2/\hbar c$ and the spin of the field particles, the character of the force is also determined by the mass of the field particle; the range of the force is inversely proportional to the mass of the field quanta. This relation can be illustrated qualitatively with the aid of Figure 12.8, which illustrates, as an example, the nucleon-nucleon force generated by the exchange of a pion.

Describing the strong force between the neutron and proton in terms of the exchange of the pion, the diagram shows that there is initially only a stationary proton and neutron. During the time of exchange *dt,* after the pion is emitted by the neutron and before it is absorbed by the proton, energy is not conserved classically, as the mass of the neutron, proton, and pion is certainly greater, by the pion mass, than the original mass of the neutron and proton. However, the Uncertainty Principle keeps us from defining the energy to an accuracy greater than $dE = \hbar/dt$. Then, for a time as short as $dt < \hbar/dE = \hbar/mc^2$, energy conservation is not violated. Traveling no faster than the speed of light, the intermediate pion can travel only as far as $a = c \cdot dt <$

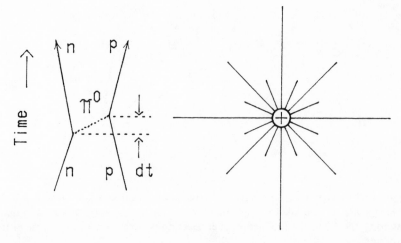

Figure 12.8 At the left, the exchange of a pion between a neutron and a proton contributes to the nuclear force between the two nucleons. Energy is not conserved during
the time interval *dt*. At the right, the lines of force emitted by a nucleon describing the
short-range nuclear force die away with distance.

$\hbar/mc$, the Compton wavelength of the pion and the range of the nuclear
force.[4]

We call these particles that exist briefly by borrowing energy under the
aegis of the Uncertainty Principle *virtual particles*. It is important to understand that the description of virtual particles in terms of the Uncertainty
Principle is not a primary argument in itself but a qualitative explanation of
the results of well-defined calculations.

Describing the forces transmitted by the exchange of massive virtual particles in terms of the force lines of Faraday, aside from the spreading out of
the lines with distance (which leads to the inverse square law for field intensities), the lines of force shown to the right in Figure 12.8 die out with distance. In $4\pi g$ lines of strong interaction force are emitted by a nucleon coupled to the pion field with a strength or charge of g, after a distance equal to
the range $a = \hbar/mc$ (about $1.4 \cdot 10^{-13}$ centimeters for nuclear forces
mediated by a pion of mass m), only about two-thirds of the lines will remain;
after a distance of $2a$, only about four-ninths will remain. In general, numerically the range of a force mediated by a field particle of mass m is $a = 200$
f/m, where m is measured in MeV/c^2 and 1 fermi (written as f) equals 10^{-13}
centimeters. For the pion mass of 139 MeV/c^2, $a = 1.4$ f, a result that accords
with our general knowledge of the range of nuclear forces. The masses of the
W and Z, the particles that mediate the weak interactions, are about 75 and
95 times, respectively, the mass of the proton (and about 500 times the mass
of the pion); hence, the range of the weak forces must be about $2.5 \cdot 10^{-16}$
centimeters, which is very small indeed—about 500 times smaller than the
size of a proton, which is the range of the pion field generated by the proton.

Even as the range of forces carried by heavy intermediate virtual particles is very short, the range of the electromagnetic force carried by massless photons is infinite. The number of lines of force from an isolated charge does not diminish with distance and is equal to the number emitted from the charge.

The Structure of the Vacuum

It is possible to consider that the vacuum is not empty at all, but rather is alive with virtual particle-antiparticle pairs of matter particles. Although this statement is somewhat metaphorical, in the usual sense of the word the vacuum is still *empty;* the rationale behind the statement is valid and important—and connected with the Uncertainty Principle in a manner similar to the use of that principle in the discussion of the range of forces.

Nominally, the creation of particles from nothing violates the principle of the conservation of energy and the conservation of matter particles. There is no violation of particle number in the production of particle-antiparticle pairs, and if the energy required to make the particles is "borrowed" for a sufficiently short time, in a manner introduced in the discussion of the ranges of forces, the Uncertainty Princple foils accounting. If the mass of a particle is m, energy equal to $2mc^2$ must be provided if a pair of such particles is to be created. Hence, from the Uncertainty Principle, the particle-antiparticle pairs can exist in the vacuum for periods of the order of Δt, where

$$\Delta t \approx \frac{\hbar}{2mc^2}$$

without violating the conservation of energy. If one probes times this small, one can "see" the pairs. Similarly, if one probes distances Δx such that

$$\Delta x = c \cdot \Delta t \approx \frac{\hbar}{2mc}$$

which is the distance the particle, moving at the speed of light, can travel in the time Δt, one can again "see" the particle-antiparticle pairs. The effective time of existence of electron-positron pairs is about $\Delta t \approx 10^{-21}$ seconds, and they extend over a region of about $\Delta x \approx 10^{-11}$ centimeters. If smaller distances and shorter times are probed, pairs of particles of greater mass will be evident. Hence, on the microscopic level, we can consider that the vacuum is full of virtual particle-antiparticle pairs, electrons and positrons, quarks and antiquarks, and so on, appearing and disappearing, all owing their transient existence to borrowed energy.

If the vacuum is so full of particles, how can anything pass through it without colliding with the particles? Surely the vacuum is transparent! This paradoxical transparency of a medium full of particles follows from the conservation of energy and momentum; a free particle cannot scatter from a virtual particle and conserve both energy and momentum. Such collisions are then forbidden, and free particles, electrons, protons, photons, and so on sail through the vacuum unimpeded.

Although the vacuum is empty to free particles, sets of interacting parti-

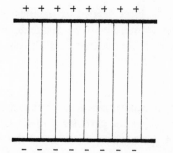 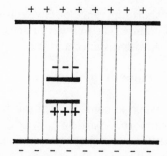

Figure 12.9 At the left is a strong electric field; at the right is the same field with an electron-positron pair introduced such that the fields of the pair cancel the external field.

cles can "see" the vacuum, as energy and momentum can be supplied by the extra particles. We illustrate the effect of virtual particles in the vacuum on more complex systems by considering the production of electron-positron pairs from the vacuum by a strong electric field—the "sparking of the vacuum." Figure 12.9 shows, to the left, a pair of plates holding positive and negative charges and the strong electric field generated by the charges. To the right, an electron-positron pair is added to a similar system of charges. To simplify the exposition, we may view the electron and positron as small flat plates holding a unit charge (rather than point charges). We make the electron and positron small enough so that the field they generate just cancels the external field. We can see from the figure that the existence of the electron-positron pair eliminates the electric field from the region between them. Because that electric field held energy, the field at the right holds less energy than the field at the left. There is less field energy with the electron-positron pair in existence than without the pair. If the energy subtracted from the field has a mass greater than the masses of the electron and positron, there is no energetic reason why an electron-positron pair will not spontaneously appear in a strong field. And that happens: an electron-positron pair will be produced—out of the vacuum—by a strong enough electric field.[5]

Such strong fields should exist at the position of the first Bohr orbit for super-heavy nuclei with a nuclear charge of about $Z = 137$ (where, for uranium, $Z = 92$). Such nuclei are not found in nature but may be produced fleetingly in the course of very high-energy collisions of heavy nuclei. Indeed, positrons have been seen in the course of such collisions which seem to be made from the vacuum by the huge electric fields produced in the course of the collisions. Other particles are produced from the vacuum by other forces in much the same way, although larger energies are required to produce heavier particles.

Charge and Mass Renormalization

We illustrate physically important effects of these virtual pairs by considering the effects of the "polarization of the vacuum" on the force between two electrons at close distances. Near the pointlike electron, the large electric fields polarize the electron-positron pairs that appear spontaneously from the vacuum so that the positron is closer to the electron than is the electron. Because this configuration tends to cancel the electron field and reduce the energy that is borrowed, the transition is more probable with the electron-positron aligned in that configuration than the reverse. The left diagram of Figure 12.10 illustrates the kind of pairs one might expect to see very close to the electron.

Of course, this polarization occurs at every distance, albeit most strongly at small distances; but to illustrate the effect on forces, we consider only the polarization at some specific radius and divide the charge produced into a positive band near the electron (from the virtual positrons) and a negative band further away (from the virtual electrons), as shown in the right diagram of Figure 12.10. Now if we almost superpose two such electrons with their "garment" of virtual electron-positron pairs, the repulsive force between the two electrons will have three contributions: one from the repulsion of the "bare" electrons, one from the repulsion of the overlapping virtual-positive sheaths, and one from the repulsion of the virtual-negative shells. In any case, the repulsion will be greater than from the bare electron as a consequence of the second and third contributions. When the two electrons are far apart, the effects of the positive and negative shells will cancel, and the polarized vacuum will not affect the force between the two electrons. However, the existence of the polarization increases the forces at small distances. The effective charge at small distances is then larger than the long-distance charge we measure—and define— as the charge of the electron. Indeed, if we attempt to calculate the charge at zero distance, we find an infinite charge! This tells us that our description of the electron as simply a charged object is inadequate at extremely small distances.

From the results of scattering experiments at high energy, we know that the electron is a pointlike object with a spatial extent no larger than $\approx 10^{-16}$ centimeters, or $\hbar/(100 \text{ GeV}/c)$, the wavelength of a particle with a momentum of 100 GeV/c. Because the density of energy held by an electric field is proportional to the square of the field strength, and the field is inversely proportional to the square of the distance from the charge, the energy density goes to infinity at the position of a point particle. Indeed, for such a pointlike

Figure 12.10 At the right is a schematic suggestion of the polarization of virtual electron-positron pairs in the field of an electron; at the left, the polarized vacuum in a given radius is divided into shells to illustrate the effects on the forces between two electrons.

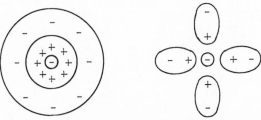

charged particle, the total energy of the accompanying electric field—and thus the mass of the electron—is infinite![6]

More sophisticated calculations also give an infinite mass to the electron, just as such a calculation gives an infinite charge. Clearly, something has been left out of the calculations; but this should not worry us unduly, as we know that we cannot at this time properly consider the other forces—gravity and the strong and weak interactions—which cannot be neglected at very small distances. However, these infinities do complicate the calculation of quantities that are physically accessible and manifestly cannot depend on the very small distance structure of the electron, which we do not now understand.

Early attempts to understand fundamental electromagnetic processes resulted in a curious state of affairs. Consider, for example, the simple Coulomb scattering of one electron by another. A calculation that considered only the exchange of one virtual photon from one electron to the other, as shown in Figure 12.11a, gave an answer that described the experimental results quite well. However, in a complete calculation, amplitudes from all processes that lead from the same initial state (here, two electrons before the scattering) to the same final state (the two electrons after the scattering) must be evaluated and added to determine the total amplitude. Then the amplitude describing the process in Figure 12.11a must be added to the amplitudes describing those in Figures 12.11b, 12.11c, and all others. But the effects of the additional processes—as shown in Figure 12.11b where the electron emits and reabsorbs a virtual photon, and in Figure 12.11c, where the electron emits a virtual photon that produces an electron-positron pair (vacuum polarization)—which should have resulted in small modifications to the scattering, actually gave infinite corrections! Again, the infinities found in the calculations were seen to come from the very small distance contributions of the electron structure. They were of the same character as the infinities found in the calculations of the mass and charge of the electron.

Figure 12.11 (a) The process that is the most important in the scattering of one electron by another. (b) The process that contributes to mass (self-energy) infinities. (c) The process that leads to charge (vacuum polarization) infinities.

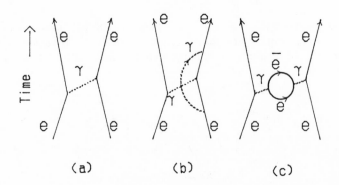

But the mass and charge of the electron are known. Hence, theoretical physicists found that it was possible to segregate the effects shown by diagrams such as Figure 12.11*b* as contributions to the mass of the electron, and to substitute the measured mass for the infinite mass in the calculations. Similarly, the contributions from processes described by diagrams such as Figure 12.11*c,* which result in infinite charge, were replaced by the measured charge. The problem of the infinities was solved by sweeping them under the rug! With this *renormalization* procedure (developed about 1948 independently by Julian Schwinger, Richard Feynman, and Sin-itiro Tomonaga), the calculations gave finite results that agreed with the experimental results with great precision.

Because not all models of elementary forces can be renormalized, the renormalizability of a description of forces is considered another constraint on guesses as to the path taken by nature: Physicists believe that only renormalizable forces are found in nature.

Summary of the Characteristics of the Forces

Assuming Feynman diagrams such as those in Figure 12.4 define all forces, we can list these forces, their associated source (matter) particles and field (force) particles (or, better, the names given to observed and conjectured associated particles), and their effective coupling strengths, as in Table 12.1. The spin of the characteristic field particle J largely determines the properties of the field under a Lorentz transformation. For example, the unit spin of the photon together with its zero mass gives us Maxwell's equations as the description of the classical electromagnetic field. For the strong interactions, the table lists the field properties of composite particles as well as the properties of the fundamental quarks and gluons. Probably only the composite particles are directly observable and, for many purposes, the description of strong interaction phenomena in terms of the fields of the composite particles is especially useful.

Table 12.1 Characteristics of the fundamental forces.

Interaction	Source Particle	Field Particle	J Field Particle	Strength $g^2/\hbar c$	Abelian?
Strong	Quark	Gluon	$1\ \hbar$	≈ 0.2	No
	(baryons)	(mesons)	$0, 1, 2, \ldots, \hbar$	≈ 1	No
Weak	Quarks and			$\frac{1}{137}$ or	No
	Leptons	W^+, W^-, Z^0	$1\ \hbar$	10^{-7}	
Electromagnetic	Charged particles	Photon	$1\ \hbar$	$\frac{1}{137}$	Yes
Gravity	Mass	Graviton	$2\ \hbar$	$\approx 10^{-40}$	No

Note: The fundamental strong interaction can be considered to be very weak at high energies or small distances (asymptotic freedom) and very strong at low energies or large distances (containment). Because of the short range of the force, the weak interaction is effectively very weak, although the fundamental weak charge is about the same as the electric charge.

Notes

1. The force will be proportional to the frequency of transfers, to the velocity v of the snowballs, and to the mass m of the ball. If the mass is m thrown with a velocity v, the momentum transfer from one exchange will be $v \cdot m$. If the number of throws per second is e and the number caught per second is e, the force between the skaters, defined as the momentum transfer per second, will be

$$F = e^2 m \cdot v$$

This model describes only repulsive forces. When stretched in an attempt to include attractive forces, the metaphor loses verisimilitude.

2. Lepton is derived from the Greek λεπτοσ, meaning light or slight; baryon from βαρυσ, heavy; and hadron from αδρυσ, taken loosely as massive or thick.

3. This is most easily seen in the system in which the electron is at rest. Here the initial energy is just the rest mass of the electron. After emitting a photon and recoiling, the total energy, equal to the rest mass of the electron plus the kinetic energy of recoil plus the energy of the photon, must increase.

4. More precisely, the force between two particles generated by the exchange of an intermediate particle of mass m can be described by the Yukawa form

$$V(r) = \frac{g^2}{r} e^{-r/a} \qquad \text{where } a = \frac{\hbar}{mc}$$

and g^2 is the coupling constant (or the square of the interaction charge). The factor of $1/r$ follows from the three-dimensional geometry of space, whereas the factor $\exp(-r/a)$ contributes the constraint on the range from the finite mass of the field quanta.

5. We can estimate the field strength required by taking the size of the square electron as d on a side where $d = \hbar/mc$, the distance between the electron and positron allowed by the Uncertainty Principle. To conserve energy, the energy from the field excluded from the volume $V = d^3$ must equal $2mc^2$, where m is the mass of the electron. Then

$$\frac{E^2 V}{8\pi} = \frac{E^2 d^3}{8\pi} = 2mc^2 \qquad \text{where } d = \frac{\hbar}{mc}$$

and

$$E = \left(16\pi m^4 c \frac{5}{\hbar^3} \right)^{1/2} \approx 8 \cdot 10^{15} \text{ volts/centimeter}$$

6. The energy dw stored in a shell of radius r and thickness dt about the electron will be equal to the volume of the shell $(4\pi r^2 dt)$ multiplied by the energy density in the electric field $(E^2/8\pi)$:

$$dw = (4\pi r^2 t) \cdot (E^2/8\pi) = \frac{2E^2 dt}{r^2}$$

From Coulomb's law, $E = e/r^2$ and

$$dw = \frac{e^2 dt}{2r^2}$$

For a shell of half the radius and half the thickness, the volume will be smaller by a factor of 8, but the energy density will be 16 times greater; the smaller shell will hold more energy.

The smaller the shell, the larger the contribution to the energy. For a pointlike electron, the total energy w stored in all shells becomes infinite according to this classical calculation, and the mass associated with that energy, $m_e = w/c^2$, is also infinite.

13

Symmetries and Conservation Laws— *CPT*

> "That's the effect of living backwards," the Queen said kindly: "It always makes one a little giddy at first."
>
> Lewis Carroll, *Through the Looking Glass*

Mirror Symmetry and Parity

Our local environment displays asymmetries. Certainly most of this asymmetry is accidental and not fundamental in nature. Most humans are right-handed, and most bears appear to be left-pawed; our hearts pump arterial blood from the left side; all life on earth uses left-handed amino acids. We feel intuitively that all these characteristics are essentially accidental; we are loathe to believe that the dominant selection by builders of right-hand threaded nuts and screws is an indication of some basic character of nature.

However, the question whether a fundamental symmetry between left-handed and right-handed systems exists is valid and important. Is there any fundamental characteristic of nature that differentiates between left-handed and right-handed systems? When Alice passes through the looking glass, she finds a world where most people are left-handed and left-hand screws are used to hold the furniture together—but perhaps this is only an accidental characteristic of this world, along with Humpty Dumpty and the Walrus and the Carpenter. Is there any observation Alice can possibly make that can tell her she is in looking-glass land? Is there any fundamental characteristic of the universe that differentiates between left-handed and right-handed screws, or between the world we know and a possible mirror image?

In Figure 13.1 Alice is looking into the mirror—or Alice-through-the-looking-glass is looking out of the mirror. A coordinate system labeled x, y, and z, which is a right-handed coordinate system according to convention, and a clock are shown together with their looking-glass counterparts. The hands of the "real" clock move in a way we call clockwise; the hands of the

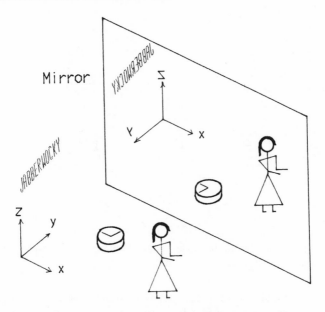

Figure 13.1 Alice looking through a looking glass.

looking-glass clock move counterclockwise. The mirror coordinate system is labeled *x, y,* and *z* in a way that is called left-handed. How can Alice tell the looking-glass world from the usual world?

We leave Lewis Carroll's classical Alice for science fiction to further examine the character of mirror symmetry. Assume that we have contacted another civilization through some means of communication like radio such that symbols (words, for example) can be transmitted, although no material devices can be transferred. We can easily determine many of the characteristics of our alien friends; perhaps they have four arms and antennas (see Figure 13.2). We can even send pictures and exchange television programs. But there are limits to this exchange of information. We have already emphasized invariances that suggest that we cannot determine the position of the other system, the orientation of that system, or the time difference between the systems. The alien system may be a billion light years from us in the direction of Polaris. North, in this system, may be west in ours; its residents may be 10,000,000 years in our future; but if the universe is invariant with respect to position, direction, and time, no measurements that we perform and compare through our communication system can allow us to determine these relations.

Can we, however, determine whether the aliens use a right-handed or left-handed coordinate system? Do the hands of their clocks move clockwise or counterclockwise? Do they use right- or left-handed nuts and bolts? Can Alice find a way to tell whether she has passed into Looking-Glass Land? We can answer these questions if we find some *fundamental* process or structure that defines a screw direction. Then both we and the aliens can compare our con-

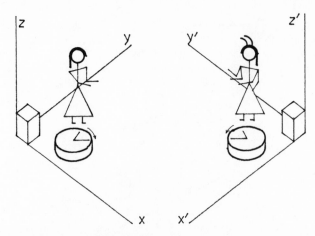

Figure 13.2 An alien from another galaxy and a human; the alien and the human use different coordinate systems.

ventions with that structure. If our screws turn in the same way as the structure and the aliens' screws turn differently, we can conclude that they prefer left-handed screws.

However, it seems that as long as we consider only phenomena induced by gravity, electromagnetism, and the strong nuclear interaction, we can find no fundamental phenomena or structure that differentiates between left-handed and right-handed coordinate systems or between the usual world and the looking-glass world. To the extent that we restrict ourselves to measurements of processes mediated by these forces, it appears that the universe is invariant with respect to mirror reflection or the choice of left-handed or right-handed coordinate systems, and we will not be able to determine which system the alien uses.

Until 1956 it was generally believed that the descriptions of all interactions were independent of the choice of coordinate system, hence that any observation of the universe was unchanged, or invariant, under reflection in a mirror. We label the operation of mirror reflection P. Then Richard Dalitz, then at Cornell University, showed that some features of the decay of a hadron named the K-meson, which decayed sometimes to three pions and sometimes to two pions, appeared to be inconsistent with the hypothesis that no observation could discriminate between left-handed and right-handed coordinate systems. In the course of efforts to understand this result, C. N. Yang and T. D. Lee, working at Brookhaven National Laboratory on Long Island during the summer of 1956, were able to show that nothing known about the weak interactions was in contradiction to the hypothesis that the weak interactions might generally proceed in such a manner as to define a screw direction and thus a coordinate system that was fundamentally different from its mirror image.

Shortly thereafter, C. S. Wu at Columbia University, convinced by the arguments of Lee and Yang, initiated a collaboration with Eric Ambler of the

National Bureau of Standards, who was an expert on the requisite low-temperature techniques. This collaboration led to an experiment that showed conclusively that the weak interactions do define a coordinate system; therefore, nature does discriminate between left- and right-handed coordinate systems.

Cobalt 60, a radioactive isotope of cobalt that emits electrons, was aligned by a magnetic field so that the spin of the nucleus was in the direction of the field. In general, nuclei have a certain angular momentum derived from the motion and intrinsic spins of the nucleons that make up the nucleus. The spins and motions of these charged objects tend to generate magnetic fields such that most nuclei act a little like small bar magnets with the bar along the axis of the nucleus spin. Hence, the external magnetic field served to line up the nucleus spin along the direction of the field.

It was then determined that the electrons emitted by the cobalt were produced predominantly opposite to the direction of the nuclear spin and opposite to the direction of the magnetic field **B**. It is useful to strip our description of conventions concerning directions of currents and directions of magnetic fields by stating that the decay electrons are emitted upward if the current electrons are moving counterclockwise. If the Co^{60} were placed on the face of a clock and the electron current were moving opposite to the direction of motion of the clock, the decay electrons would move away from the face of the clock. We can then use the experiment to define counterclockwise and, of course, clockwise.

Hence, nature does define a coordinate system, and we can use the results of such a measurement of Co^{60} decays to determine whether we are in a looking-glass world or to determine the coordinate system used by the aliens with whom we are talking. In particular, we can arrange with the aliens to perform the same experiment we do and compare results. In Figure 13.3 we see the

Figure 13.3 A human and an alien observing Co^{60} decays so as to compare coordinate system conventions.

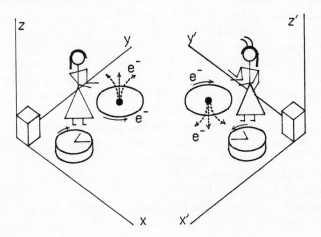

human and the alien each performing the Co^{60} decay experiment, which can be used by each to define a coordinate system. If the moving electrons that provide the magnetic field that aligns the nucleus travel opposite to the direction of the hands of a clock, and the decay electrons travel upward, a right-handed coordinate system is in use; if the electrons travel downward, a left-handed coordinate system is in use—the clock convention is different, and we are in a looking-glass world. In the particular case shown in Figure 13.3, we see that the alien concludes that her coordinate system differs from ours as a reflection in a mirror.

The Conservation of Parity

The symmetry with respect to mirror reflection P of all the observations of nature that depend on the strong interactions, the electromagnetic interactions, or gravity results in a conservation law—*the conservation of parity*. Parity is a more subtle concept than energy and momentum, and it is important only in the microscopic areas of atoms, nuclei, and elementary particles. We approach the idea of parity by noting that if the forces of nature do not differentiate between coordinate systems, the elementary structures formed as a consequence of those forces cannot differentiate between coordinate systems; that is, no such elementary structure or process can define the direction of a screw. (Elementary processes can produce screw-defining structures, but they must produce left-handed and right-handed structures with equal probability, thus not differentiating between coordinate systems.)

From our knowledge of the quantum character of nature, we know that microscopic structures can be described as standing-wave patterns of de Broglie waves. Moreover, the physically realizable information concerning the structure is found in the intensity of the wave. Our requirement that the structure cannot define a coordinate system is equivalent to stating that the intensity pattern of the wave is invariant under mirror reflection.

It is useful to consider parallel classical and quantum systems in illustrating the conservation of the parity symmetry by symmetric forces. Hence, we follow the simple, if artificial, de Broglie wave of a particle in a one-dimensional box[1] together with similar vibrational modes of a violin string.

Figure 13.4 shows amplitudes and intensities of the de Broglie wave[1] or, equally, the vibrating violin string. The second harmonic state (*a*) invariant under reflection about the center of the box—or center of the violin string—has even parity, whereas the state (*b*), the first harmonic of the violin string, which changes sign upon reflection, has odd parity. In either case, the intensities are symmetric about the center. The forces on the particle, provided by the ends of the box, and the forces (tension) on the string, provided by the constraints at the ends of the string, are symmetric. A change in the interaction or force on the particle described by the de Broglie wave will change the wave. Similarly, a change in the distribution of tension along the string will change the wave-form of the vibrating string. However, if the additional

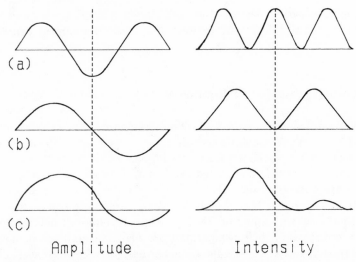

Figure 13.4 Amplitudes of de Broglie waves representing a particle in a box, or the modes of vibration of a violin string: (*a*) is a state of even parity, (*b*) of odd parity, and (*c*) of mixed parity. The intensities are shown to the right.

interaction is symmetric, the parity of the state cannot be changed by the interaction. For example, we could add springs to each end of the box repelling the particle. Or we could attach fine wires from each end of the string to points one-fourth the distance from the end and increase the tension on the center of the string through tension on the wires. In either case, the shape of the wave would be changed by the additional interaction, but the parity would not change.

If the additional perturbing forces are not symmetric, those forces will generate states that are of mixed parity (parity will not be conserved), and the intensities of these mixed states *will not be symmetric.* If the spring at one end of the box is stronger than the other, or if the tension on one of the guy wires is greater than the other, the de Broglie wave or violin amplitude will be changed as suggested in Figure 13.4*c*. That state is largely a mixture of the fundamental or ground state and the first harmonic or first excited state of the system. Considering the particle state, if the energy of the state is measured, the result will sometimes be the ground state energy, thereby indicating that the parity of the state is even, and will sometimes be the first excited state, which has odd parity. The parity of the state changes and is not conserved because of parity nonconserving forces. Parity nonconserving forces will differentiate between left and right in these simple one-dimensional models.

For more realistic, and much more complex, quantum states in three dimensions, parity is conserved by forces that are mirror symmetric. We know that gravity, the strong interactions, and the electromagnetic interactions conserve parity, whereas the weak interaction does not—and weak interaction effects differentiate between left- and right-handed coordinate systems.

Charge Conjugation *C—CP* Invariance

In 1928 the British physicist P.A.M. Dirac developed the first wave equation for spin $\hbar/2$ particles which was consistent with the Special Theory of Relativity. A number of striking results were implied by these equations. In particular, for each particle there must exist an antiparticle with opposite electrical properties—opposite charge and opposite charge distributions, currents, and magnetic fields. Somewhat later, similar results were found for bosons, particles with spins of $0\hbar$, $1\hbar$, $2\hbar$, ... , although neutral bosons could in some cases be their own antiparticle: there is no antiphoton, or rather the photon and antiphoton are the same particle.

A further property of the entire set of theories was the result that antimatter was seen to follow exactly the same laws as matter as far as electromagnetic forces were concerned. Later it was seen that matter and antimatter behave in the same way under the strong forces—and under gravity. Perhaps this was the case for all interactions, including the weak interactions. Then, a universe composed of antimatter where antihydrogen gas, made up of atoms in which a positron or antielectron circled a negatively charged antiproton in a Bohr orbit, was used to fill antiballoons made of antirubber and held by antichildren, would be just like ours.

Any interaction between universe and antiuniverse would be catastrophic, however. Because the conservation laws that operate so as to guarantee the stability of matter—that is, the conservation of leptons and quarks—do not ensure the stability of mixtures of particles and antiparticles, the matter in such mixtures will annihilate to produce mesons (largely) with large kinetic energies; some of that kinetic energy will be dissipated in a manner that will eventually be seen as heat before the mesons disintegrate to form leptons with kinetic energy. That kinetic energy will also be transformed to heat before the heavier leptons decay to the lightest lepton, the electron. And these transformations would occur with lightning speed to be largely completed in a few millionths of a second.

The physical interactions between the two universes must then be limited. Friendships must be distant: any affair between girl and antiboy must be platonic. But if the symmetry between matter and antimatter were exact, and we were to make contact with some extragalactic civilization (such as that of the aliens introduced earlier by, say, radio so that we could exchange information but not material), we could not determine whether they were anticreatures living in an antiuniverse or merely creatures like ourselves. A universe constructed of antimatter would be identical to ours: any fundamental observations (which do not, of course, include those of the length of the antenna on the aliens' heads) would be the same in the two universes.

However, the experimental and theoretical work connected with the discovery of the nonconservation of parity by the weak interactions, which was discussed in the last section, also showed that the weak interactions do distinguish between matter and antimatter. The weak interactions are therefore not invariant under the operation, labeled *C,* called *charge conjugation,* which exchanges particles and antiparticles. In particular, it was shown that leptons and antileptons are produced differently by the weak interactions. In the limit where they are produced at the speed of light, the leptons' intrinsic angular momentum or spin is directed opposite to their direction of motion, whereas the spin of the antileptons is directed along their direction of motion. A neutrino moving toward us is spinning clockwise; an antineutrino moving toward us will be spinning counterclockwise. This behavior is shown in Figure 13.5, which also shows the view in a mirror.

We can use this information concerning the orientation of lepton spins to determine the relative particle-antiparticle character of our alien friends. They could observe if their neutrinos and other leptons are spinning clockwise like ours, or counterclockwise. From this information they could determine whether their local environment is composed of particles or antiparticles. (Of course, like the label *alien,* the terms *particle* and *antiparticle* are our nomenclature. Our friends would call us aliens and our material antimatter if it differed from theirs.)

However, in order to implement this exchange of information we would have to teach the aliens what clockwise and counterclockwise mean, which means that we would have to determine our and the aliens' relative coordinate system conventions. Do they use a left-handed or right-handed coordinate system? Earlier, we showed that they could determine the relative convention by performing the cobalt-60 alignment experiment, or they could observe the relation between the spin and direction of a neutrino.

But perhaps they do not have Co^{60} but anti-Co^{60}; perhaps they live in an antiuniverse and would therefore observe antineutrinos if they followed our

Figure 13.5 Leptons and antileptons, and their mirror images (to the right).

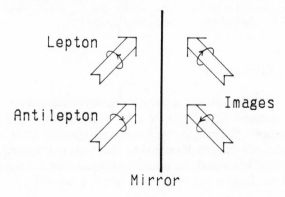

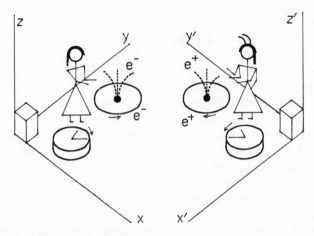

Figure 13.6 A human and an antialien aligning their Co^{60} sources according to their different coordinate systems and observing the decays.

prescription. Just as a neutrino behaves like an antineutrino viewed in a mirror, anticobalt, in a magnetic field produced by a current of antielectrons (positrons) in an anticopper wire, emits antielectrons in a manner that is indistinguishable from the corresponding particle experiment viewed in a mirror. As shown in Figure 13.6, antialiens performing the Co^{60} experiment with anti-Co^{60} in a universe of antimatter will obtain results that will be described in the same manner as our results if they use a left-handed coordinate system.

The weak interactions are not invariant with respect to the mirror inversion operation, which we designate P for parity. Nor are weak interactions invariant with respect to the exchange of particles and antiparticles—the operation we call C for charge conjugation. However, to the extent that the description we use here is correct, it appears that the weak interactions are invariant with respect to the combination of operations CP. If we substitute antiparticles for particles and observe the interactions in a mirror, we cannot distinguish between these observations and the direct observations of particle interactions. The other forces, gravity, electromagnetism, and the strong interactions, seem to be invariant under the C and P interactions separately—and are then obviously invariant under CP.

Violation of CP Invariance

Certainly, to a high degree of accuracy the weak interactions, like all other interactions, seemed until the 1960s to be invariant with respect to the combined operations CP. Then, in 1964, James Cronin and Val Fitch of Princeton University, working at Brookhaven National Laboratory, showed that violations of CP invariance do occur. The arguments concerning the interpretation of the original experiment are indirect; we will not consider them here but will instead look at the results of a closely related experiment that allows a more transparent analysis. If we run either a beam of K^+ mesons into copper or a beam of K^- (antiparticles to the K^+) mesons into an anti-copper target, we will produce a beam of K_L^0 mesons (which are half particle state and half antiparticle state). We now know that the K_L^0 particle decays slightly more often into positive muons and positive electrons than into negative muons and electrons. This result violates *CP* invariance. We can show that it does so by using the results of such an experiment measuring this charge ratio to determine whether our friend, the extragalactic alien, is an antialien living and experimenting in a universe of antimatter that she describes using a left-handed coordinate system, or whether she uses a right-handed coordinate system and is constructed of the same kind of matter as we are.

We proceed in this quest by asking the alien to set up a beam of K_L^0 mesons and measure the curvature of the decay products, the electrons and muons, in a magnetic field derived from an electron current through a copper coil. The two setups, ours and the alien's, are shown in Figure 13.7. We take a K^+ beam from a particle accelerator facility and run the beam into a copper target to produce K_L^0 mesons. The antialien, nominally following our directions, runs a beam of anti-K^+ mesons, or K^- mesons, into an anticopper target to produce K_L^0 mesons, which are the same as ours. We identify the dominant leptons from the K-decays as positive muons and electrons by passing the leptons through the magnetic field we generate by an electron current moving counterclockwise. The alien attempts to copy our procedures, but her electron beam is actually an antielectron or positive current, and her version of counterclockwise is opposite to ours because she uses a left-handed coordinate system. Of course, the main decay products of her K_L^0 beam are positive electrons and muons, the same as ours. But a little study of the relative

Figure 13.7 A human and an antialien examining the decays from K_L^0 mesons. The observations show them that they are related as antibeings in systems of antimatter.

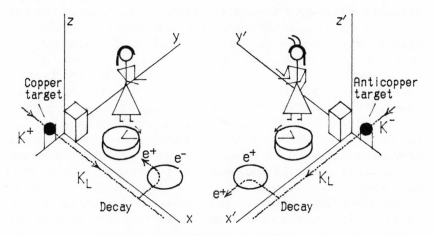

charges and directions of motion of the charges shows that the K_L^0 decay products in the alien's universe will be bent toward the negative y-direction, whereas the same decay products will be bent toward the positive y-direction in our universe. The difference is obvious and can be communicated. We know now that the alien is an antibeing using a left-handed coordinate system. And we know that not everything in the universe is invariant under the combination of mirror inversion P and charge conjugation C, the change of particle to antiparticle; not everything is left unchanged by CP.

Mach's Principle

In the last part of the nineteenth century Ernst Mach, in his consideration of mechanics, emphasized the importance of the relative meaning of space and velocity and considered the problem of the nominally definite, nonrelative meaning of acceleration. He concluded, implicitly, that acceleration must also be relative—perhaps relative to the mean position of the entire mass of the universe—if we understood it correctly.

Mach's specific concerns have been expanded by others to consider further symmetries. Just as Mach expected that the laws of nature, hence observations of nature, should not depend on the position, velocity, orientation, or acceleration of the coordinate system, nature should not differentiate between left-handed and right-handed geometric coordinate systems; observations of nature should be invariant under a mirror inversion or P transformation. Of course, the Co^{60} experiment dashed that hoped-for symmetry. It seemed that nature did define a screw direction. But the discovery of CP invariance, which followed almost immediately, largely preserved Mach's principle. The universe was not left-handed; the universe did not distinguish between particles and antiparticles; the universe treated left-handed particles and right-handed antiparticles identically.

But this retrieval of Mach's principle through differences between particles and antiparticles was destroyed by the observations of the decays of the K_L^0 particles. Let us examine the consequences of this exhibition of CP violation in a way that may illuminate the deep philosophical problems that arise from this connection between CP violation and the extended Mach's principle. If in empty space (an empty universe?) we prepare a spherical oven with no built-in asymmetries and produce matter and antimatter from the interaction of the radiant energy from the oven, we will eventually produce more fast left-handed electrons than right-handed positrons. Somehow, nature fundamentally differentiates between matter and antimatter, and, contrary to Mach's principle that no physical process can differentiate between different geometric coordinate systems, nature does differentiate between left-handed and right-handed coordinate systems. *God is not completely ambidextrous.*

Our local system is, of course, constructed largely of particles rather than antiparticles. The predominance of matter over antimatter in cosmic rays leads us to believe that our galaxy is made up completely of matter. Most astrophysicists guess that the entire universe is constructed of matter, not of half matter and half antimatter, and physicists are attempting to understand this asymmetry as a consequence of the effects of the *CP* violation acting in the early (first microsecond!) history of the universe.

Time Reversal *T*

Because the descriptions of many fundamental processes have been found to be invariant with respect to *P*, a reversal of all coordinates, we might ask whether the descriptions of these processes are not also invariant with respect to a change in the direction of time—*time reversal.* Aside from the tendency of the universe to move toward a more probable situation or toward an increase in entropy, is there any fundamental process that determines the direction of time? We have noted that the collision of two billiard balls is (almost, neglecting friction) invariant with respect to time reversal; the elastic collision fits the conservation laws just as well if time is reversed as in the original course of events. If we were to observe a motion picture of the collision, we could not tell whether the film were being run backward or forward by observing the event.

Until quite recently, physicists believed that all interactions were invariant under time reversal and, aside from the Second Law of Thermodynamics and considerations of the increase in probability, no fundamental interaction could differentiate between past and future. The experimental evidence in this matter was never highly precise, and there is now some indirect evidence that shows definitely that not all interactions are invariant under time reversal. As mentioned earlier, the decays of the K_L^0 mesons are not invariant under the combined operations *CP*. However, there are theoretical reasons that are taken to suggest that all interactions are invariant under the combinations of the three interactions *C*, *P*, and *T*, where *T* symbolizes the operation that reverses time; quantum field theory as it is now understood requires invariance under *CPT*. Then, if the description of some interaction is changed under *CP*, it must also be changed under *T* if the observation is to be invariant under *CPT*. The result of this argument, that the K_L^0 decays violate time reversal invariance, has been confirmed directly, but those arguments are not transparent.

Lacking transparent indications of *T* invariance violation, let us examine a *T*-violating phenomena that must occur on some level but has not yet been seen experimentally: the electric dipole of the proton. Because we know that the weak interaction violates *T* invariance (through the evidence from the K_L^0 decays), we know that the neutron must have an electric dipole moment, although on some models of T violation the moment must be very small. Figure 13.8 presents a schematic diagram of the neutron with an electric dipole distribution. At the right we see the same neutron under time reversal; the spin is reversed but the electric charge distribution is unchanged. Clearly

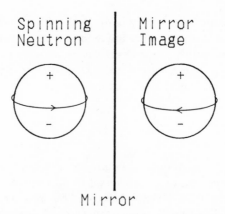

Figure 13.8 A spinning neutron with electric dipole moment together with the same state under time reversal.

there is a fundamental difference between the two pictures. If we consider the direction of the dipole moment to be the direction from the negative to positive charge, on the left side of the diagram, the dipole moment is in the direction of the spin; in the other time-reversed diagram (to the right), the dipole moment is opposed to the spin direction. If T invariance were to hold, there could be no difference between the two states, and thus there could be no neutron electric dipole moment.

It seems that nature is not invariant with respect to CP or T. Physicists have an occasional nightmare to the effect that there really is no fundamental order in the universe and that the little order we see is accidental. Before Johannes Kepler, men thought the planets moved about the sun in perfect circles. This is almost true; the orbits are very nearly circles. Then why are the circles imperfect? The orbits are almost circular for the least fundamental of reasons: frictional forces such as tides or the resistance of residual gases in space affect elliptical orbits in a way that brings them to a circle. Perhaps the symmetries we see, broken to some extent as CP is, are not the result of the Grand Design but only consequences of trivial accidents. Perhaps the elegance we expect in the laws of the universe is simply not there. Einstein said, "God is subtle but not malicious"—that is, He would not play tricks on us. Was Einstein right?

Summary of the *C, P,* and *T* Invariance Properties of the Fundamental Forces

We will now review the character of C, P, and T invariances. Invariance of interactions with respect to space inversion restricts observables to those which do not differentiate between a left-handed and right-handed coordinate system. Time reversal invariance allows only observables that do not depend on the direction of time, and invariance under charge conjugation restricts observables to those which remain unchanged when all particles are changed

Table 13.1 Transformation properties of the various fundamental interactions

Interactions	P	C	CP	T	CPT
Strong	+	+	+	+	+
Weak	−	−	+	+	+
Weak*	−	−	−	−	+
Electromagnetism	+	+	+	+	+
Gravity	+	+	+	+	+

Note: Invariance is denoted by +, noninvariance by −. Weak* includes the *CP*-violating force in the weak interaction.

to antiparticles. These invariances can be demonstrated in a particularly simple way. Consider a motion picture of a fundamental process, perhaps an elementary particle interaction in the presence of electric and magnetic fields resulting from charge and current distributions shown in the scene. If the interactions are invariant under space inversion, it will not be possible from a consideration of the projected scene to determine if the film has been reversed in the projector or projected by reflection in a mirror. If the interaction is invariant under time reversal and if entropy is not changed in the process, it will not be possible to tell if the film is run backward, and if the interactions are invariant under charge conjugation, it will not be possible to state whether the picture is that of our universe or an antiuniverse where every particle is replaced by its antiparticle.

These three invariances are probably not independent. In the framework of local field theory (the only way we know how to describe fields), invariance under a proper Lorentz transformation (that is, invariance of the description of nature with respect to a change from one inertial system to any other, but without space or time inversion) leads to the invariance of all interactions under the combined operations *CPT*. The equality of the masses and lifetimes of particles and antiparticles follows from this theorem (developed by G. Luders and W. Pauli). Although no violations of *CPT* invariance have been found, the experimental tests of *CPT* invariance have been limited in completeness and in sensitivity.

We now have enough information to list the transformation properties of all the interactions in Table 13.1. We have mentioned that there is a non-*CP*-invariant force that is probably—but may not be—specifically related to the weak interaction. The effective strength of this force is smaller, by a factor of perhaps 10^{-3}, than the primary weak interaction and can be considered a small perturbation on most weak interaction processes. We can then list the weak interactions with and without the addition of the *CP*-violating part of the total force.

Magnetic Monopoles

It is instructive to consider a simple, classical (though possibly hypothetical) system of particles and their interactions, which illustrates the use of such a basic analysis of invariances under C, P, and T. Figure 13.9 shows the trajectory of a particle holding an electric charge of $+e$ in the field of a large, isolated plate holding magnetic charges (monopole charges) g. We can differentiate between the "real" event to the left and the mirror image of the event to the right. The experimental configuration of monopoles and charges in the diagram differentiates between left-handed and right-handed coordinate systems, between the real world and the mirrored world; hence, parity is not conserved by the interactions that define the experiment. It is easy to see that the system is not invariant under the change of the electric charge to the anti-electric charge C_e or the exchange of the magnetic charge with the antimagnetic charge C_m. Each change would reverse the direction of the force on the moving charge. But the scene is unchanged if both electric and magnetic charges are reversed: $C = C_e \cdot C_m$. Similarly, the scene is not invariant under time reversal T; if we were to observe a motion picture of the electric charge passing through the plates, we could tell if the film were being run forward or backward. However, the scene (and thus the total interaction) is invariant under CPT: if we reversed the sign of the electric and magnetic charges and photographed the scene by reflection in a mirror—and then ran the film backward—everything would look normal.

There are still further subtleties. If the scenerio in Figure 13.9 is to have any meaning, we must have an operational way to define the sign of g (or, more properly, the relative signs of g and e). We can do this, for example, by considering the direction of the force on an element g in a field produced by a current of e^+ as in the lower left side of the figure. There, if the force is directed upward, the sign of g is positive. However, this definition is reversed in the right-hand mirror world: with that definition of the sign of g, the sign changes on reflection in a mirror (or reflection through the origin). Then we will have CP violation or time reversal violation only if there is another way of defining the sign of g apart from a definition solely in terms of electromag-

Figure 13.9 An electron moving through the magnetic field produced by a capacitor plate holding a magnetic pole charge seen directly and in a mirror.

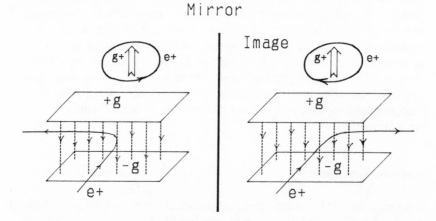

netic properties. As of 1986 we do not know if free monopoles exist, although there are compelling theoretical reasons for believing that they do.

If there are monopoles, and modern unified theories of elementary particles suggest that very heavy monopoles may exist, we are in a position to gain some insight into the quantization of electric charge. Dirac has pointed out that if monopoles exist (with pole strengths measured in the same units as electric charge), the product of any magnetic pole strength P and electric charge Q must be such that the product $PQ = n\hbar c/2$, where n is an integer.[2] Hence, if magnetic poles exist, a quantization of electric charge—and the magnetic pole strength—is required such that if g is the unit pole strength, then

$$eg = \frac{n}{2} \hbar c \qquad \text{where } n = 1, 2, 3, \ldots$$

Knowing that $e^2 = \frac{1}{137} \hbar c$, we see that the smallest possible value of the unit pole strength will be such that $g^2 = \frac{137}{4} \hbar c$; the unit pole strength must be much larger than the unit charge. Although the existence of a *Dirac monopole* requires the quantization of the electric charge, it does not shed any light on the value we observe, $e^2/\hbar c = \frac{1}{137}$.

The Conservation of Charge and Particle Numbers

We have made the point that the existence of conservation laws can usually, if not always, be related to the existence of symmetries. We observe that electric charge is conserved—the net charge, or the amount of positive charge minus the amount of negative charge, in an isolated system will be conserved. We observe that similar laws seem to hold for other quantities; the color charge appears to be conserved, the net number of quarks (quarks minus antiquarks) in an isolated system seems to be conserved, and the net number of each of the three families of leptons (electron, muon, and tau) appear to be conserved. It is important, indeed essential, that we learn to associate these conservation laws with other aspects of nature; and it is attractive, moreover, to attempt to connect the conservation of these quantities with a set of symmetries or invariances.

Although these quantities, charge and particle numbers, seem to be similar in that each conservation rule concerns the conservation of some integral number where individual particles can have either negative or positive values of the number, we have been rather successful only in relating the conservation of charge to an observed invariance of nature. The conservation of charge seems to be intimately related to the gauge invariance of electromagnetism. For simplicity, we might consider only a piece of gauge invariance; we know of no observable effect that depends on electrostatic potential.

We may illustrate what is meant by the invariance with respect to electrostatic potential by examining observations of nature as made in two laboratories placed in the electric fields found between the very large charged (capacitor) plates shown in Figure 13.10. The electric fields in the two laboratories are identical (and equal to $4\pi\sigma$, where σ is the charge density on the plates). The two laboratories are equivalent except that more energy will be required to bring a positive charge from the bottom plate to the higher laboratory than to the lower laboratory. We then say that the high laboratory is at a higher electrical potential. No known experiment can be performed in these laboratories which will allow the observers in the laboratories—our old friends from relativity studies, Oliver and Olivia—to determine which of the two laboratories is higher, or is at the higher potential.

According to Figure 13.10, both observers in their different laboratories are supplied with equal positive charges. Because the electric field is the same in the two laboratories, the force on the two charges is the same: only their absolute energy is different, but that is not measurable by any device we know. However, if the charges could be annihilated singly and the energy of annihilation could be measured, the energy seen by Oliver would be greater than that seen by Olivia at a lower potential (just as more energy was required to bring the charge Q to its position than the charge Q'), and that difference in energy could be used to measure the position or potential of the laboratories. Nature would not then be invariant with respect to electric potential. Conversely, if nature observes this invariance, we are able to understand our observation of the conservation of electric charge in terms of that invariance.

It is important to note that the annihilation of a pair of opposite charges is not forbidden by these arguments, just as no energy is required to bring positive-negative pairs of charges from the bottom plate to either laboratory.

Formally, the invariance of phenomena with respect to potential is related to a constraint on the description of charged particles in an electromagnetic field called *gauge invariance*. The concepts of gauge invariance have now been extended to other fields, so that observations of the conservation of

Figure 13.10   Laboratories at different electrostatic potentials.

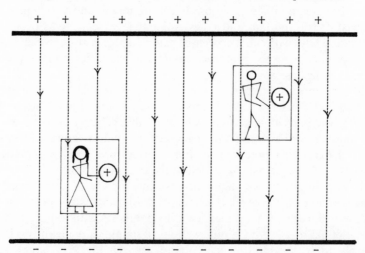

other particles might eventually be connected with the gauge properties of these fields in a coherent and unified picture of nature.

Notes

1. The wave functions describe states of a particle in a one-dimensional box of length L as shown in Figure 13.4. The amplitudes have the form

$$\psi = \frac{\sqrt{2}}{L} \sin (n\pi x/L)$$

where for (a), $n = 3$, and for (b), $n = 2$.

2. Although Dirac's proof of the relation between pole strength, charge, and Planck's constant is not simple, we can illustrate the character of the relation by a simple estimate. Here we consider that poles exist and one isolated plate of Figure 13.9 holds a pole density of σ_g poles per unit area. By analogy with the calculation of the electric field from such a plate holding an electric charge density (Chapter 4) and the symmetry between the electric field $\mathbf{E}$ and the magnetic field $\mathbf{B}c$, we find that the magnetic field near the plate is

$$cB = 2\pi\sigma_g$$

The electric charge moving with a velocity v, as shown in Figure 13.9, will move in a circle of radius r if the centrifugal force on the charged particle, mv^2/r, is equal to the force on the moving charge generated by the magnetic field, Bev

$$Bev = \frac{mv^2}{r} \quad \text{or} \quad Be = \frac{(mvr)}{r^2}$$

From the quantization of angular momentum, mvr is equal to $n\hbar$, where n is an integer. Substituting the pole density expression for B,

$$\frac{(2\pi\sigma_g)}{c} \cdot e = \frac{n\hbar}{r^2} \quad \text{or} \quad 2(\pi r^2\sigma_g) \cdot e = n\hbar c$$

This relation requires that the magnetic pole strength enclosed by the orbit of the electron be quantized. Setting that pole strength to be the smallest value, one single pole with a unit pole strength g, we have $\pi r^2\sigma_g = g$ and

$$g \cdot e = \frac{n}{2} \hbar c$$

Although our estimate is crude, it serves to illustrate the connection between the quantization of angular momentum and the quantization of charge and pole strength. Quantum mechanics requires a quantization of charge—if monopoles exist.

14

The Strong Interactions

Three quarks for Muster Mark!

James Joyce, *Finnegans Wake*

As of 1955, most physicists believed that nature had constructed the universe of nuclei made up of neutrons and protons held together with nuclear forces—the strong interaction forces—and that these forces between neutrons and protons were generated by the exchange of pions (or pi-mesons) between the nucleons (the generic label for neutrons and protons). Although other baryons, sources of the force and relatives of neutron and proton, and other mesons, carriers of the force and relatives of the pions, were visible on the horizon, there was an implicit hope that they would not prove to be important. However, during the next decade, experimental discoveries served to increase explosively the size of the sets of baryons and mesons. To make a universe from three building blocks is satisfactorily economical, but to use one hundred and more elements!—surely God would not be so untidy. There must be something more fundamental.

In the two decades following the introduction of the quark concept by Murray Gell-Mann and George Zweig in 1964, physicists constructed a simpler picture of fewer fundamental elements. The baryons and mesons (together, hadrons) were understood as compounds made up of quarks, a smaller set of more nearly fundamental particles, where the quarks were held together by forces generated through the exchange of gluons, the field quanta of the strong interactions. The quarks were source or matter particles, the gluons field or force particles, in a description of the strong forces parallel to that of electromagnetic forces between charges generated by the exchange of photons.

Although we might describe the strong interactions in terms of these revealed quarks and gluons, we would lose much of the spirit of modern particle physics by not discussing the path that led to this point. In science, the search is often as important as the discovery, as the character of past searches serves to guide future research. Hence, we begin by discussing the search.

Symmetries of the Strong Interactions—Charge Independence

Beginning with the pioneering work of Eugene Wigner and others, searches for an understanding of the strong interactions—and later, for an understanding of other forces—centered on analyses of the symmetries exhibited by the forces. Typically, observed patterns of particle behavior led to definitions of symmetries that in turn led to deep understandings of obscure fundamental structures.

Charge independence is the first (and simplest) important symmetry to come under such analysis, and it is a paradigm of the class of symmetries basic to present and future understandings of particles and fields. Hence, we study charge independence and $SU(2)$, the symbol for that symmetry, as a comparatively accessible example of a broad class of concepts as well as for its intrinsic importance in classifying particles and forces.

The discovery that the positive charge of atoms was contained in a very small nucleus of the atom led immediately to the question as to what held the charge together. It was easy to calculate that the repulsive forces generated by the charges were enormous on the scale of atomic energies. Indeed, for the gold nuclei used by Ernest Rutherford to scatter alpha particles in his discovery of the nucleus, the energy required to bind the nuclear charge together could not be less than 1 billion electron volts (1000 MeV),[1] very much larger than the energies of a few electron volts that bound atoms together to form molecules. It seemed that nature must provide another, specifically nuclear, force to provide the very strong binding required.

The discovery of isotopes by the mass spectroscopy of F. W. Aston in 1919 laid to rest the objections to the hypothesis of Prout that the mass of all atoms could be expressed as an integral number (which we write as A) of proton masses. But the X-ray spectra of H. G. J. Mosely, measured in 1913 (Mosely was killed at Gallipoli at the age of 27), augmented chemical evidence such as the periodic table of Mendeleev, which demonstrated that the number of units of positive charge Z was much less than the number of proton masses of a nucleus (for all save the lightest nuclei, Z is about one-half A). Hence, there must be neutral elements of the mass of the proton, and nuclei must be made up of about equal numbers of these protons and "neutrons." The existence of these neutral protons or neutrons was demonstrated by James Chadwick in 1932. In the next decade it became clear that except for the lack of charge, the neutron was very much like the proton.

The mass of the electron is considered to be largely due to the mass of the energy stored in the electrostatic field generated by the electron charge. Unfortunately, calculations of the electromagnetic mass, sophisticated or naive, lead to values of infinity unless corrections are introduced that cut off the contribution to the mass of the field at very small distances. Whatever the character of this "cutoff," the electromagnetic mass is proportional to the square of the electric charge, which, in the natural units of $\hbar c$, is equal to $\frac{1}{137}$ (that is, $e^2/\hbar c = \frac{1}{137}$).

We have described the forces between nucleons in terms of a nuclear force charge g, held by neutrons and protons, which couples to pions (or pi-mesons), considered quanta of the strong interaction field, rather as the electric charge e couples to photons, the quanta of the electromagnetic field. By analyses of various processes—the scattering of π-mesons from nucleons, nucleon-nucleon scattering, and the production of mesons from the interaction of very high-energy photons (gamma rays) with nucleons—the square of the strong interaction charge coupling the nucleon to the meson field is measured to be about 14.5 in the same units used to define the electric charge strength ($g^2/\hbar c$ = 14.5). We might then expect the mass associated with the strong interaction field to be on the order of 14.5 · 137 = 1850 times the mass of the electron—a number (accidentally) close to the measured ratio of the masses of the proton and electron. But if the accuracy is fortuitous, the result does suggest that the mass of the proton and neutron must be largely derived from the strong interaction field and then from the strong interaction charges of the proton and neutron.

Moreover, the mass of the proton is almost exactly equal to the mass of the neutron; the neutron is heavier by only about 0.15%, and this small difference is about the magnitude that might be attributed to different electromagnetic energies stemming from the different electric charge and current structures of the two particles. It would then seem that the neutron and proton have the same strong interaction charge and identical strong interaction properties. As far as the strong interactions are concerned, the neutron and proton appear to differ only by the labels associated with their different electric charges; one nucleon, called the proton, is labeled +, wheras the other is labeled 0 and called the neutron. Any binary label—red-green, boy-girl, or whatever—would serve as well as plus and zero charge.

This strong interaction symmetry of the neutron and proton, expressed as the *charge independence* of nuclear forces, has interesting and subtle consequences characteristic of a broad class of symmetries critically important to our understanding of forces. Hence, we will now consider charge independence, as the simplest member of this class, in some detail.

If the strong interactions do not differentiate between a neutron and proton, we might expect that a state (or particle) that is part neutron and part proton[2] must be affected in the same way as a proton or neutron by the strong interactions. An abstract coordinate system where the proton amplitude P is described as the "up" direction and the neutron amplitude N is "down" is shown in Figure 14.1. The other directions in this three-dimensional abstract *I-spin* space define different ratios of N and P with different phase differences between the amplitudes. If the strong interactions are charge independent, a new choice of proton amplitude P' and neutron amplitude N' will not be detectable by inspection of any strong interaction phenomena. This invariance is akin to the invariance of natural phenomena with respect to the orientation of a coordinate system. As a consequence of the rotational invariance of nature, the description of nature is independent of the direction chosen as the z-direction; as a consequence of charge independence, the strong interaction phenomena are independent of the direction in I-spin space chosen as the proton.

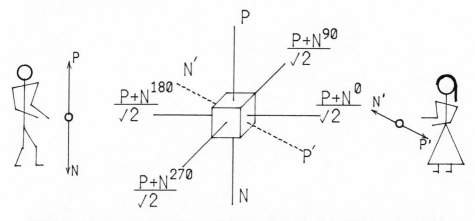

Figure 14.1 The diagram in the center shows the relation of Olivia's proton and neutron wave amplitudes, P' and N', to proton and neutron amplitudes in Oliver's universe, P and N. The axes of the three-dimensional abstract space are labeled with the values of the "rotated" proton and neutron in terms of P and N, the amplitudes in Oliver's universe. The superscript to N is the phase between the N and P amplitudes.

To illustrate this invariance with respect to orientation in the abstract I-spin space, consider a universe (perhaps ours) where a nuclear physicist, Oliver, measures properties of nuclei where only the strong interactions count. In Oliver's universe, the P and N states are defined as shown in Figure 14.1. He will fill his notebook with the results of his experiments. In another universe, which nature has provided with different rules, the proton and neutron are defined differently, as P' and N', in a manner described by the rotation shown in Figure 14.1. Here another nuclear physicist, Olivia, is filling her notebook with the results of her measurements. When the two notebooks are compared later, (somehow!) the results are identical. The strong interactions are invariant with respect to the orientation of the coordinate system in the abstract I-spin space of Figure 14.1. Olivia's proton and neutron are each combinations of Oliver's and vice versa, but the strong interactions cannot discriminate between the two systems.

As charge independence is defined, no matter how Nature might position the arrow, the universe She would generate would follow the same strong interaction rules. Indeed, we might think that She set the index arrow arbitrarily in constructing our universe and then assigned a positive electric charge to the proton She defined and zero charge to the neutron.

To illustrate the consequences of charge independence in a homely way, we might hire a very small observer and ask him to examine a set of neutrons and protons using only a strong interaction meter, as suggested by Figure 14.2. Because a proton and neutron differ only by a rotation of 180° in the I-spin coordinate system, and the strong interactions are invariant with respect to such rotations, our observer would find that he would not be able to differentiate between the neutron and proton with only this tool. Of course, if he were then to pull an electric field meter out of his pocket, he would see that the proton was charged and the neutron was not. The symmetry holds

Figure 14.2 A small observer with a strong interaction meter attempting, unsuccessfully, to differentiate between a neutron and a proton.

only for certain varieties of observations—observations of the effects of the strong interactions.

Because our small colleague with the strong interaction meter cannot tell a neutron from a proton, he must be equally inept in differentiating between the members of the different sets of two, three, or more nucleons arranged in the same configuration that can be constructed from neutrons and protons. Figure 14.3 shows such different sets of nucleons viewed by our puzzled helper. If these were sets of red and green children's blocks rather than nucleons marked with plus and zero charges, a color-blind observer would judge that the different members of each set of blocks were equivalent, and we might then expect our charge-blind helper to come to a similar conclusion concerning the sets of nucleons. Although this is almost the case, quantum rules hold in the submicroscopic world of nucleons, and there are nonclassical effects that would allow our small observer to separate each of these sets into smaller subsets of identical elements.

In describing the properties of sets of neutrons and protons under the *exchange* of label (or charge), it is necessary to keep in mind the wave nature

Figure 14.3 At the left are two sets of two nucleons in identical configurations together with multiplets; at the right are two sets of three nucleons in identical configurations.

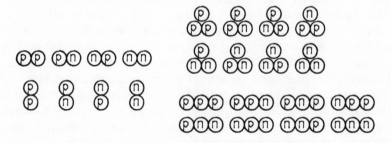

of the particles. The interchange of the labels of two particles—seen identically by relevant forces—of a set must at most change the sign of the wave describing the entire set. Because two exchanges must leave matters as they were, one change can result only in a multiplication of the de Broglie wave by $+1$ or -1. States are odd or even under the interchange of two of the constituents. Odd and even states differ, and the small observer in his submicroscopic quantum land can notice this difference. Nature is equally discriminating and, in general, treats the different subsets differently, holding them together with forces of different strength and thus giving them different masses.

Consequently, our observer does not find that the four two-nucleon sets are equivalent. He finds that three are identical but one is different. Obviously, even under interchange (as switching labels changes nothing), the *pp* and *nn* states are the same. Yet the *np* and *pn* are neither odd nor even under interchange, but change into each other. However, combinations can be constructed that are properly

$$\text{even: } \sqrt{\tfrac{1}{2}}\,(np + pn) \qquad \text{or} \qquad \text{odd: } \sqrt{\tfrac{1}{2}}\,(np - pn)$$

(The $\sqrt{\tfrac{1}{2}}$ sets the total intensity—the square of the wave amplitude—to one particle.) Now the small observer cannot differentiate between the even $np+pn$ combination and the even *nn* and *pp* states, but the odd state is different. The quartet breaks down into an even triplet of three equivalent states and an odd singlet that is different. Although two nucleons can be arranged in many ways—indeed, in an infinite number of standing waves or orbits—for each configuration, there will be either three states identical except for the electric charge (and effects of the charge), or one singlet state. If there are three states, one must have zero charge (as it is made up of two neutrons), one must have a charge of $+1$ (as it is made up of a neutron and proton), and one must have a charge of $+2$ (as a two-proton state). Any singlet (set of one) must have a charge of $+1$, just as it must be made up of a neutron and proton. Nature has set nuclear forces so that the most strongly bound set of two nucleons is such a singlet, the deuteron. Two nucleons in a triplet state are not strongly enough tied together by nuclear forces to form a bound state.

Larger sets of nucleons can also be broken down into characteristic subsets with different properties under the exchange of the labels of two nucleons. Although the systematic analysis of such separations into subsets or *multiplets* requires some facility with tensor manipulations,[3] the multiplets can usually be described simply. Sets of three nucleons form quartets, even under interchange of any two nucleons, and two different doublets. Moreover, all configurations of three nucleons will be seen by the small observer as equivalent quartets or doublets. Larger sets also have characteristic patterns of multiplets with members indistinguishable under strong interactions. For example, sets of four nucleons form quintet patterns, three different kinds of triplet patterns, and two different singlets. The mathematical symbol for this symmetry under the exchange of two labels is $SU(2)$.

Even as the $SU(2)$ symmetry leads to multiplet patterns in nuclei, a study of the sets of observed patterns serves to reveal the underlying $SU(2)$ symmetry. Assume that our small friend lives in a universe in which (quasi) fundamental particles seem to come in charge-independent quartets and doublets. After studying the patterns of particles, he might conclude that the particles follow an $SU(2)$ symmetry applied to sets of three basic entities. With this knowledge, he might guess that his nominally fundamental particles are actually compounds of three truly basic particles of two kinds, one with no charge and one with a positive charge. And if he does not want to call them quarks, he might give them the name "neutron" and "proton"! In our universe, we must also raise the question as to whether certain particles long considered fundamental, such as the neutron and proton, are not actually compounds of more truly fundamental particles, such as quarks. And we search for an answer to the question by looking at patterns.

Even as the neutron and proton constitute a doublet, two particles equivalent with respect to the strong interactions, the three pions π^+, π^0, and π^- (where the superscript gives the electric charge), form such a charge-independent triplet. The three pions have almost the same mass (about ⅐ of the mass of the nucleon, or 270 times the mass of the electron) and again seem to be identical in their strong interaction properties. We can presume that our small observer, armed only with his strong interaction meter, could not differentiate among the three pions. Looking further, we find that all strongly interacting elementary particles belong to families or multiplets of 1, 2, 3, or more particles, where the members of the multiplet have different electric charges but are not distinguishable through the strong interactions.

Charge-Independence Selection Rules

The existence of the symmetry requiring that no measurement—by the small observer or by anyone else—using only the strong interactions can differentiate between members of a multiplet leads to interesting *selection rules* for the interactions of hadrons through the strong interactions. We may gain insight into such selection rules by considering a possible decay of a triplet meson state A^0 to two singlet meson states B^0 and C^0. In our notation the subscript represents the electric charge; hence, the states we consider are all neutral. We presume that the mass of the A-particle is greater than the sum of the masses of the C- and B-particles. Then energy and momentum would be conserved in the decay, and there would seem to be no reason that the transition would not occur. However, the charge-independence symmetry forbids the decay!

$$\begin{matrix} A^+ \\ A^0 \\ A^- \end{matrix} \quad \rightarrow \quad B^0 + C^0$$

We note that the sibling states A^+ and A^- cannot decay to states B and C, since charge could not be conserved. Then if the A^0 particle decayed through the strong interactions (but not A^+ or A^-), the small observer could differentiate between the members of the A-triplet. The contradiction is removed only if the decay $A^0 \rightarrow B^0 + C^0$ is forbidden.

Although the consequences of symmetries such as charge independence in forbidding reactions and decays are interesting, the illumination of fundamental symmetries through the observation that some transitions are forbidden is much more important.[4]

Relative Decay Rates

Aside from forbidding decays, the charge-independence symmetry of strong interactions leads naturally to relations among the rates of decay, or lifetimes, of unstable particles. We illustrate relations imposed by charge independence of the strong forces by examining the decay of members of an I-spin quartet of baryons called the Δ to a pion and nucleon. The Δ states were discovered experimentally by Enrico Fermi and his colleagues about 1951.

Using the basic rule that the strong interactions do not differentiate among the different electric charge states of an I-spin multiplet, we may undertake an idealized experiment (similar to real experiments conducted by Fermi) in which we select an equal number of the four different charge states of the Δ and observe their decays into nucleons and pions. If the strong interactions do not discriminate between the four Δ states, each must decay at the same rate—in a given time, the probability of each decaying must be the same. Moreover, the set of four must decay to equal numbers of neutrons and protons, and equal numbers of π^+, π^0, and π^- mesons, or else the strong interactions would discriminate among the members of the doublet nucleon set or the triplet pion set.

As an illustration, assume that a large set of Δ particles was produced such that there was no charge bias in the sample—that is, initially there were equal numbers of each of the four charge states. Upon observing 12,000 decays in some short time, we would expect *on the average* to observe 3000 Δ^{++} decays, 3000 Δ^+ decays, 3000 Δ^0 decays, and 3000 Δ^- decays. Among the decay products there must be about 6000 neutrons and 6000 protons and 4000 π^+, 4000 π^0, and 4000 π^- mesons. From these conditions, imposed by charge independence, it follows that there will be 3000 $\Delta^{++} \rightarrow p + \pi^+$ decays, 2000 $\Delta^+ \rightarrow p + \pi^0$ decays, 1000 $\Delta^+ \rightarrow n + \pi^+$ decays, 1000 $\Delta^0 \rightarrow p + \pi^-$ decays, 2000 $\Delta^0 \rightarrow n + \pi^0$ decays, and 3000 $\Delta^- \rightarrow n + \pi^-$ decays. The decay ratios are expressed simply in Table 14.1.

In summary, charge independence requires that all the Δ states decay at the same rate but that the Δ^{++} and Δ^- decay only to a nucleon and a pion. However, the Δ^0 must decay one-third of the time to a proton plus a π^- and two-thirds of the time to a neutron and a π^0, and the Δ^+ must decay two-thirds of the time to a proton and π^0 and one-third of the time to a neutron and π^+.

These decay ratios are observed and serve to add to the evidence that the

Table 14.1 Decay ratios

Reaction	Relative Rate
$\Delta^{++} \rightarrow p + \pi^+$	1
$\Delta^+ \ \ \rightarrow p + \pi^0$	⅔
$\Delta^+ \ \ \rightarrow n + \pi$	⅓
$\Delta^0 \ \ \rightarrow p + \pi^-$	⅓
$\Delta^0 \ \ \rightarrow n + \pi^0$	⅔
$\Delta^- \ \ \rightarrow n + \pi$	1

strong interactions are charge independent; that the neutron and proton form a nucleon doublet; and that the three pions constitute a triplet of particles that are identical as viewed by strong forces.

SU(3)—Hypercharge and Quarks

For some time after C. F. Powell and his colleagues discovered the pion in 1948 in the course of their analyses of tracks in special photographic emulsions exposed to very high-energy cosmic rays, many physicists believed that they were quite close to an almost complete understanding of the character of the strong interactions. Neutrons and protons were thought to be the only particles carrying the strong interaction charge, and the pion was thought to be the only quanta of the strong interaction field. This simple picture, first described by Hidekei Yukawa a decade earlier, was thought to be essentially correct, and there was a near consensus of belief that if the mathematical problems of handling large charges could be solved, almost everything would fall into place.

However, at almost the same time, G. D. Rochester and C. C. Butler at the University of Manchester, England, found evidence of new particles in *V*-shaped droplet tracks that appeared in cloud chambers exposed to high-energy cosmic rays. They gave the particles whose decays formed the strange tracks the phenomenological name *V*-particles. Later, the several varieties of particles of the kind whose decays caused the strange *V*-shaped tracks were called "strange particles."

These strange particles, produced in collisions between very high-energy cosmic-ray protons with nuclei of nitrogen and oxygen in the air, were found to interact with matter strongly (as with the strong interactions). However, the particles decayed to nucleons and pions (and to other particles) with lifetimes that were about 10^{13} times longer than the characteristic lifetimes for strong interaction decays such as that of the Δ. This enormous difference in lifetimes—the difference between a day and the age of the universe—had to be fundamental and important. Indeed, the decay rates were of a magnitude to be expected if the strong interactions took no part at all and the decay took place only through the weak interactions responsible for beta decay. Furthermore, these strange particles were never produced singly in the strong interactions of the production process but instead in association with other strange particles.

Perhaps the simplest description of the phenomena was contained in the postulate that the strongly interacting particles could carry an additive quantity, like a charge, and that this quantity was conserved in the strong interactions but not by the weak interactions. The name *hypercharge* or *strangeness* was given to this hypothetical quantity. (The slight difference in definition of hypercharge and strangeness is not important to us.)

From this view, the strange particles could be produced in pairs through the strong interactions of high-energy hadrons where one strange particle would have a positive strangeness and the other a negative strangeness; ordinary matter was considered to have zero strangeness. Because the strangeness charge was conserved in the strong interactions, these particles, carrying a strange charge, could not decay quickly through the strong interactions but only very slowly through the comparatively feeble weak interactions that did not conserve strangeness.

Having observed $SU(2)$ patterns or multiplets of elementary particles where members of a multiplet with different charge behaved the same with respect to the strong interactions, it seemed reasonable to search for larger patterns or *supermultiplets* of particles with different electrical charge and different hypercharge acted upon equally by the strong interactions. Did supermultiplets exist such that members of the set, although differing in charge and hypercharge, could not be separated by our small observer armed only with his strong interaction meter?

Members of an $SU(2)$ multiplet, indistinguishable by the strong interactions, differ only in their electrical charge; hence, the particles of a multiplet can be displayed symbolically as points on a line separated by one unit of length representing a unit of charge. Then the neutron-proton doublet can be represented by two points on a charge axis; the π^-, π^0, and π^+, as three points; and the Δ^{++}, Δ^+, Δ^0, and Δ^- by four equally spaced points in a line. But points on a line cannot display particles with different electrical charges and different hypercharges; a two-dimensional plot is required. Attempts to assemble such charge-hypercharge supermultiplet patterns were complicated by what seemed to be strong hypercharge forces. If there was a symmetry with respect to strong interactions such that members of a supermultiplet could not be distinguished by strong interactions alone, that symmetry was badly broken by what appeared to be forces associated with the hypercharge; sets of particles with different hypercharge had quite different masses. It was as if the symmetry of red and green blocks of the same size were broken by having the green blocks made of much heavier wood. However, through the exercise of skill and ingenuity, ways were found to separate effects resulting from hypercharge, and two-dimensional patterns were found.

Figure 14.4 shows the most important meson octet and baryon octet and decuplet patterns where the points on a horizontal line represent particles that are members of $SU(2)$ multiplets and are very much alike except for their different electrical charges. The vertical separations show differences in

Figure 14.4 The lightest meson octet is shown at the left, the basic baryon octet in the center, and the baryon decuplet at the right. The electric charge of the particles is placed as a superscript and the hypercharge as a subscript. The η^0 and π^0, and the Σ^0 and Λ^0, occupy the same places on the diagram.

strangeness or hypercharge. The precise definition of the coordinates is otherwise not important.

With these patterns established by experiment and analysis, Gell-Mann in the United States and Yuval Ne'mann, an Israeli army officer on leave to study in England, found the meaning of the patterns. The patterns were characteristic of an $SU(3)$ symmetry, a symmetry that could have its origin in the invariance of forces with respect to the interchange of three labels—hence a symmetry that might follow if all hadrons (mesons and baryons) were made up of three basic particles.

But was there a basic set of three particles—ones more deeply fundamental than any of the familiar particles found in nature? In order that compounds of these triplets should fit observed particles in a satisfactory manner, it would seem that the particles must be spin ½ fermions with fractional charge! The values of the charge must be one-third and two-thirds of the charges of the proton or electron, and *nothing like that had ever been seen.* But with boldness and insight, Gell-Mann and, independently, George Zweig, suggested in 1964 that three real particles with fractional charge—as seen in Figure 14.5—might exist as the fundamental building blocks of the strongly interacting particles, the hadrons. From this view all the known particles were compounds of the basic triplets and their antiparticles. Gell-Mann, a sometime student of the Gaelic language and of James Joyce, named the hypothetical particles *quarks* after a line from a sonnetlike sequence from Joyce's *Finnegans Wake,* "Three quarks for Muster Mark!" Zweig called the particles "aces," but quarks they became.

Just as two points on a line, the simplest nontrivial symmetry in one dimension, represent the neutron and proton that make up the fundamental nucleon doublet, so do three points arranged in a triangle as shown in Figure 14.5, the simplest nontrivial symmetry in two dimensions, represent the members of the fundamental multiplet of triplet quark states labeled[5] *u, d,* and *s.* Just as symmetry under the exchange of two particles, the neutron and proton, was the basis for the construction of a one-dimensional nuclear sym-

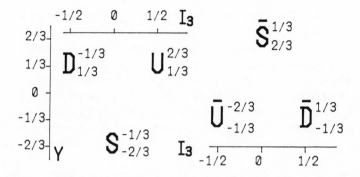

Figure 14.5 The *u*-, *d*-, and *s*-quarks are shown on a two-dimensional manifold with the antiquark triplet to the right. The superscripts are the particle charge Q, the subscripts the hypercharge Y. The orthogonal coordinates[7] Y and $I_3 = Q - Y/2$ are useful technically.

metry $SU(2)$, symmetry under the exchange of three particles, the *u*,- *d*,- and *s*-quarks, would constitute the basis of a two-dimensional symmetry,[6] $SU(3)$. Hence, the conjecture of interest is the hypothesis that the strong interactions are (almost) independent of the label of the quark, just as we know that nuclear interactions are independent of the label of the nucleon. We call the quark labels "flavors," and we presume that the assumption that the strong forces are flavor-independent will be a useful approximation.

Aside from the flavor independence of the strong forces, in order to account for the properties of hadrons, quarks must have certain properties:

1. The quarks must have fractional charge—the *u*-quark a charge of ⅔ and the *d*-quark and *s*-quark charges of −⅓, with the respective antiquarks having charges of opposite sign.
2. The three quarks have to be fermions with a spin of $\hbar/2$.
3. Each of the quarks must be conserved by the strong interactions—for each variety, *u*, *d*, and *s*, the number of quarks minus the number of antiquarks cannot be changed by the strong interactions.
4. Although the weak interactions can change one kind of quark to another, the total number of quarks (minus antiquarks) is conserved absolutely.

The conservation of particular quarks accounted for the conservation of strangeness; the conservation of quarks as a whole accounted for the conservation of baryons and the stability of matter made up of the particular baryons, the neutron and proton.

Meson Octets

With the concept of quark building blocks in hand, we can consider the supermultiplets that might be constructed from the blocks and relate those supermultiplets to sets found in nature. Three quarks and three antiquarks are represented by points in the coordinate system of Figure 14.6. Points on the diagram to the right show the nine various possible quark-antiquark combinations. In a manner similar to the reduction of the four combinations of two nucleons to singlet and triplet $SU(2)$ multiplets, in view of the phases of the de Broglie amplitudes, the set of nine quark-antiquark states is split to an octet that is even under the exchange of two labels, and a singlet that is odd under label exchange. If the symmetry were exact, the masses of the eight members of an octet would be the same, but that mass could differ from the mass of a singlet state.

One might expect that the mass of such quark compounds would be a little less than the sum of the masses of the individual quarks and then provide some indication of the quark mass. Because the binding energies are so large, this is not the case. Although all bound systems must be lighter than the sum of the free parts by an amount $dm = E/c^2$, where E is the energy required to break up the compound, the relative magnitudes of the binding mass dm and the mass of the parts is much larger for quark systems than for chemical or nuclear bindings. For chemical systems the ratio of binding energy to mass is typically on the order of 10^{-9}; the water and carbon dioxide from the burning of gasoline in an engine weigh about one-billionth less than the gasoline and oxygen consumed. Nuclei have masses about 1% less than the sum of the masses of the neutrons and protons from which they are constituted. But the interactions that hold quarks together to form mesons and baryons are even stronger, and the masses of the mesons and baryons must be much less than the masses of the free constituent quarks—indeed, we now believe that an infinite energy would be required to free a quark, hence no

Figure 14.6 Construction of a nonet from a set of (three) quarks and antiquarks. The breakdown of the nonet to octet and singlet multiplets with definite symmetries is shown below. The $d\bar{d}$, $u\bar{u}$, and $s\bar{s}$ occupy the same places in the diagram, and the two central members of the octet and the singlet state lie on the same coordinates.

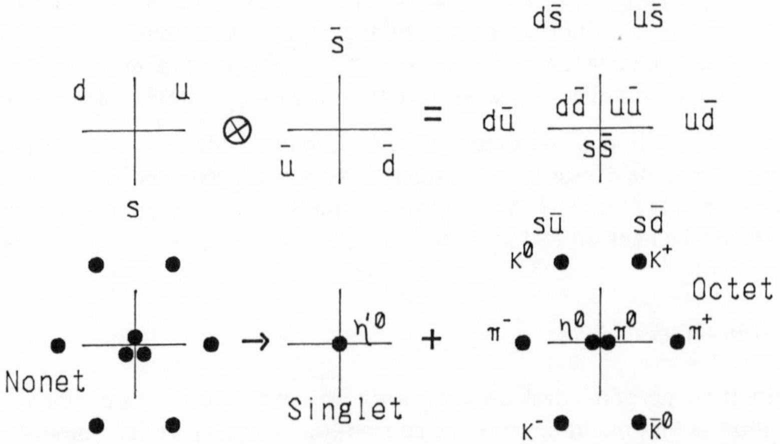

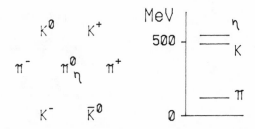

Figure 14.7 At the left, the light meson octet is shown on a $Y-I_3$ diagram. At the right, the masses of the octet particles are shown on an energy-level diagram.

free quarks exist. Nevertheless, the differences in the masses of compounds made up of different kinds of quarks do seem to follow differences in the intrinsic masses of the quarks.

We have noted that compounds of a quark and antiquark form only octets and singlets. Remarkably, we see such octet and singlet mesons in nature—and only singlets and octets. The lightest mesons can be grouped into two octets and two singlets such that for one octet and singlet, the quark and anti-quark have their spins in opposite directions, adding to zero, and for the other, the quark and antiquark spins are aligned, adding to $1\hbar$.

The $SU(3)$ symmetry is surely inexact. The members of the octets are not identical, since they have quite different masses. The points plotted on the Y-I_3 coordinate system—Figure 14.7—represent the particles making up the lightest meson octet. An energy-level diagram at the right shows the masses of the different states. The masses differ; the symmetry must be broken by some asymmetric effect.

Broken symmetries are common in physics, and if the symmetry breaking is not too large and has a sufficiently simple character, the imperfect symmetry can be useful in classifying information and as a guide to the structure of nature. It seems that this $SU(3)$ symmetry is broken in an especially simple manner; the strong interaction forces appear to be the same for all quarks, but the s-quark is intrinsically heavier than the u-quark and d-quark, which have almost the same mass. [The more nearly exact $SU(2)$ charge-independence symmetry follows from invariance under the exchange of the u-quark and d-quark.]

Baryon Multiplets

Although quark-antiquark compounds describe meson states satisfactorily, the baryons such as the neutron and proton—fermions with spin ½—must be made up of three quarks, with antibaryons constructed of three antiquarks. (Compounds of two quarks are not found in nature.) If the three quarks were children's blocks with different letters—u, d, and s—engraved on them, for any configuration of the blocks, 27 sets could be constructed representing different permutations of the letters. Then baryons would come in sets of 27 if

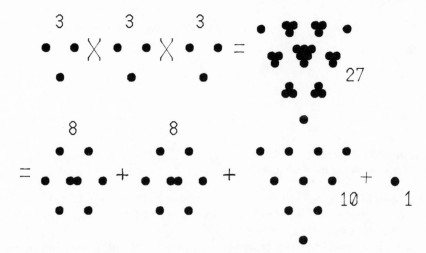

Figure 14.8 The 27 baryon states constructed from combinations of three quarks are shown; below is the breakdown of the 27-plet into two octets, a decuplet and a singlet, each with a different symmetry.

the forces were independent of flavor. When the phases of the de Broglie amplitudes are properly taken into account, different supermultiplet subsets emerge; there is a singlet such that the amplitude changes sign upon any exchange of two quarks, a decuplet such that the amplitude does not change sign upon any exchange, and two different octets such that the amplitude changes sign upon one exchange but not the other. These multiplets are displayed in Figure 14.8.

The neutron and proton are members of the lightest baryon supermultiplet, an octet made up of three quarks disposed such that the total spin is $\frac{1}{2}\hbar$. Figure 14.9 shows the positions of the members of this lightest baryon octet on a Y-I_3 diagram as well as an energy-level diagram showing the masses of the different particles. Another especially interesting set of closely related

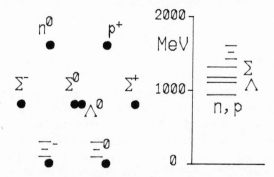

Figure 14.9 The baryon octet and an energy-level diagram.

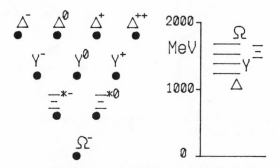

Figure 14.10 The baryon decuplet and energy-level diagram.

states form a decuplet, although with the three quarks aligned so that the particle spin is $\frac{3}{2}\hbar$. Figure 14.10 presents the quark configuration of these states as well as an energy-level diagram showing the particle masses.

The decuplet energy-level diagram illustrates with special clarity the dependence of the mass of the particles making up the decuplet on the number of s-quarks, as well as the independence of the mass on the exchange of u- and d-quarks. As one goes from the Δ to the Y^* to the X^* to the Ω, the number of strange quarks goes from 0 to 1 to 2 and to 3 for the Ω^-. Likewise, the mass increases in increments of almost exactly 145 MeV/c^2, slightly more than the mass of a pion, as one goes in steps from the Δ to the Ω. The pattern is very much as if the mass of the s-quark were 145 MeV/c^2 greater than the masses of the u-quark and d-quarks. A more detailed inquiry shows that one cannot make quite so precise a statement of the mass differences, but that one can consider that there is a real quark mass difference of about that magnitude: the s-quark is appreciably heavier than the u- and d-quarks.

The entire systematics suggests that these quark mass differences are intrinsic and do not stem from any difference in the effective strong charges carried by the quarks. The strong interaction forces appear to be flavor independent, although the masses of the different flavor quarks differ.

As one moves horizontally from left to right on the Y-I_3 diagrams of Figures 14.9 and 14.10, one exchanges d-quarks for u-quarks. The mass increases slightly with such exchanges, thereby suggesting that the u-quark is slightly heavier (a few MeV/c^2) than the d-quark.

In spite of the success of the quark model in explaining properties of mesons and baryons, the character of the decuplet indicates a grave inadequacy in the model. The particles represented by the corners of the decuplet triangle shown in Figure 14.10 are compounds of three identical quarks: the Δ^- is a *ddd* state, the Δ^{++} is a *uuu* state, and the Ω^- is an *sss* state. Because all decuplet particles have spin $\frac{3}{2}\hbar$, the three spin $\frac{1}{2}$ quarks must be aligned; their spins must be in the same direction. Quarks are fermions, hence the Pauli Exclusion Principle requires that no two identical quarks occupy the same state. Because the spin directions are the same for the three identical

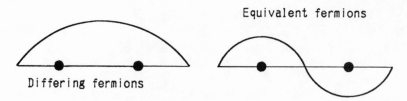

Figure 14.11 At the left, a half wave de Broglie amplitude with no nodes. At the right, a full wave with central node.

quarks making up these states, the space wave functions (or orbits) must be different. In particular, the de Broglie wave describing the quarks must change sign upon the interchange of any two identical quarks. But if the wave function is positive at the position of one particle and negative at the position of the second—so the signs reverse upon exchange of the particles—the wave function must go to zero somewhere between the two, as suggested by Figure 14.11; there must be a node in the space wave function. Hence, there must be nodes in the standing waves that describe the quarks bound together to form the decuplet states.

The least energetic state of a particle in a box is that state which corresponds to the standing wave with the longest wavelength (recalling that the momentum of a particle is inversely proportional to the wavelength). The fewer nodes, the longer the wavelength, the lower the momentum, and the lower the energy. Hence, the least energetic state can be expected to have no nodes in the wave function (save at the boundaries of the box). We can consider the decuplet states as boxes holding the three quarks. But the decuplet in question is the lowest mass (or energy) state of its category; hence, there can be no nodes in the decuplet space wave function. A critical contradiction exists. Although the $SU(3)$ model explains a hundred things and more correctly, with one failure we know that this simple model, describing hadrons as compounds of quarks carrying charge and hypercharge, is wrong. We must change the model—but not too much!

Color

A paradox exists. The quark picture works too well to give up, but there are serious problems with the conventional quark model; the restriction of all hadron states to $q\bar{q}$ meson states and qqq baryon states seems excessively arbitrary, and the symmetry of the low-mass baryon states violates the Exclusion Principle and thus our understanding of the fundamental character of fermions. The paradox is resolved by the introduction of a new quark property called *color*.

In essence, we postulate three different kinds of strong interaction charge (as compared to one electromagnetic charge), and each quark carries one of these three charges. This is very like saying that there are three times as many quarks, inasmuch as each flavor, *u, d,* or *s,* can have one of the three different strong charges.

We need names for the three charges. Physicists have come to use the names of colors for the different charges and the term "color" as a generic name for this class of charges. We choose to call the three colors red, green, and blue for the primary colors, and call the negative or anticolors carried by the antiquarks simply antired, antigreen, and antiblue (though the actual complementary colors would be cyan, magenta, and yellow). Although there are some relations—more poetic than scientific—between "color forces" and "spectral colors," we must be careful to differentiate between the two; a red quark will scare no microscopic bull, and the blue of a blue quark has nothing to do with Yale blue.

With more kinds of quarks, we can solve the symmetry problem. For any space, spin direction, and flavor, there are three kinds of quarks: red, green, and blue (that is, carrying red, green, or blue charges). Moreover, we assume that the color charges are exactly equal so that the forces are independent of the particular color of the quark. There is then a symmetry under a threefold color exchange that we label $SU(3)_c$. The problem of the decuplet symmetry is then removed by giving the otherwise identical three quarks different color charges in an appropriate manner. Symbolically, the ddd state (for example) is then a $d_g d_r d_b$ state, where the subscripts indicate the color charge held by the particle. With different color charges, the three d-quarks are no longer identical fermions, and the Pauli Exclusion Principle is modified.

The multiplet structure of three quarks under the exchange of three color labels is the same as the structure under the exchange of three flavors. A small color-blind observer will be able to separate the 27 color combinations into a decuplet that is even upon exchange of colors; two octets that are even under one exchange and odd under another; and a singlet that is odd under any exchange. If the flavor decuplet states that are even under the exchange of flavor, spin, and space coordinates are to be odd under total exchange and satisfy the Exclusion Principle, they must hold a color singlet combination.

We extend our already rather fanciful use of the term color by calling the singlet color set white. The singlet color combinations for real colors actually do combine to give white, whereas the other representations have members that in terms of real color would turn out to be various garish hues.

For the meson states, we must consider combinations of color and anti-color again in an $SU(3)$ format. The nine combinations of three colors and three anticolors (or complementary colors) form an $SU(3)_c$ octet, and a singlet that is odd under label interchange. Again, we find that the singlet color-state assignment fits the observations we have made. This combination, expressed fancifully in real colors, is of a primary color and its complement adding to white. Again, singlets are white.

What evidence do we have that some of the many particles we observe do not belong to other color representations? Because the other representations have high multiplicities, for every member of a definite space, spin, flavor, category, there would be many color states. If the nucleon belonged to a color octet instead of a singlet, we would expect to see eight similar particles, eight nucleons with different colors. Indeed, if the color symmetry were exact (as we believe it is), the eight different nucleons, carrying different color-charge

combinations, would all have the same mass. Because they would be different, however, the Pauli Exclusion Principle would not hold for the differently
colored nucleons, and they could be distinguished from each other even if the
color per se were difficult or impossible to detect directly.[7] Even if we were
color-blind, the use of the Pauli Principle would tell us that the particles were
different from each other.

Hence, the universe we observe is made up only of singlet color states—
or only of white particles. Why? Although from the color symmetry we can
be sure that the color forces act in the same manner for each member of a
color multiplet, the color forces can be quite different for the different multiplets. We know that same-sign electric charges repel, and opposite sign
charges attract. The three color charges and their anticolors follow more complicated rules. In particular, our observations suggest that the color forces are
attractive only for the color singlets. For the qqq color octets and color decu-
plets, and for the $q\bar{q}$ color octet and the triplets, sextets, and other representations constructed from combinations of quarks other than qqq or $q\bar{q}$ (which
would produce particles of fractional charge), the forces are effectively repulsive. As a consequence, sets of quarks are bound together only in $SU(3)_c$ singlets, and only qqq and $q\bar{q}$ combinations can form singlets. One cannot make
white from two or four primary colors; hence, one cannot have singlet color
states from two-quark or four-quark combinations. (Actually, combinations
of baryons and mesons will also constitute color singlets, and such hadron
"atoms" may exist.)

In summary, the introduction of color solves the symmetry problem of
the baryon multiplets and presents a logically consistent explanation for the
lack of existence of quark combinations other than the qqq of barons and the
$q\bar{q}$ structure of mesons. We will find that there are other strong interaction
phenomena that provide evidence for color. Moreover, the weak interaction
systematics provide reasons for the introduction of color that are at least as
compelling as the strong interactions.

Charm, Top, and Bottom—The Six Conserved Flavors

The three quarks u, d, and s are different, and differentiable, particles conserved separately by the strong interactions. Early suggestions that this set of
basic, different, conserved particles should be extended from three quarks to
four—or six or more—was given substance when it was noted that the existence of a fourth quark, labeled c for charm, would act so as to suppress certain weak decays to the observed level (Chapter 15).

The introduction of a fourth c-quark also adds an aesthetically attractive
symmetry to the flavor systematics; the s-quark and c-quark form a charge
doublet similar to the u-d doublet. We now know of two more quarks, a fifth
quark named b and are confident of the existence of a sixth quark t, which
are considered to form a third doublet family as shown in Table 14.2. We
cannot be certain that there are not even more families of quarks, albeit with
very large masses.

Although the flavor independence of color forces appears to extend to the

Table 14.2   Quark families

Quark	Electric Charge
d	$-\frac{1}{3}$
u	$+\frac{2}{3}$
s	$-\frac{1}{3}$
c	$+\frac{2}{3}$
b	$-\frac{1}{3}$
t	$+\frac{2}{3}$

heavy c,- b,- and t-quarks, the $SU(6)_f$ flavor symmetry we might expect upon the exchange of six quarks is badly broken by the very different masses associated with the different flavor quarks. Although the mass of particles that are never found free is not reasily determined or even defined, we know that the masses of the u- and d-quarks are nearly equal and small. Depending on the definition used, the masses will be between a few MeV/c^2 and, perhaps, 300 MeV/c^2, one-third the mass of the proton. The s-quark must be heavier by an amount on the order of 150 MeV/c^2, and the c-quark mass is about 1500 MeV/c^2, or about 1.5 times the mass of a proton. The b-quark has a mass of about 5000 MeV/c^2 (heavier than a helium atom), and the t-quark mass may be about 40,000 MeV/c^2, or near the mass of a calcium atom. As a consequence of the large symmetry breaking, only the u- and d-quarks can be interchanged and leave a state nearly unchanged (hence the equivalence of the neutron and proton and the success of $SU(2)$ nuclear symmetry). Although we believe that the forces on the other quarks are not different from the forces on the light quarks, the great differences in quark masses mask ordinary symmetry effects, and $SU(6)$ super-supermultiplets cannot be expected to be evident.

Partons

We have been considering a picture of baryons and mesons as quark compounds very much as a molecule is a compound of atoms or a nucleus is a compound of neutrons and protons. Considerable evidence for this picture has been accumulated from inspection of the symmetries evident in the sets of mesons and baryons uncovered experimentally. However, quite different kinds of observations led to quite independent conclusions to the effect that the hadrons were compounds of more nearly fundamental particles; only later were the quarks identified certainly with the pointlike *partons* seen in the high-energy electron and neutrino collisions we discuss now.

Recall that Rutherford established the existence of a very small nucleus in a very much larger atom by bombarding the atom with alpha particles and observing large-angle collisions of the alpha particles from the (relatively) pointlike nucleus as a scatterer. Physicists from the Massachusetts Institute of Technology and Stanford University conducted the same kind of experiment using electrons instead of alpha particles and bombarding nucleons (neutrons and protons) instead of atoms. Instead of observers looking with a microscope at flashes produced by scattered alpha particles striking a zinc sulfide sheet, the MIT-Stanford group observed flashes produced in scintillating plastic using light-sensitive vacuum tubes to record the passage of scattered electrons. And instead of an investment of perhaps a thousand pounds Sterling and a man-year of labor for Rutherford's experiment, the modern experiment cost several million dollars and required the expenditure of more than 20 man-years of effort. But the results were similar and of similar importance. The patterns of scattered electrons showed conclusively that the electrons were scattered from pointlike constituents in the nucleons. The nucleons could not then be considered fundamental but must be compounds made up of some more fundamental entity.

Electrons with energies of the order of 15 GeV/c^2 were used in the experiment, which means that the effective mass of the electron was about the same as that of an oxygen nucleus. For such high-energy collisions, the forces holding the partons in place could be neglected, and the partons acted very much as free particles. Indeed, one could consider that the electrons were striking free partons rather than the nucleons. Then, from the analysis of the scattering of the electrons, the charges of the partons could be determined, and the motion of the parton in the nucleon could be deduced. Through such measurements—and other, better indicators—it became clear that the partons were fractionally charged quarks; the two levels of inquiry combined.

It is especially interesting to consider the kinetic energy distribution of the quarks making up a neutron or proton in a coordinate system such that the electron is stationary and the nucleons are moving with very large energies toward the electron. In such a frame of reference, the total nucleon energy E—which is very much greater than the rest mass energy of a nucleon; $E >> M_n c^2$—is almost exactly the sum of the energies of the constituents. From measurements of the scattering of electrons—and especially neutrinos—from the quarks, it is possible to construct the energy distribution of the quarks. This distribution is shown in Figure 14.12, where the proportion of the total nucleon energy carried by quarks with a definite proportional energy is plotted against the fraction x of quark energy to total nucleon energy. For example, the shaded area in Figure 14.12 is proportional to the amount of energy carried by all quarks that have a quark energy between 0.5 and 0.6 of the nucleon energy.

The ratio of the area under the curve of Figure 14.12 to the total enclosed area is equal to the proportion of the total nucleon energy carried by the quarks on the average. That ratio is about 50%—the three constituent quarks carry about 45% of the energy, and antiquarks, which have a transient existence in the nucleus, carry another 5% of the energy. If only half of the energy of the nucleon is carried by quarks, what carries the rest? We conclude that the strong interaction field carries that energy in the same manner that the electromagnetic field carries energy. Because the electrons used in the electron recoil experiments do not "see" the field, the strong or color field quanta, which we call "gluons," must not be charged. Neutrinos, which carry no electric charge, scatter off quarks as a consequence of the weak charges carried by

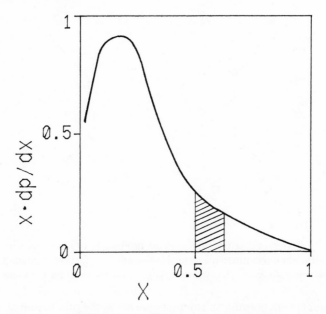

Figure 14.12 The probability dp/dx of a quark having a fractional energy between x and $x + dx$ of the nucleon energy, multiplied by that fraction, plotted against x. The shaded area is then proportional to the portion of energy probably carried by quarks in the nucleon carrying between 5/10 and 6/10 of the nucleon energy.

neutrino and quark. Because the neutrino scattering experiments give the same distributions as the electron scattering measurements, the neutrinos evidently do not "see" the field either, and the gluons must not carry a weak charge.

The Production of Quark-Antiquark Pairs in High-Energy Electron-Positron Collisions—The Charge of the Quark

If quarks are pointlike, truly elementary particles, as suggested by the electron-scattering experiments, very high-energy electron-positron collisions should produce quark-antiquark pairs just as such collisions produce lepton-antilepton pairs. However, we cannot expect to see free quarks as quarks seem to manifest themselves outside the nucleon interior as showers of hadrons. But whatever the final form of the quark, we should expect that the total probability of interaction would reflect the probabilities of the initial processes.

In the course of electron-positron collisions at high energies, muon pairs are produced profusely, and such pairs can be detected easily and measured accurately. The process is described by the Feynman diagram in Figure 14.13a, where a virtual photon is produced by the annihilation of the electron-positron pair, and the photon then generates the μ^+-μ^- pair. At high energies the probability of the reaction will be proportional to the product of the square of the charge of the electron and the square of the charge of the

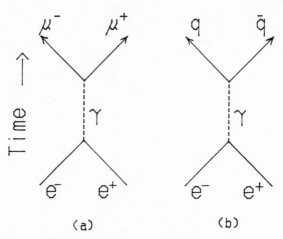

Figure 14.13   Feynman diagram (*a*) shows the production of muon pairs by electron-positron annihilation; Feynman diagram (*b*) shows the production of quark pairs from electron-positron collisions. The symbol γ (gamma) is the label for a photon.

muon. If quarks are pointlike elementary particles, the production of any particular variety of quark can be calculated using the same rules. For example, a pair of a *u*-quark colored blue and its antiquark should be produced ⁴⁄₉ as often as the muon pair, just as the *u*-quark electric charge is ⅔ of the muon charge. If we consider *u*-quarks of all three colors, the production of *u*-quark pairs will proceed with $3 \times ⁴⁄₉ = ¹²⁄₉$ times the probability of muon pair production. Adding *d*-quarks and *s*-quarks, each with a charge of ⅓ and each with three possible colors, gives a production of hadrons through the quark-anti-

Figure 14.14   The ratio *R* of hadron production of muon pair production, as calculated from the color-flavor quark model, plotted as a function of the energy of the electron-positron collisions. (Resonance contributions have been subtracted.) The discontinuities represent the energy thresholds for the production of the $c\text{-}\bar{c}$ and $b\text{-}\bar{b}$ pairs. The experimental measurements of *R* are well represented by the curve.

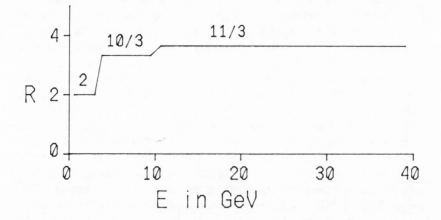

quark production about two times the muon production; the ratio R of the probability of the production of hadrons to the production of muons would be such that $R \approx 2$. At energies above 4 GeV, where pairs of charmed mesons can be produced, the ratio will be increased to $R \approx 3\frac{1}{3}$. Above 10 GeV, the contribution of the b-quark, with a charge of $\frac{1}{3}$, would add another $\frac{1}{3}$ through the production of b-mesons. Above the threshold at 80 GeV for the production of a pair of t-quarks, with a charge of $\frac{2}{3}$, the ratio R should increase by $\frac{4}{3}$ and reach 5. Figure 14.14 shows the measured values of the ratios of hadron production in electron-positron collisions to the production of muon pairs. The thresholds for the production of the different quark pairs are shown together with an estimate of the theoretical production based on the simple arguments made here.

Although the model is quite simple, the adequate agreement between theory and experiment does establish the basic conclusion that quarks have the fractional charges we have postulated and that there are three colors of quarks. If colors were not included, the theoretical result would be reduced by a factor of 3 in disagreement with the experiments; if the nucleons were made up from integrally charged quarks (according to a particular model), one would expect about twice the production of hadrons—and twice the value of R—that is observed.

The Charmonium Spectrum and Quark-Quark Forces

Although the color-quark model provides a remarkably good fit to observations of the strong interactions of baryons and mesons, free quarks have never been observed. It appears that the forces that bind quarks to each other are so strong that single quarks never escape the communal bonds. The experimental limits on the existence of free quarks are stringent. Less than 1 in a billion collisions of high-energy protons on nuclei result in the production of free quarks. Moreover, few free quarks seem to have survived from the Creation; there is less than 1 free quark to 10^{19} nucleons, and less than 1 quark in 100 micrograms of matter. The results of an interesting and subtle experiment at low temperatures have been interpreted to suggest that free quarks are found in superconducting niobium spheres, but this interpretation has not been generally accepted.

Just as the masses of the meson and baryon states are strongly affected by the magnitude and character of the forces between the quarks that make up these particles, an analysis of the spectrum (or distribution) of masses of such hadron states can be expected to provide information on the character of the forces. Because two-body systems are simpler than three-body systems, it is easier to analyze meson spectra than baryon spectra. And for technical reasons, the spectra of mesons made up of heavy quarks are more easily understood than the spectra of lighter mesons.

It is then especially attractive to consider the energy-level spectrum of states of the c-$\bar{c}$ system of a charm quark and its antiquark, called charmonium. We can discuss these states in a manner reminiscent of the spectroscopy of the hydrogen atom (Chapter 11). Just as the distribution and the different states of the hydrogen atom tell us of the character of the forces binding the proton and electron, we might expect to learn about the forces that hold the quarks together from an analysis of the charmonium states.

The spectrum of the lower (lighter) states of the charmonium (c-$\bar{c}$) system is shown in Figure 14.15. The mass of the state and the orbital angular momentum of the quark-antiquark pair are indicated by the position of the state in the figures. (By a century-old convention, we label states with zero orbital angular momentum as S, and states with one unit ($\hbar$) of angular momentum as P.) The heavy solid horizontal lines show the energies of spin-triplet states such that the quark and antiquark spins are aligned to give a total spin of $1\hbar$; the dotted lines show the masses of spin-singlet states where the quark and antiquark have their spins opposed. The total angular momentum—quark spins added vectorially to the orbital angular momentum—is shown below in the diagram. The difference between the masses, hence the forces between the quarks, for parallel and antiparallel quark spins is illustrated by the difference between the spin-triplet and spin-singlet S-states with no orbital angular momentum. It seems that the quarks have color-magnetic moments rather as electrons (and bar magnets) carry regular (electric current-induced) magnetic fields. The quark and antiquark color magnetic fields act to attract each other when the spins are opposed and repel each other when the spins are aligned. Color magnetism also affects the three P-states; it gives them different energies, as the quark-antiquark spins are aligned differently with respect to the color-magnetic fields set up by the moving color charges generated when the quark and antiquark rotate about their common center.

Because the distribution of the energies of the states depends on the vari-

Figure 14.15 The charmonium spectrum. The energy is given in MeV; the mass of a helium atom is about 4000 MeV. The dotted lines show gamma-ray transitions that have been observed in parallel to photon transitions seen in the hydrogen atom, although at energies 10 million times greater.

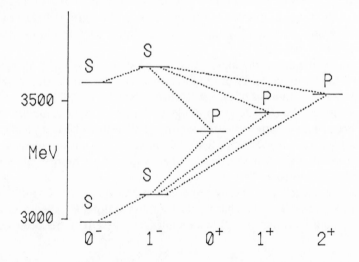

ation of force, or potential energy, with distance between the particles, we can gain some insight into the variation of the quark-antiquark force with separation distance by comparing the charmonium energy-level distributions with the known energy-level distributions from two different forces.

If the force falls off inversely with the square of the separation distance (and the potential energy falls off inversely with the distance), a hydrogen-type spectrum of levels is obtained. That potential, and the spectrum of the lower states, is plotted in Figure 14.16a. This potential does not contain particles; if the electron in a hydrogen atom has enough energy, it can escape the proton. Note that the P-state (with an orbital angular momentum of $1\hbar$) has the same energy as the second S-state (with zero orbital angular momentum).

A second well-understood force law is that which describes the ideal harmonic oscillator. Here the force increases linearly with the separation distance, and the potential energy increases as the square of the distance (as for a spring that stretches forever without breaking). This is a contained potential; the particle held by the force can never escape. The potential and the spectrum of low-lying states of the harmonic oscillator potential are shown in Figure 14.16b. Aside from theoretical calculations, such energy-level spectra are seen in nature, as in the vibrational spectra of diatomic molecules such as nitrogen. For this potential, the P-state is halfway between the lowest 1S state and the 2S state.

Figure 14.16c shows a potential that increases linearly with separation distance (corresponding to a constant force, independent of separation) and the corresponding spectra. This is a containing potential; an infinite energy is required to free the particles from one another, but they are not held quite so closely as for the oscillator potential. The P-state is seen to be about two-thirds the distance from the 1S and 2S states.

If we consider just the charmonium triplet states, as shown in Figure 14.15, and take the average energy of the three P-states (thus neglecting color-magnetic effects), this mean P-state energy falls about two-thirds the distance

Figure 14.16 (a) An electrostatic potential between two charges such as for the hydrogen atom together with the associated spectrum. (b) A harmonic oscillator potential together with the associated spectrum. (c) A constant force potential together with the associated spectrum.

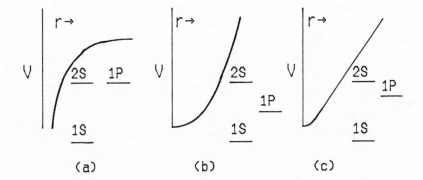

between the $1S$ and $2S$ states very much as for the constant force potential that contains particles. Hence, we suspect that the force between quark and antiquark is nearly constant with separation. Such a force requires an infinite energy to separate the particles completely and thus contains quarks, a conclusion that is consistent with the observation that free quarks are not found.

The Containment of Quarks—Long-Range Color Forces

It seems that quarks are *contained*. Free quarks are produced extremely rarely if at all in the course of the interactions of hadrons, even at very high energies, and we have seen no remnants of quarks that might have been present at the Creation—nor have we seen free gluons. It is easy to separate charged particles of an atom and ionize the atom, and it is easy to produce a photon by the acceleration of a charge in the atom—we do see light! But we have never been able to ionize a nucleon by stripping off a quark, and no apparatus has ever detected (or seen) a free gluon. Although there are similarities between electromagnetism and the strong interactions, there are also deep differences between the forces. Electromagnetic forces between charges weaken as the charges are separated. It now seems that the color forces holding quarks together are independent of that separation. These striking differences seem to stem from deep differences between the field quanta. The photon is neutral—the source of no force field—but gluons, which carry neither electric charge nor weak charge, appear to carry the strong color charge.

We can gain some insight into the effects of the gluon charge by comparing the electrostatic forces between an electron and antielectron (or positron) and the color forces between a quark and an antiquark. At the upper left of Figure 14.17 are electric field lines linking the electron and positron; at the upper right are the color field lines linking a quark and an antiquark. In the lower diagrams the same kind of field lines are shown with the quarks further apart, and then still further apart where a second quark-antiquark pair has been produced by the energy of the field.

We have noted previously that there is energy stored in the electric field. (The energy stored in a unit volume is just $E^2/8\pi$ in scientific units.) When the distance between an electron and a positron with opposite unit electric charges is doubled, additional field energy is produced (supplied by the energy required to separate the charges against the attractive forces holding them together). If the distance is again doubled, the field energy would increase again, but the second increase would be only one-half as great as the first. To move the two charges infinitely far from one another would add only twice as much field energy as the first move: the energy required to separate the two charges completely is only twice that required to double the distance between them.

The diagrams showing the color field lines linking the quark and antiquark are quite different. The field lines can be considered to represent trajectories of field quanta. Because the gluons carry color charges, the lines are

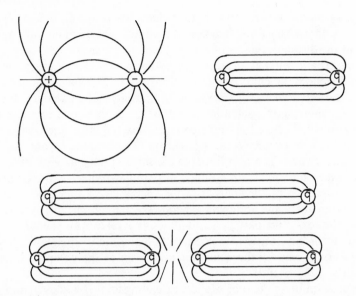

Figure 14.17 At the upper left, the electric field lines connect an electron and positron. At the upper right, the color field lines connect a quark and an antiquark in a singlet (white) state. Below these are the color field lines for quarks at a greater separation. At the bottom is the production of mesons by the breaking of the force line tube.

rather like lines of color current. Parallel electric currents attract each other through the magnetic fields generated by the currents. In much the same way, parallel color currents attract each other through color magnetism generated by the currents. There are then color-magnetic forces between the gluon currents that act to pull the field lines together. Like the electric field, the energy in the gluon field is determined by the square of the field strength and then the square of the density of field lines. It seems that the color-magnetic forces between gluon currents hold the color lines together as shown in Figure 14.17, so that the cross-sectional area of the tube of lines is independent of the distance between the source particles, the quark and antiquark. It follows that the density of lines of force in the tube is independent of the tube length. The energy density is then independent of the tube length, and the field energy is simply proportional to the volume of the tube. Doubling the distance between quark and antiquark doubles the volume of the tube and doubles the field energy. Increasing the distance by a factor of 4 increases the volume by that factor and then increases the field energy, again by a factor of 4. To separate the particles by an infinite distance requires infinite energy. Hence, the quark and antiquark cannot be separated; quarks are contained.

Similar arguments, differing only in detail, hold for the separation of a quark from the other two quarks of a baryon and for the freeing of a gluon. Quarks and gluons are contained by color forces as a consequence of the color charge held by the gluons.

Because the force between the color charges is constant, independent of distance, whereas the force between the electric charges falls off with the square of the distance, there will always be some distance where even a very small color charge will have a greater effect than a large electric charge. Hence, at large distances the color charge is effectively very strong. Figure 14.16*a*, showing the potential energy of systems of opposite electric charges, and Figure 14.16*c*, showing the potential energy of a system of two opposite color charges plotted as a function of the distance between the charges, suggest the different character of the electromagnetic and color forces. (Again, the force between the particles is proportional to the slope of the potential, very much like the force on a small cart rolling on a similar profile in a gravitational field.) As suggested by the analysis of the energy levels of charmonium, the force between quark and antiquark generated by the color field is independent of the separation of the particles, and although it takes a finite energy to free a particle from the electric field potential, no amount of energy will free the particle (quark) held by the color potential even as the potential increases indefinitely with distance.

The description of the tube of force extending forever was oversimplified, because when sufficient field energy to create new quark-antiquark pairs is accumulated, the tube will break as shown in the lowest diagram of Figure 14.17. So when a quark is hit very hard by a neutrino, electron, or another quark in a very high-energy collision, the quark will leave a string of quark-antiquark pairs behind in the form of mesons.

Asymptotic Freedom of Quarks—Short-Range Color Forces

In the course of discussing the effects of vacuum polarization by an electric charge, we showed that the forces between charged particles are greater at very short distances as a consequence of the effects of the polarization charges. The same effect occurs for color charges except that the vacuum polarization forces are six times as great even as a polarization of the six different kinds (or flavors) of quark-antiquarks takes place.

The diagram at the left of Figure 14.18 is meant to suggest something of the character of the vacuum polarization of virtual quark-antiquark pairs by the color field of an isolated quark. Specifically, a quark with a blue color charge repels a blue charge and attracts an antiblue charge polarizing the virtual pairs (as suggested by the rather fanciful diagram). Now if we were to (almost) superimpose the diagrams for two such blue charged quarks (to simulate two such quarks very close together) added to the repulsion q_b-q_b, we would have repulsions from the virtual pairs u_b-u_b and from $\bar{u}_b$-$\bar{u}_b$ as well as for the other flavors. The total repulsion would be greater than the simple repulsion of the two primary blue charges as a consequence of the extra repulsions from the vacuum polarizations. From the vacuum polarization, the forces (repulsive or attractive) between quarks are increased.

However, the gluon fields about the two blue charges attract each other

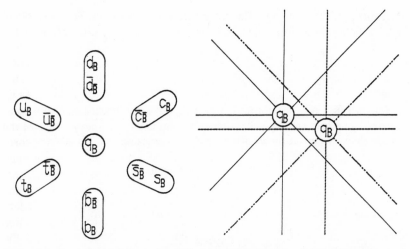

Figure 14.18 At the left is a schematic of the polarization of the vacuum by a color-charge showing the contributions of the six different color flavors. The diagram at the right suggests how the attraction of the field lines from the two color charges may provide an attractive force between the quarks, thereby reducing the regular repulsion.

and add an attractive force to the basic repulsion that is even greater than the added repulsion from the vacuum polarization. The diagram at the right of Figure 14.18 is meant to show the lines of force from two blue charges close together. Again, the lines can be viewed as trajectories of gluons and then as color currents that generate a color magnetism. Because similar currents attract, just as parallel electric currents attract, the (solid) lines of force from one charge attract the (dashed) lines from the second charge, and this attraction forces the charges together. The force is stronger than the repulsive vacuum polarization force from the six flavors and acts such that the total force between two quarks becomes quite small when they are very close. This small force is suggested by the flat potential at very small distances shown in Figure 14.16c.

The weakness or freedom of force in the limit of small distances translates into a freedom of force at very large energies—just as, quantum mechanically, distance and momenta are complementary and small distances correspond to large momenta and then large energies. Thus, at very high energies, quarks act almost as free particles unbound by any forces. Similarly, quarks closely bound in hadrons can be described as almost free particles at the short distances defined by the hadron dimensions, although an enormous amount of energy will not shake them loose. This freedom from forces in the asymptotic limit of very high energies or very short distances is called *asymptotic freedom.*

In summary, the color forces holding quarks together are extremely strong at large distances on the order of 10^{-13} centimeters, the size of the nucleons and mesons. Conversely, the color forces are very weak at small distances of less than 10^{-14} centimeters. From the complementary nature of distance and momentum, and then energy, the color forces are effectively strong for low-energy particle interactions and weak for very high-energy interactions.

Notes

1. Although the exact value of the energy depends on the distribution of the charge throughout the nucleus, one can easily estimate the magnitude of the energy: the potential energy of a charge $+e$ on the surface of a set of Z positive charges held in a sphere of radius r is

$$E = \frac{Ze^2}{r}$$

Then the total potential energy of the Z charges will be on the order of

$$E_{\text{tot}} = ZE = \frac{Z^2 e^2}{r}$$

Taking Rutherford's estimate of the size of the gold nucleus as $r = 10^{-12}$ centimeters and the charge of the nucleus as $Z = 79$, and putting in Millikan's value of the fundamental unit of electric charge e (and converting units), we find that $E_{\text{tot}} \approx 1000$ MeV.

2. When the relative phase of the neutron and proton are properly considered, the rotation can be defined by two angles θ and ϕ such that the particle amplitudes of our researcher friend Olivia are described in terms of her colleague Oliver's states by the relation

$$p' = p \cos \theta + n \sin \theta \underline{/\phi} \qquad \text{and} \qquad n' = -p \sin \theta \underline{/-\phi} + n \cos \theta$$

where ϕ is the phase difference between the neutron and proton amplitudes.

3. The wave function of the three-nucleon quartet, even under any n-p label exchange, can be described as

$$nnn, \qquad \sqrt{\tfrac{1}{3}} \, (nnp + npn + pnn), \qquad \sqrt{\tfrac{1}{3}} \, (ppn + pnp + npp), \qquad ppp$$

and two different doublets, even under one exchange and odd under another, can be written as

$$\sqrt{\tfrac{2}{3}} \, pnn - \sqrt{\tfrac{1}{6}} \, n(pn + n), \qquad \sqrt{\tfrac{2}{3}} \, npp - \sqrt{\tfrac{1}{6}} \, p(np + pn)$$

and

$$\sqrt{\tfrac{1}{2}} n(pn - np), \qquad \sqrt{\tfrac{1}{2}} p(np - pn)$$

where, as usual, the symbols n and p stand for the de Broglie amplitudes of the neutron and proton. Using the quantum mechanical rule that the product of two substates (such as npn and pnp) is 0 if they are different and 1 if they are identical, the square of each member of the quartet or doublets is 1 (normalization) and the product of any two different states is 0 (orthogonality). In common language, each wave function describes one particle, and the particles are completely different.

4. General selection rules derived from charge independence take a simple form. A state a, a member of a multiplet with multiplicity m_a, can decay into two states c and d, members of multiplets with multiplicities m_c and m_d, only if

$$|m_d - m_c| \le m_a \le m_d + m_c$$

5. The mathematical or logical similarities between the description of spin $\hbar/2$ particles in space, where spin-up and spin-down states are not differentiated by space-symmetric forces, and the two-quark states acted upon equally by the strong interactions led to the names "up" and "down" for the members of the u-d quark doublet. The s-quark was named after the observed property "strangeness" later linked to the postulated quark. Now we have "charm," a sibling to the s-quark (with no excuse for the name) and a third doublet, the t- and d-quarks, "top" and "bottom," in parallel to "up" and "down."

6. Usually quarks and hadron compounds of quarks are shown in geometric diagrams where a particle composed of a number $N(s)$ s-quarks minus anti-s-quarks, $N(u)$ u-quarks minus anti-u-quarks, and $N(d)$ d-quarks minus anti-d-quarks is described as a point on the y-axis or the abscissa with the value $-2N(s)/3 + [N(u) + N(d)]/3$ (called "hypercharge" and labeled Y) and at a position on the x-axis or ordinate, $[N(u) - N(d)]/2$, labeled I_3. Hence, in this coordinate system the coordinates (or I_3, Y values) of the three quarks are $u(\frac{1}{2}, \frac{1}{3})$, $d(-\frac{1}{2}, \frac{1}{3})$, and $s(0, -\frac{2}{3})$.

7. For example, the alpha particle, made up of two neutrons with spins in opposite direction and two protons with spins opposed, is the lightest nucleus with all four nucleons in the simplest orbit (without nodes in the wave function). As a consequence, the alpha particle is especially strongly bound. If there were 8 different colors of nucleons—as a color octet— 32 different nucleons could occupy the same lowest orbit, and the sulfur nucleus of 16 protons and 16 neutrons would be much more strongly bound than it is.

15

The Weak Interactions

> Creator of all and giver of their sustenance.
>
> Hymn to the Sun, Egypt, c. 1400 B.C.

Looking back, we can see that the beginning of our understanding of the heat of the sun stems from the discovery of radioactivity by Henri Becquerel in 1895. In the decade following, researches by Pierre and Marie Curie in France, Ernest Rutherford in Canada and then England, and others led to a separation of the radiation from uranium and certain other heavy elements into three classes, named alpha, beta, and gamma rays. Eventually it was determined that alpha rays were positively charged helium nuclei, gamma rays were neutral electromagnetic radiation, and beta rays were negative electrons traveling near the speed of light.

We know now that alpha rays are connected most closely with the strong interactions and gamma rays with the electromagnetic forces. But the beta rays are generated by a different, specific force called the weak interaction. At characteristic nuclear energies, the weak interaction is a hundred-billion times weaker than electromagnetic forces. Nevertheless, it is the weak interactions that drive the great engines of sun and stars—and are even a prime mover in supernovae explosions.

Lepton Helicities—God Is Left-Handed!

Leptons and quarks carry the weak interaction charge. Both are spin ½ particles (in standard units of $\hbar$, Planck's constant divided by 2π), and they are probably closely related. Quarks carry a strong interaction charge, whereas leptons—collectively, electrons, muons, and tau-leptons, together with their associated neutrinos and antiparticles—do not; quarks carry fractional electric charges, whereas leptons carry integral charges, including zero for the neutrinos. Although quarks take part in the weak interactions very much as leptons do, it is lepton phenomena, uncomplicated by strong interaction effects, that provide an ideal probe of the weak interactions. Consequently, it is convenient to outline the character of the weak interactions by a study of leptons.

280

All leptons (and quarks) are spin ½ fermions. To consider their properties adequately, we review the description of spin or angular momentum required by quantum mechanics. The total angular momentum of a particle described as "having spin ½" is properly $\sqrt{\tfrac{3}{4}}\ \hbar$. From the Uncertainty Principle, the direction of the spin axis of the particle can be known only within an angular uncertainty $d\theta$, where $d\theta$ is an angle about equal to half of the arc of a circle. However, a measurement of the component of spin in a given direction can result in but one of two possible results, $+\tfrac{1}{2}$ or $-\tfrac{1}{2}$ (in units of $\hbar$). There are then just two well-defined "average" directions of spin with respect to a given fiducial direction. Evidently, the lepton carries only one bit of directional information, and the mean spin can be directed only "forward" or "backward" —or "up" or "down"; the particle is either spinning counterclockwise or clockwise about that fiducial direction. Although the exact direction of the spin is always uncertain, the average direction is well defined and customarily called the spin direction.

Any direction such as a spin direction is relative to another direction. If some special direction, such as the direction of motion of a particle, is defined, the spin direction can be defined only as forward or backward. Conversely, if the (mean) direction of the spin is determined, the orientation of any other direction, such as the direction of motion of a particle, can be determined only within a twofold distinction with respect to the spin direction of motion; the direction can be forward or backward. If the direction of motion of a particle is known exactly, the true direction of the spin of the particle will be uncertain; if the direction of the spin is known exactly, the direction of motion will be uncertain. The direction of spin and the direction of motion are complementary quantities related by the Uncertainty Principle.

In this context, "forward" and "backward" directions of motion are used imprecisely but helpfully as labels for probability distributions of directions that are largely forward and directions that are largely backward. The character of the distributions is almost defined by simple conditions; the sum of the two distributions must represent no preferred direction—isotropy; and the distributions corresponding to forward and backward must be identical, although reversed.[1] Figure 15.1 shows operationally defined distributions of the direction of motion of spin ½ fermions traveling "forward" and "backward" with respect to their direction.

If the direction of the particle is known, the spin is forward or backward. If the spin direction is known, the particle travels forward[2] or backward (in the sense of the distributions of Figure 15.1). If the spin and motion are in the same direction, the particle is said to have positive *helicity*. If the spin and direction of motion are opposed, the particle has negative helicity. Right-handed nuts spinning clockwise as they move forwards on a threaded bolt have positive helicity; left-handed nuts spinning counterclockwise have negative helicity. If nature does not differentiate between screw directions, between left-handed and right-handed coordinate systems, then no natural process will define a screw direction, and leptons spinning clockwise must be produced equally with leptons spinning counterclockwise. Positive and neg-

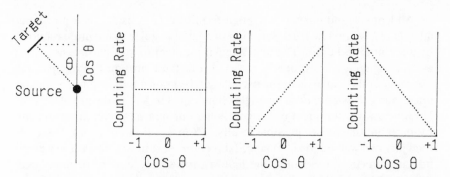

Figure 15.1 At the left, a counter at a distance *r* from a source of spin 1/2 fermions (perhaps electrons from the decay of an aligned nucleus) counts the particles passing through it. The counter is at an angle Θ with respect to a direction of quantization— the direction of the particle spin. The three graphs at the right show counting rates plotted against different values of the projection of the position of the counter on the axis (which is proportional to cos Θ). The right-hand graph shows the distribution for particles emitted in the "backward" direction; the "forward" distribution is shown in the center graph, and an isotropic distribution is shown in the left-hand graph.

ative helicities will be equally probable. But the "Old One," in Einstein's words, had different ideas and constructed His universe with particles of a definite helicity. This handedness or *chirality* (from the Greek χειρ, hand) is a striking characteristic of the weak interactions.

Because much of the phenomenology of the weak interactions can be defined in terms of rules describing the helicities of the particles involved in the interactions, it is useful to classify the possible lepton helicity-direction relations. To illustrate the relations, we might consider, as an example, one typical weak interaction decay, the beta decay of a hadron (usually a nucleus) to an electron, an antineutrino, and a final-state hadron. There are 16 different decays; the electron can be emitted up or down (forward or backward) with the spin up or down, and the neutrino can be emitted up or down with the spin up or down. Figure 15.2 shows a complete set of such possibilities in a manner introduced pedagogically by Harry Lipkin. Nature will choose only a subset of these. The early history of modern studies of the weak interactions was devoted to determining nature's selection of helicities.

The Electron Helicity

C. S. Wu, Eric Ambler, and their colleagues performed the first important "modern" experiment on beta decay by measuring the asymmetry of electrons emitted in the beta decay of polarized Co^{60}, a radioactive isotope of cobalt that decays to an electron, an antineutrino, and a final nuclear state of a nickel isotope. Using high magnetic fields and temperatures near absolute zero, Wu and Ambler polarized the cobalt nuclei so that the nuclear spin was

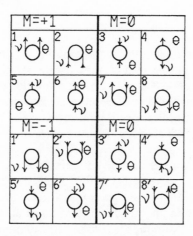

Figure 15.2 Lipkin diagrams showing all possible allowed beta-decay configurations of an electron and an antineutrino emitted by a nucleus. The direction of quantization is up, and the direction of helicities of the leptons are shown by the arrow heads; the direction of emission is shown by the direction of the lines. The sector headings M indicate the total spin in the up direction.

in the direction of the magnetic field. The spin of the Co^{60} nucleus was known to be equal to $5\hbar$, and the spin of the residual nickel nuclear state was $4\hbar$. Hence, by conservation of angular momentum, the total spin in the direction of quantization must not change, and the spin of both neutrino and electron—with spins of $\frac{1}{2}\hbar$—must be in the direction of the Co^{60} spin (and then the magnetic field) as indicated by Figure 15.3. Thus, the transition could be represented by diagrams 1, 5, 2, or 6. Because the electrons were found to be emitted preferentially opposite to the direction of the field (downward in Figure 15.3), the helicity of the electrons must be negative, and the decay must be described by either diagram 5 or 2. These diagrams describe electrons spinning counterclockwise; electrons are left-handed!

Figure 15.3 Angular momentum configuration and possible Lipkin diagrams for the beta decay of the polarized Co^{60}.

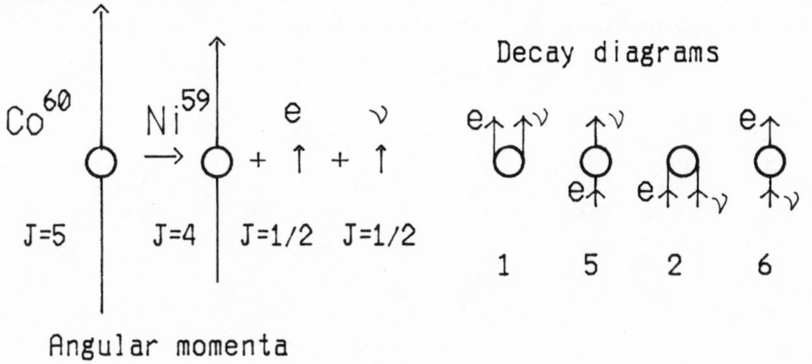

The Neutrino Helicity

The Co^{60} experiment, together with other measurements leading to the same conclusions, tell us that the electron is emitted with negative helicity, but we do not know the neutrino helicity; we do not know whether nature follows the pattern of Lipkin diagram 5 or 2. Because the neutrino is illusive, such a measurement must be made indirectly and ingeniously. The ingenuity was provided by Maurice Goldhaber and his associates in 1958 at Brookhaven National Laboratory.

Because the neutrino itself cannot be detected easily, inasmuch as its interaction with matter is so weak that it will usually pass through any detector untouched, the neutrino helicity is best measured indirectly through measurements of the momenta and angular momenta of all the other particles taking part in the decay. Assuming the conservation of momentum and angular momentum, any missing momentum and angular momentum must be assigned to the neutrino. If we know both the direction of the neutrino and the angular momentum of the neutrino, we can determine its helicity.

But does nature provide decays that allow all the required measurements? Yes, but the set of possibilities is curiously sparse. We know of only one decay transition that allows a direct measurement of the neutrino helicity. That transition is shown by the energy-level diagram of Figure 15.4. A radioactive isotope of the rare-earth element europium, which has zero spin, forms an excited state of the samarium nucleus by capturing an electron from the cloud of electrons about the nucleus and emitting a neutrino. That samarium state then emits a gamma ray—a photon with spin 1—returning to its spin-zero ground state. Like leptons, the spin of zero-mass particles such as the photon is directed only forward or backward; hence, the spin component of the photon in its direction of motion is $+1$ or -1 ($\hbar$).

Because both initial and final nuclear states have spin zero, the sum of the components of the spins of the photon and neutrino in any direction must

Figure 15.4 At the left are indicated the energies of the states of the original europium nucleus and the final samarium nucleus, together with the transitions studied in the experiment that established the helicity of the neutrino. At the right are the relations between the helicities and the directions of the momenta.

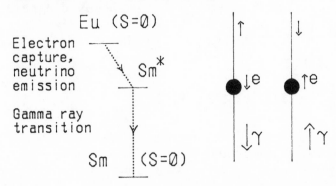

equal the spin component of the captured electron. It follows that the spin of the emitted gamma and the spin of the emitted neutrino must be opposed. The direction of the spin of the photon can be determined by passing the gamma-ray photons through polarized iron that acts as a polarimeter, passing photons spinning (circularly polarized) in the direction of the magnetic field more easily than those with opposite spins (polarizations). Schematically, we can consider that the iron can be magnetized to pass either photons with their spin down or photons with spin up; hence, the direction of the spin of the photons could be determined by measuring the transmission of the photons through the magnetized iron. If photons could be selected from events such that the neutrino was emitted upward, the helicity of the neutrino could be determined by measuring the polarization of the photon; if the photon spin was down, the neutrino spin would have to be up, and the neutrino would have positive helicity. If the photon spin was up, the neutrino spin would have to be down and the neutrino would have negative helicity.

Particles are emitted equally in all directions in the decay of a spin-zero state, as no direction is defined by the state.[3] From the isotropic emission of the neutrinos, the experimenters had to select those decays where the neutrino was emitted upward (in Figure 15.4). This selection was made by exploiting the recoil of the samarium nucleus left after the neutrino emission.

If a tuning fork is set in vibration and emits sound, an identical tuning fork nearby will absorb some of the sound in resonance, begin vibrating, and reemit the sound. Atoms and nuclei act similarly. If the excited samarium nucleus emits electromagnetic waves of a particular frequency as a high-energy gamma-ray photon of definite energy, another (identical) samarium nucleus can absorb and reemit the photon. If the first (emitting) tuning fork is moving away from the second (absorbing) tuning fork with an appreciable velocity, the frequency of the sound at the absorbing fork will be reduced by a Doppler effect, and no resonance absorption and reradiation will occur. Similarly, although the emitting fork is fixed, if the absorbing fork moves away, that fork will "hear" a lower note, and again there will be no resonance absorption of the sound.

Much the same holds for samarium nuclei. If it can be held motionless, a second samarium nucleus will absorb and reradiate a photon—an electromagnetic vibration—emitted by the first samarium nucleus which was held motionless. However, because the high-energy photon carries considerable momentum, upon emitting the photon the first nucleus recoils, and the frequency of the photon reaching the second nucleus is reduced. Also, upon absorbing the photon, the absorbing nucleus recoils, so that the required frequency must be increased if resonance absorption is to occur. Consequently, there will be no resonance transfer of radiation from one samarium nucleus to another if they are both initially stationary, as they are not held in position tightly enough by atomic forces. However, if the first nucleus is moving toward the second with a velocity that will compensate for the two recoils, the second nucleus will absorb and reemit the photon.

Remarkably, for the europium-samarium transitions, the downward recoil of the samarium nucleus that follows from the emission upward of the neutrino is just sufficient so that the photon emitted downward by the excited samarium will scatter resonantly—that is, be absorbed and readmitted—from a target of samarium metal. Exploiting this feature of the decay, the Brookhaven group set up their apparatus in a manner suggested by Figure 15.5 so that they would measure only the polarization of photons scattered into their magnetized iron polarimeter by a samarium scatterer, thus guaranteeing that the neutrino, which they could never see directly, had actually been emitted upward. Then, by measuring the transmission of the magnetized iron, they found that the resonantly scattered photons were polarized upward. Hence, the neutrinos moving upward had spins pointing down, and neutrinos must have negative helicity. Neutrinos then spin counterclockwise like electrons; neutrinos are also left-handed!

An accumulation of evidence from many other observations leads to the conclusion that all leptons have negative helicity in the limit of their emission at the speed of light. All antileptons have positive helicity in the same limit. Leptons are left-handed, anti-leptons right-handed. Nature differentiates not only between left-handed and right-handed coordinate systems but also between particles and antiparticles.

If the decays to the leptons and antileptons are not constrained, the absolute value of the helicity for emissions at smaller velocities v is found to be equal to β, where $\beta = v/c$. Hence, for an electron emitted at a velocity $\beta = 0.5$, the helicity will be equal to $-\frac{1}{2}$. Broadly speaking, the intensity or probability of emission of the lepton with "natural" helicity is proportional[4] to the factor $(1 + \beta)/2$, and the intensity of emission a lepton with "unnatural" helicity is $(1 - \beta)/2$. (The intensity of emission is also proportional to the square of the weak charge and proportional to the momentum of the emitted particle.) Stating that the helicity of the emitted lepton is $-\frac{1}{2}$ is then the same as saying that 75% of the emitted electrons have negative helicity, and 25%

Figure 15.5 A schematic representation of the design of the experiment that determined the helicity of the neutrino.

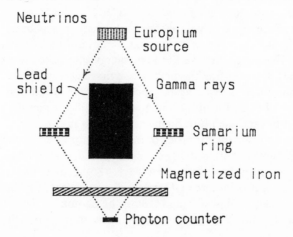

have positive helicity. (If half have each helicity, the electrons are unpolarized.) Because neutrinos are generally believed to have zero rest mass and always travel at the speed of light, neutrinos always have 100% negative helicity and antineutrinos always have 100% positive helicity.

The establishment of the negative helicity of the electron and the positive helicity of the antineutrino limits the possible beta-decay transitions to Lipkin diagrams 5 and 7, and 5′ and 7′, which are just 5 and 7 rotated by 180° (see Figure 15.2). Diagrams 5 and 5′ show the lepton and antilepton spins parallel, adding to a total angular momentum of 1. Such transitions are known as G-T (for Gamow-Teller) transitions. Diagrams 7 and 7′, where the lepton spins are opposed, thus giving a total spin of 0, show F (for Fermi) transitions.

The Dependence of Decay Intensities on Helicity

Nature is subtle; things that are very much the same can behave quite differently. Electrons and muons are siblings acted upon by the same forces. Yet the muon is much heavier than the electron—we do not know why—and sometimes acts very differently from the electron because of that greater mass. The pion decays ever to muons and (almost) never to electrons. We understand this now to be a consequence of the helicity rules and the differences in masses of the particles.

The rules concerning the decay probabilities of states of definite helicity explain some striking features of meson decays. The charged pi-mesons π^+ and π^- have a spin of zero and decay to leptons (electrons and muons) and antileptons (antineutrinos). To conserve momentum, the charged lepton and the antineutrino are emitted in opposite directions. To conserve angular momentum, their spins must be directed in opposite directions. Because the antineutrino rest mass is zero, it always travels at the speed of light. Therefore, the helicity of the antineutrino emitted in the π^- decay must be positive and, to conserve angular momentum, the helicity of the electron or μ^- must also be positive, as shown in Figure 15.6. But that positive helicity is the

Figure 15.6 At the left are the helicities of final-state leptons produced in the decay of a pion to a muon μ^- and an antineutrino. At the right is the decay to an electron e^- and an antineutrino in a similar fashion.

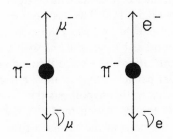

"unnatural" helicity for the electron or μ^-, and the intensity of emission will be proportional to $(1 - \beta)/2$. If the velocity of the emitted lepton is near the speed of light and $\beta \approx 1$, the decay will be strongly inhibited.

The mass of the pion is about 139 MeV/c^2; the masses of the electron and muon are about 0.5 MeV/c^2 and 105 MeV/c^2, respectively. The neutrino masses are very small if not zero. The difference between the masses of the π-meson and the masses of the decay products is available for the kinetic energies of those particles. For the $\pi \rightarrow \mu + \nu$ decay, 34 MeV/c^2 is available; the decay energy for the transition $\pi \rightarrow e + \nu$ is about 139 MeV/c^2. For the $e + \nu$ decay, the very light electron and the neutrino practically share the energy so the electron is emitted with an energy of about 69.5 MeV, which is about 136 times the electron rest-mass energy—and a momentum of 69.5 MeV/c. A simple calculation shows that β_e is 0.999974 and $1 - \beta_e = 2.95 \cdot 10^{-5}$. The decay must be strongly inhibited. In contrast, the muon is omitted relatively slowly with an energy of about 5 MeV, which is much smaller than the muon rest mass, and the muon momentum is 30 MeV/c. The muon velocity is $\beta_\mu = 0.274$, $1 - \beta_\mu = 0.726$, and the muon decay is not much inhibited. With the greater momentum of the electron, there is an added factor of about $69.5/30 = 2.3$ in favor of the electron transition. In spite of the fact that the weak interaction forces responsible for the decay act equally on the muon and electron, as a consequence of the helicity rules only about 1 pion in 10,000 decays to an electron, and the rest decay to muons.

Similarly, charged K-mesons with a spin of zero and a mass of 493 MeV/c^2 decay to muons plus neutrinos, but only rarely do they decay to electrons plus neutrinos—although again, the forces responsible for the decays are the same for muons and electrons.

The Universal Fermi Interaction and the Intermediate Vector Boson

The electric charge of the electron is exactly the same size as the charge of the proton—to an accuracy of 1 part in 10^{40}! The size of the electric charge of all particles is exactly the same. Surely Nature must give us an equally universal weak charge—or is Her scheme more subtle?

At a time before it was necessary to consider seriously any hadron decays other than those of the neutron and proton (usually as constituents of nuclei), it was attractive to consider that the weak interactions dealt equally with the neutron-proton system, the electron-neutrino system, and the muon-neutrino system, as suggested by Figure 15.7. The dashed lines are meant to imply that the transitions from the pair of particles at one corner of the triangle to the pair of particles at any other corner proceed with equal strength. The solid lines show that these transitions take place through an intermediate particle, and the coupling to that particle—that is, the weak charge—is the same for each member of the pairs. In this hypothesis the charge is universal.

This particle, called the W, which mediates the weak interaction must decay freely to any of the pairs at the corners of the triangle of Figure 15.7.

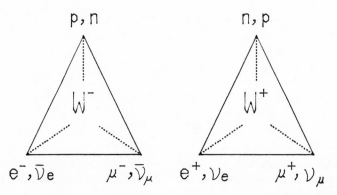

Figure 15.7 The triangular diagrams show relations for weak interaction transitions. The dotted lines show the transitions of the projected intermediate vector bosons, the W^+ and W^-, to the particle pairs at the corners. (For a free W, these represent different possible decays of the W.) The solid lines making up the edges of the triangles represent transitions from one pair of particles to another through the intermediate W state.

In particular, as shown in Figure 15.8, the W^- must decay to a left-handed electron and a right-handed antineutrino, each with spin $\frac{1}{2}\,\hbar$; hence, the conservation of angular momentum requires the W to be a boson with spin 1. The W is called the *intermediate vector boson* (vector, used here as an adjective, signifies spin 1).

A typical reaction, such as that representing the decay of a neutron to a proton, an electron, and an antineutrino, can be described in a two-stage form:

$$n \rightarrow p + e^- + \bar{\nu} \quad \text{or} \quad n \rightarrow p + W^- \quad \text{and} \quad W^- \rightarrow e^- + \bar{\nu}$$

Feynman diagrams representing such transitions are shown in Figure 15.9.

We now know that there are also forces mediated by the transfer of a neutral intermediate particle, usually called Z^0. This transfer plays no part in the leptonic decays of hadrons, and it is then convenient to consider neutral intermediate particles later.

Just as the electromagnetic interactions can be considered the interaction between two electric currents (charges that are in general moving) mediated by the exchange of a spin-1 boson, the photon, coupled to the electric charge,

Figure 15.8 An intermediate vector boson at rest decays to an electron and an antineutrino. The arrowheads show the direction of the lepton spins. Conservation of angular momentum and the known helicities of the electron and antineutrino require that the W have a spin component of 1 ($\hbar$) in the direction of emission of the antineutrino.

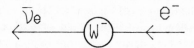

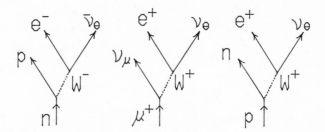

Figure 15.9   Feynman diagrams showing weak decay transitions. The rate of decay is proportional to the product of the square of the weak interaction charges acting at each of the two vertices.

so can the weak interaction be considered an interaction between two weak currents (weak charges that may be moving) mediated by a spin-1 boson, the *W*, coupled to the weak interaction charge. From the similarities between the descriptions of the two forces, it is attractive to presume that a universal weak charge strength exists with the same value for any interaction in the same way that a universal electric charge strength exists that has the same value no matter what particle carries the charge. This assumption, that the coupling between each set of particles is equivalent in form and in strength (that the weak charge is "universal"), was first stated clearly by Fermi in his early description of beta decay. It is called the *Universal Fermi Interaction* (UFI).

The Weak Decays of Quarks—Breakdown of UFI— The Cabibbo Model

Although the UFI was extremely attractive aesthetically and seemed to constitute an approximately valid description of the decays of nucleons (in nuclei) and muons, it was noticed early that the neutron decays (to a proton, an electron, and an antineutrino) seemed to indicate a slightly smaller nuclear weak charge than the weak charge measured in the muon decay (to an electron, a neutrino, and an antineutrino). Although the difference between the nucleon and muon effective weak charge was only about 3%, the theory required that the equivalence be *exact;* any discrepancy would destroy the theory. Moreover, measurements of decays of strange particles (baryons with hypercharge values different from that of the nucleon, and mesons with hypercharge values different from that of the pion), such that the hypercharge values changed, indicated that the effective weak interaction charge was much weaker for such transitions than might be expected from the assumption of a universal charge. It was clearly not possible to add strange particles to this description of nature by simply adding one or more corners describing strange particles to the triangles of Figure 15.7. If the attractive idea of a UFI was to be retained, a new formulation would have to be found.

For a long time, the beta-decay transitions of nuclei had been understood in terms of the beta decay of individual nucleons—neutrons and protons—

that made up the nucleus. We may now consider the decays of hadrons similarly as a consequence of the decays of individual quarks that make up the hadrons. It is not the neutron and proton that should occupy the corner of the UFI triangles of Figure 15.7, but rather quark states. Moreover, the quark states defined by the weak interactions turn out to be different from the states defined by the strong interactions in a manner that explains the small effective charge deduced from strange particle decays, while retaining the universal character of the weak decays and the universal strength of the weak interaction charge.

After correcting for the energies of the decays and the relation of the quarks to other quarks in the hadrons (requiring a subtle analysis), it seemed that the quark transitions $u \to d$ and $d \to u$ were about 93% as probable as lepton transitions such as $\mu \to e$, and quark transitions of the type $s \to u$ and $u \to s$ were about 7% as probable as the lepton rates. Such transitions are shown in Figure 15.10. (The lifetimes for hypercharge-changing u-s transitions are typically about 13 times as long as the u-d transitions after corrections for different energy releases.) The decay rates were expected to be proportional to the square of the weak interaction charges carried by the quarks; hence, the quark charges seemed smaller than the lepton weak charges. These differences, interesting in themselves, seemed to violate the attractive concept of the universal weak charge (UFI).

In 1962, when Nicolo Cabibbo, then working at CERN, the central European particle research laboratory in Geneva, showed that it was possible to salvage the UFI by expressing the weak charge (or current) in a manner that behaved in a specific way with respect to the SU_3 hadron symmetry, quarks had not yet been considered. It is most convenient to consider Cabibbo's ideas in terms of the characteristics of weak charges held by quarks. We will first consider only the three light quarks, u, d, and s; later we will extend the model to consider four quarks, and we will note the modifications required for the six quarks we now know.

According to Cabibbo's description of the weak interaction, the weak interaction defined or chose different quarks than the strong interaction—but different in a simple, rational way. The weak quarks are combinations of the strong quarks; the strong quarks are combinations of the weak quarks. This

Figure 15.10 Feynman diagrams suggesting typical weak interaction transitions of quarks. The transition probability is proportional to $g_q^2 g^2$ where g_q is the quark weak charge coupling the quark to the W, and g is the lepton weak charge.

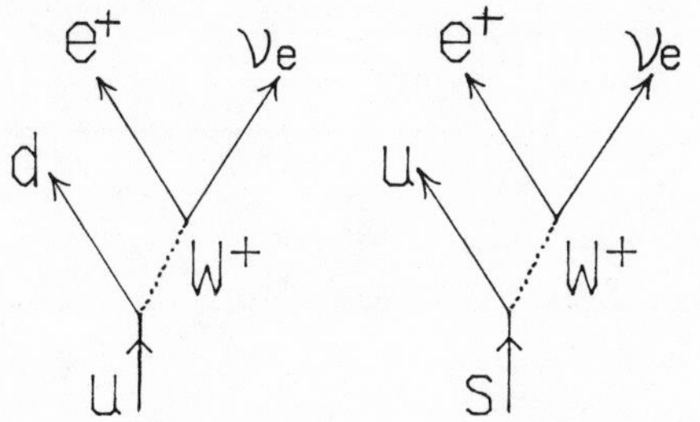

difference can be seen with the aid of Lewis Carroll by asking our researcher friends Oliver and Olivia to drink enough of Alice's potion so that they shrink to the size of particles. We provide Oliver with a strong interaction meter and Olivia with a weak interaction meter and ask them to examine the two sets of quarks shown at the left in Figure 15.11 (and to write down the results in their notebooks).

When they conduct their examinations of the particles using their meters, Oliver finds that the left-hand set is simply d-, u-, and s-quarks, as labeled in the figure. But in a series of measurements Olivia finds that the quark labeled d is an s-quark about 7% of the time and a d-quark 93% of the time. Her measurements on the quark labeled s are similarly complicated; 7% of the time she finds that the quark is a d-quark, and 93% of her measurements read s. But when they move over to the second set (labeled with the subscript W), Olivia finds that her measurements of the quarks agree with the labels and Oliver's results are aberrant. He finds that the quark labeled d_W is a d-quark 93% of the time but an s-quark in 7% of his measurements. Similarly, the s_w quark is measured to be a d-quark in 7% of his determinations and an s-quark in 93% of the measurements.

Later (after drinking the antidote to the shrinking potion), they compare notes and realize that if they take the square roots of the numbers 0.07 and 0.93 that they have found, so as to compare amplitudes rather than intensities, their results differ in just the same manner as their geographic surveys (described in Chapter 3) differ when they use different definitions of north. Olivia's (weak) quarks are related to Oliver's (strong) quarks as vectors in different coordinate systems. The weak quark amplitudes are "rotated" (in an abstract space) from the strong quark amplitudes by an angle θ_c, now called the Cabibbo angle. According to this description, the weak quarks (d_w and s_w) are related to the strong quarks (d and s) rather as distances north and east defined according to Oliver's maps, which use the geographic pole as north,

Figure 15.11   At the left Oliver and Olivia inspect quarks using weak and strong interaction meters. A geometric description of the relations they find between weak and strong charge $-1/3$ quarks is shown to the right.

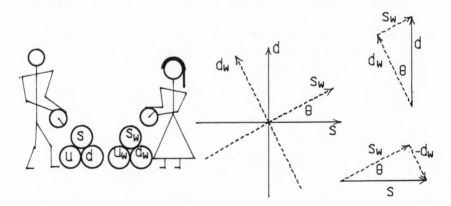

are related to distances north and east defined by Olivia's maps, which use magnetic compass coordinates (as discussed in Chapter 3). The diagrams at the right in Figure 15.11 show the relations graphically.[5] (We use a convention such that $u = u_w$.)

With this description of weak quarks, the quark decay strengths are simply described and consistent with UFI. The UFI transitions are between u_w and d_w, and these occur with the full strength of the lepton transitions; the weak charges of u_w and d_w are the same as the weak charges of the leptons. Then the other transitions, between s_w and u_w (and the transition between d_w and s_w, which must be carried by an electrically neutral weak charge or current) do not occur at all, as if the s_w held no weak charge.

Let us expand on the meaning of the relations between the weak and strong quarks. Because the effective strong interaction charge is greater than the weak interaction charge by a factor of about 10,000,000, the quarks that make up matter, which is held together by the strong interactions, are the strong quarks (u, d, s, c, t, and b). These strong quarks are combinations of weak quarks (and vice versa) according to the transformation relations shown at the right in Figure 15.11. Explicitly, the d-quark is about 93% d_w and 7% s_w, and the s-quark is 7% s_w and 93% d_w where the amplitudes are squared to determine the intensities; in this description, the u-quark and u_w are the same. But the weak interactions know only weak quarks, and weak transitions occur between the weak quark elements of the strong quarks.

We may better apprehend the meaning of these statements by considering the relations between the (hypothetical, in part) weak transition probabilities for the decays of (strong) quarks, where we presume that the transitions are energetically possible and that there are no energy differences leading to changes in the transition probabilities:

1. $\quad d = d_w \cos \theta + s_w \sin \theta \rightarrow u_w + e^- + \bar{\nu}$
2. $\quad s = -d_w \sin \theta + s_w \cos \theta \rightarrow u + e^- + \bar{\nu}$

The initial quarks d and s are each part d_w and part s_w, and these parts decay to the u-quark (which is also the u_w-quark); according to the rules expressed by Figure 15.12, only the d_w part will make the transition. Hence, the transi-

Figure 15.12 The relative strength of different quark and antiquark transitions for strong quarks is shown at the left; for weak quarks, at the right.

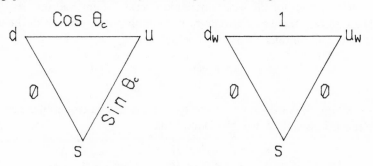

tion probability or decay rate is reduced from the rate that might be expected from a pure d_w decay or a lepton decay by the fraction of time the d (or s) quark is found as the d_w quark. And that probability is just $\cos^2 \theta \approx 0.93$ (or, for s, $\sin^2 \theta \approx 0.07$). Hence, the relative probabilities for (1) and (2) will be proportional to $\cos^2 \theta_c$ and $\sin^2 \theta_c$ respectively, and decay (1) will take place about $0.93/0.07 = 13$ times faster than (2). According to this model, the invalid UFI diagram of Figure 15.7 should be redrawn as Figure 15.13.

In summary, just as the weak interactions conserve the various categories of leptons (for instance, the number of electron-leptons is conserved, the number of muon-leptons is conserved, and the number of tau-leptons is conserved), so do the weak interactions conserve the number of light, weak quarks; the number of $u_w + d_w$ quarks is conserved. Moreover, the u_w and d_w quarks are coupled to the weak charge with the same universal strength—they carry the same weak charge as the muon or electron. In this model, the s_w quark does not take part in the weak interactions. However, the weak quarks are superpositions of strong quark states, and vice versa. A state of a strong d-quark is, for example, part of the time a d_w-quark and part of the time an s_w-quark. Hence, although the weak interactions conserve quarks (the number of quarks minus the number of antiquarks is unchanged), the flavor of the quarks is mixed up. And, as a consequence of the isolation of the s_w-quark from the interactions, the strong quarks appear to interact with a weaker charge than leptons. The transitions ($u \rightarrow s, s \rightarrow u$) are weaker than might be anticipated by a factor of $\sin^2 \theta_c$, and the ($u \rightarrow d, d \rightarrow u$) transitions are weaker by a factor of $\cos^2 \theta_c$. If the Cabibbo angle were 0, the $s \rightarrow u$ and $u \rightarrow s$ transitions would be forbidden—the weak interactions would conserve strangeness or hypercharge just as the strong interactions and weak hadron transitions would proceed with the same strength as lepton transitions such as the muon decay.

Determination of the Cabibbo Angle

From the Cabibbo model of the weak interaction decays of quarks, one can calculate the probability of any quark transition if one knows the Cabibbo angle. Then, with an understanding of the quark structure of hadrons deduced from analyses of the symmetries of hadron states, one can calculate the transition rates for the weak interaction decays of any hadron.

The Cabibbo angle can be determined simply by an analysis of the comparative transition rates for the two decays:

$$\pi^+(u, \overline{d}) \rightarrow \mu^+ + \nu_\mu$$
$$K^+(u, \overline{s}) \rightarrow \mu^+ + \nu_\mu$$

where the quark constitution of the hadrons is shown in parentheses. The basic transitions can be described simply as

$$\pi^+ \rightarrow u + \overline{d} \rightarrow W^+ \rightarrow \mu^+ + \nu_\mu$$
$$K^+ \rightarrow u + \overline{s} \rightarrow W^+ \rightarrow \mu^+ + \nu_\mu$$

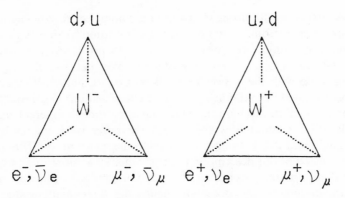

Figure 15.13 The set of universal relations (UFI) for weak interaction charged currents involving quarks and leptons.

as shown by Figure 15.14. The quarks making up the mesons combine to form the W, which then decays to the lepton states.

If other factors are equal, the ratio of the decay rates Γ_π and Γ_K (measured in decays per second) will be simply the ratio of the intensities of the d_w quarks held by the d-quark of the pion and by the s-quark of the K-meson:

$$\frac{\Gamma_K}{\Gamma_\pi} = \frac{\sin^2 \theta_c}{\cos^2 \theta_c} = \tan^2 \theta_c$$

Each transition rate is also proportional to the momentum of the emitted muon and the helicity factor $(1 - \beta)/2$, where $\beta = v/c$ and v is the velocity of the muon. These factors are different for the two decays as a consequence of the different energies of the final products from π and K decays, which follow from the different masses of the π and K-meson. However, these kinematic differences can be evaluated in a straightforward way and

$$\tan^2 \theta_c = 0.075 \quad \text{and} \quad \theta_c = 15°$$

The differences between the quarks defined by the weak interactions and

Figure 15.14 Feynman diagrams for (a) the decay $\pi^+ \rightarrow \mu^+ + \nu$ and (b) the decay $K^+ \rightarrow \mu^+ + \nu$.

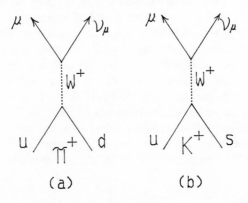

the quarks defined by the strong interactions are unexpected. Where does the Cabibbo angle come from? Who needs this complication? Although there are some interesting conjectures, physicists really do not know just how the Cabibbo angle fits into an overall view of forces. We guess that the angle has something to do with the answer to a different quark puzzle. Why are the quark masses different? Who asked for that complication? We suspect that the strong interaction mechanism, which gave the quarks different masses, also "twisted" the amplitudes about from the purer weak quarks by the Cabibbo angle, and that that angle is related to the quark-mass differences.

The extension of this description to the six quarks we now know requires a picture that is more complicated in detail but similar in principle to the three-quark discussion and does not change the three-quark results very much.

Charm and the Weak Neutral Currents

It seems aesthetically unattractive to have weak interaction forces transmitted by electrically charged intermediate particles, the W^+ and W^-, and not by a neutral counterpart to the charged W's. The "Old One" had not skipped zero in distributing charges to other sets of particles, such as the set of pions; could He have changed His ways in designing the weak forces and left us with charged currents (as charged W's) and no neutral currents?

Considering Nature's delight in symmetry, we might expect Her to have constructed a neutral sibling of the charged W's, a W^0, which would couple equally to all lepton-antilepton and quark-antiquark pairs with a charge strength $\sqrt{\frac{1}{2}}$ smaller[6] than the weak coupling of the W^+ and W^-. If the relation to the charged current forces is to hold, the transitions must be to left-handed particles and right-handed antiparticles. This W^0 particle would decay with almost equal probability to each of the pairs of particles shown. Because there are twice as many such pairs available as for charged W decays, and the total rate of decay of each sibling must be the same, the rate for the decay of the neutral W to any specific pair must be one-half of that for the charged decay to a pair. Hence, the square of the effective W^0 weak charge must be one-half of the weak charge of the W^+ and W^-.

The diagram at the left of Figure 15.15 shows possible transitions of such a hypothetical neutral current interaction. But we know, from the example of the photon, that other kinds of decays are also possible. For example, at the right a similar diagram shows a particle with photonlike couplings to charged particle-antiparticle pairs, where left-handed and right-handed particles are produced with equal frequency. We will find that nature does provide a neutral intermediate boson; however, that particle is neither the W^0 nor the G^0 of Figure 15.15 but rather a kind of mixture called the Z^0.

In the form that the weak neutral current is described by the diagram, a transition such as the decay of the neutral ρ^0 meson, which is partially made up of a u-$\bar{u}$ quark-antiquark pair, to an electron-positron pair through the transition

$$\rho^0(u\bar{u}) \rightarrow e^- + e^+$$

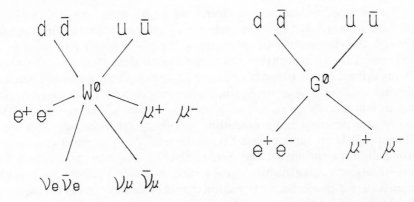

Figure 15.15 The diagram at the left shows interactions mediated by a possible W^0 intermediate boson. The W-boson would decay to each set of left-handed particles and right-handed antiparticles with equal frequency. At the right, a similar diagram shows decays through a hypothetical G^0 intermediate boson to charged particles and antiparticles. The G, like the photon, would decay equally to left- and right-handed charged particles and antiparticles.

will take place through the weak interactions and an intermediate Z^0 intermediate boson as shown by the Feynman diagram of Figure 15.16a. Here the quark-antiquark pair annihilate to produce a virtual Z that decays into the electron-positron pair. However, a similar diagram (Figure 15.16b) shows the same process mediated by electromagnetic forces and an intermediate photon rather than the Z. Because the effective electromagnetic couplings are very much stronger than the weak forces, the electromagnetic transition will dominate by a factor of about 10 million, and any weak interaction effects are hidden. In general, most possible neutral weak transitions are masked in this way by electromagnetic forces and are difficult to observe.

Figure 15.16 In (a) $\rho \to e^+ + e^-$ is mediated by weak forces; in (b) the same transition is mediated by electromagnetic forces; and (c) is the diagram for the transition $\rho \to \nu + \bar{\nu}$. For convenience, we describe the ρ-meson as a compound of $u - \bar{u}$, although it is one-half $d - \bar{d}$.

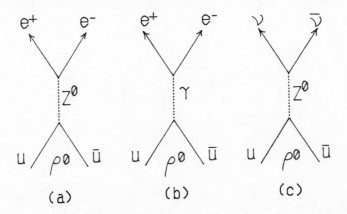

To see weak neutral currents clearly, we might better look at transitions (if there are any?) that can take place through the weak interactions but not through the electromagnetic interactions. We might search for decays to a neutrino plus an antineutrino, which would occur if the weak current has the forms defined by the diagram of Figure 15.15 but which would not happen through electromagnetic forces, since the neutrinos carry no charge. According to this description of the weak interactions, the ρ^0 can be expected to decay to a neutrino and antineutrino in exactly the same way, and with almost exactly the same probability, as the weak decay to the electron-positron pair. But neutrinos are very hard to detect. It would be easier to look for hypercharge-changing hadron decays such as a decay where an s-quark changes to a d-quark. Such a transition cannot occur through electromagnetic forces that do not change quark flavors. We might then expect that the weak neutral currents would mediate the transition

$$K^-(\overline{u}s) \rightarrow \pi^-(\overline{u}d) + e^+ + e^-$$

where the quark constitution of the mesons (in parentheses) shows that the meson transition takes place as a consequence of an $s \rightarrow d$ quark transition, as shown in Figure 15.17.

The reaction must take place through a transition $d_w \rightarrow d_w$ between the d_w parts of both the s-quark in the K-meson and the d-quark in the pion.[7] Calculating the effective charge, we should expect the transition to take place at a rate about the same as that of other hypercharge-changing decays or about $\frac{1}{15}$ as fast as ordinary weak decays. In fact, the decay is not seen at all, and the experimental limit is about a million times less than the calculated rate. Measurements of other hypercharge-changing weak neutral current transitions give similar results. If neutral weak currents exist, as our confidence in the

Figure 15.17 A possible weak decay transition $K^- \rightarrow \pi^- + e^+ + e^-$.

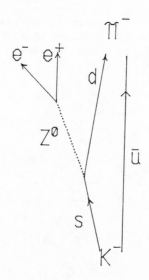

good taste of nature demands, the three-quark model provides no natural explanation for the lack of hypercharge-changing weak decays.

At this point, physicists recognized that another quark with appropriate properties was required. Before any direct evidence for its existence was uncoverd, this quark was postulated and given the flavor name "charm" and the label c.

The problem of the lonely s_w, seemingly unconnected to the weak interactions and the problem of the nonobservance of weak neutral currents, was solved by the discovery of the predicted fourth charmed quark, the c-quark found in the discovery of the J/Ψ charmonium c-$\bar{c}$ state. This quark couples to the charged and neutral weak currents with the s-quark as a c-s duo similar to the u-d quark pair. Such couplings are shown in Figure 15.18. If the charged W's couple to the s_w-quark, it is natural to assume that the neutral Z^0 also couples to the $s_w \bar{s}_w$. But this assumption radically changes the expectations for the neutral hypercharge-changing transitions. We noted earlier that the missing $s \rightarrow d$ neutral transition should take place through the d_w parts of the s and d. Now, with the additional coupling of charged and neutral currents to the s_w, we can also expect a contribution from the s_w parts of the s and d. But the contributions, $d_w \rightarrow d_w$ and $s_w \rightarrow s_w$, to the neutral hypercharge-changing decay amplitudes are equal with opposite signs and cancel.[8] The amplitude for the transition is then zero, consistent with the lack of observation of such a decay.

By the introduction of the fourth quark, the aesthetically unattractive idea of having an s_w quark that does not take part in the weak interactions is expunged, and the lack of hypercharge-changing neutral decays is accounted for. There is no longer a barrier to a belief in a neutral weak current.

With the later discovery of two more quarks, the t with a charge of $+\frac{2}{3}$

Figure 15.18 The coupling of the intermediate bosons to quark and lepton states with the c_w and s_w included. (To keep the diagram simple, the couplings to the t and b quarks and to the tau-lepton are omitted.)

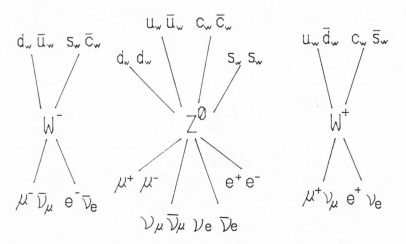

(like the u and c) and the b with a charge of $-\frac{1}{3}$ (like the d and s), the description of weak interaction has been extended to a fifth and sixth quark. The t_w and b_w presumably couple to the W and Z intermediate bosons just as do the other pairs. The precise relation of the t and b quarks to the weak quarks is not yet known, but it is clear that these quarks do not mix with the other quarks very much; that is,

$$b_w \approx b \quad \text{and} \quad t_w = t$$

Neutrino Interactions and the Helicity of the Quark

Neutrinos, carrying neither strong interaction charges nor electromagnetic charges, are an especially useful probe of matter, and a study of neutrino interactions can illuminate some properties of the weak interactions. The character of neutrino interactions with the fundamental constituents of stable matter, quarks and electrons, can be largely understood from the description of the weak interactions, which has been outlined. We now know of three different neutrinos together with their antiparticles or antineutrinos, each associated with one of the three separately conserved sets of leptons. The three families are as follows:

$$e^-, e^+, \nu_e, \bar{\nu}_e \qquad m_e = 0.511 \text{ MeV}/c^2 \qquad m_\nu < 15 \text{ eV}/c^2$$
$$\mu^-, \mu^+, \nu_\mu, \bar{\nu}_\mu \qquad m_\mu = 105 \text{ MeV}/c^2 \qquad m_\nu < 0.6 \text{ MeV}/c^2$$
$$\tau^-, \tau^+, \nu_\tau, \bar{\nu}_\tau \qquad m_\tau = 1785 \text{ MeV}/c^2 \qquad m_\nu < 250 \text{ MeV}/c^2$$

For the most part, we assume that the neutrino masses are zero. Although the direct experimental limits are not all extremely restrictive, it is difficult to fit massive neutrinos into our understanding of cosmology. From standard cosmological considerations, the universe should be full of neutrinos, and if any kind of neutrinos have a mass greater than 50 eV/c^2, the total mass of the neutrinos in the universe would be extremely large, and their mutual gravitational attraction would hold back the expansion of the universe to a rate below that which is observed.

According to conventional views, neutrinos interact with matter through their coupling with the intermediate bosons as suggested by Figure 15.15. Feynman diagrams for important interactions of neutrinos through this charged weak current are shown in Figure 15.19.

The interaction of neutrinos with matter through neutral currents, shown in Figure 15.20, displays both similarities and differences with respect to the charged interactions. In particular, the interactions with quarks through the charged weak current always lead to the production of a charged lepton, whereas no such lepton is produced by neutral current interactions. Because such leptons can be identified experimentally, the signature of a charged lepton in a final state indicates a charged weak interaction, whereas the lack of such a charged lepton in the interaction products shows that the neutral current mediates the reaction. Also, although a neutrino can scatter from an elec-

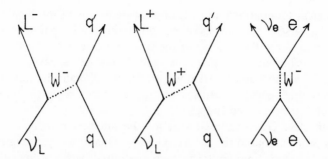

Figure 15.19 The amplitudes for neutrino and antineutrino scattering from quarks and electrons. The symbol L stands for leptons generically ($L = e, \mu, \tau$), and the symbols q and q' stand for quarks such as, d, $\bar{d}$, u, $\bar{u}$, s, $\bar{s}$, and c, $\bar{c}$.

tron through a neutral current interaction, no such scattering can take place if only charged currents operate.

Although the probability of neutrino interactions is intrinsically small—or, equally, the effective interaction areas (or cross sections) are small—with the intense beams of neutrinos available experimentally, extensive measurements have been made of neutrino interactions. Nuclear reactors produce large fluxes of low-energy $\bar{\nu}_e$ electron-antineutrinos with energies on the order of 1 MeV. High-energy electron and muon neutrinos and antineutrinos are produced by the decays of high-energy mesons produced in the course of the interactions of proton beams of high-energy accelerators with matter. These neutrinos have energies that range typically from 1 GeV (1000 MeV) to 200 GeV.

By and large, it is easy to produce large intensities of rather pure beams of muon neutrinos through the decays of positive pions and K-mesons produced at accelerators, and it is almost as easy to produce muon antineutrinos through the decay of negative mesons. It is much more difficult to produce large fluxes of high-energy electron neutrinos and antineutrinos, and tau-neutrinos have not been produced in a controlled fashion.

The interaction of high-energy neutrinos displays patterns that rest on

Figure 15.20 The interactions of neutrinos and antineutrinos with quarks or electrons through weak neutral currents.

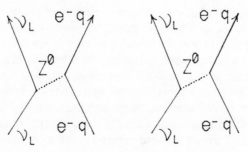

simple logical foundations and assumptions concerning the helicity of the quarks. Conversely, an analysis of the interactions of neutrinos with quarks serves to determine the helicity of the quark. As we need not differentiate among flavors (or colors) of quarks, we will simply use the designations q and $\bar{q}$ for quark and anti-quark. Also, we will implicitly consider only the reactions of very high-energy neutrinos, as the salient features of the interactions take simpler forms at high energies.

The effective cross-sectional area for the interaction of neutrinos with quarks is proportional to the neutrino energy. Qualitatively, as compared to lower-energy interactions, more high-energy initial particles enter the interaction region per second and more high-energy final interaction particles leave the region per second; hence, the overall interaction probability is greater at high energies. According to detailed calculations, if the interaction of the neutrinos with quarks and electrons takes place at a point, the cross section will increase linearly with the neutrino energy. At energies such that the wavelength of the neutrino is of the magnitude of the range of the weak interactions, a distance equal to $\hbar/m_c$ where m_W is the W-mass of about 75 GeV/c^2, the linear relation will break down, and the cross section will stop rising and remain at about the size of the interaction region. Although it must fail at neutrino energies near 1000 GeV, the direct proportionality of the cross section with energy extends to very low energies. Hence, the interaction cross section for a neutrino with an energy of 1 MeV is 100,000 times smaller than the cross section for a neutrino with an energy of 100 GeV produced from an accelerator beam. The mean path for interaction of a 100-GeV neutrino in water will be about 10,000 miles; the interaction length for a 1-MeV neutrino from a nuclear reaction will be about 1,000,000,000 miles! Nevertheless, low-energy neutrino interactions have been detected, and high-energy interactions have been studied in detail.

The description of quark-neutrino scattering takes an especially simple form in the center-of-mass system, where the two particles approach the collision point from opposite directions with equal momentum. For energies such that they travel at the speed of light, only left-handed leptons and right-handed antileptons take part in weak interactions moderated by the charged W-currents. Do the same rules hold for quarks? The helicity rules and the quark helicity assignments determine the angular distributions in quark-neutrino scattering; therefore, measurements of such scattering serve to determine the quark helicities.

Perhaps the quark is left-handed. An inspection of Figure 15.21 shows the left-handed quark and (left-handed) neutrino spinning in opposite directions before their collision; hence, the initial total spin is zero. After the collision, for both forward and backward scattering (or scattering in any other direction) the total spin of the neutrino and quark remains zero, and angular momentum is conserved. Therefore, the helicity rules do not discriminate among scatterings in different directions, and the scattering is isotropic.

The angular distribution will differ for the scattering of antineutrinos by quarks. Figure 15.22 shows antineutrino and quark spins before the collision,

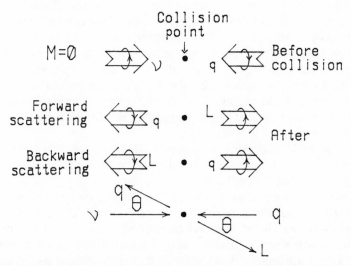

Figure 15.21 Momentum and helicity diagrams for the interactions of left-handed quarks with neutrinos and antineutrinos before and after scattering. The momenta of quark and neutrino are taken as equal and opposite in this center-of-mass Lorentz frame. The diagram below defines the angle Θ of scattering.

and for forward and backward scattering after the collision. Because the anti-neutrinos are right-handed, the total spin is initially 1 $\hbar$ directed clockwise facing to the right. After the collision, the spins of the particles still add to the same value for forward scattering, but the total spin is reversed for backward scattering. Then, the conservation of angular momentum together with the helicity rules forbid backward scattering, and the scattering of antineutrinos from left-handed quarks is directed strongly forward.[9]

Using the same arguments and Figure 15.22, one sees that the results are reversed if only right-handed quarks interact through the weak interactions. The scattering of neutrinos by quarks is strongly forward, and backward scattering is forbidden, whereas antineutrinos and quarks scatter isotropically.

Figure 15.22 Momentum and helicity diagrams for the interaction of right-handed quarks with neutrinos and antineutrinos before and after scattering.

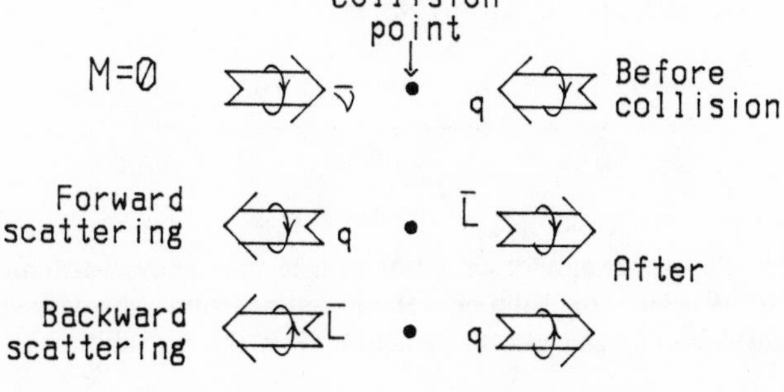

Although the angular distributions for the scattering of neutrinos from quarks are important in determining the properties of the weak interactions, how can they be measured when the neutrino is so elusive and no quark has ever been seen—or, perhaps, cannot be seen? Although a free quark is not produced in the scattering, the scattered quark is manifest in a shower of hadrons, and the direction and energy of that shower can be measured to determine the direction and energy of the scattered quark. And for the quark-neutrino interactions mediated by the charged W, the direction and momentum of the resultant lepton is easily measured.

The angular distributions are defined in the center-of-mass system where the two particles, before and after the collision, travel with equal and opposite momentum so that the collision is in some sense at rest. But the measurements are made in laboratories where the neutrino is moving and the hadron containing the quark is at rest. In that system the particle going forward in the center-of-mass system will have (almost) all the energy of the incoming neutrino, and the particle going exactly backward will have (almost) zero energy. Hence, the angular distribution in the center-of-mass system becomes an energy distribution in the laboratory system. Figure 15.23 shows the distributions observed for neutrino and antineutrino interactions. An inspection of the distributions shows that the backward scattering of antineutrinos is almost nonexistent, whereas the backward and forward scattering of neutrinos is about the same. These are just the results expected if the quarks are left-handed, and they establish that only left-handed quarks take part in the weak interactions mediated by the W-bosons, as is the case for left-handed leptons. If right-handed quarks were to take part, the two curves would be reversed.

The Neutral Weak Current and the Z^0

The netural weak interaction moderated by the Z^0 boson shows qualitative differences from the charged interaction. In particular, the measured hadron energy distributions from the neutrino-quark interactions are a mixture of the

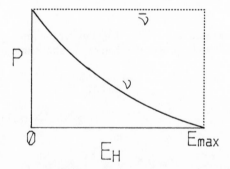

Figure 15.23 Experimentally observed hadron recoil energy distributions from the interaction of neutrinos (the solid curve) and antineutrinos (the dashed curve) with quarks through the charged weak interaction; P is the probability, E_H the hadron energy.

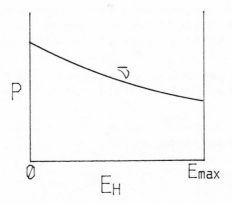

Figure 15.24 Experimentally observed hadron energy distributions from the interaction of antineutrinos with quarks through the neutral current interaction.

two different forms for neutrino interactions and for antineutrino interactions through the charged weak currents. This shows that for neutral currents, both helicities of quarks take part in the interactions—right-handed as well as left-handed quarks. This dual interaction is understood as a property of the Z^0, which is only partially a neutral sibling of the charged W's but is also partially a singular state closely related to the photon. Roughly speaking, the W^0 part interacts only with left-handed quarks like the charged W's, whereas the singular part interacts, like the photon, equally with both left- and right-handed charged leptons and quarks. The ratio of the two parts is defined by an angle θ, the so-called weak angle. The ratio of interactions of neutrinos with right-handed and left-handed quarks is a measure of that angle.

Figure 15.24 shows the results of measurements of neutral current interactions. It indicates that the right-handed quarks interact about one-half as strongly as the left-handed quarks.

Notes

1. Angular distributions are described conveniently in a form that expresses the probability dP of finding the state in an angular region $d \cos \theta$ as a function of the cosine of the angle $\cos \theta$:

$$\frac{dP}{d \cos \theta} = A + B \cos \theta + C \cos^2 \theta + \cdots$$

where θ is the angle between the direction of motion and the spin-quantization direction. For spin zero particles, only A is nonzero; for spin ½ particles, A and B may have nonzero values. In general, just as a particle of spin j may have but $2j + 1$ different defined directions, the description of the distribution of directions of such a particle must not require more than $2j + 1$ numbers, A, B, C, ...

2. Particularly for low-energy leptons, the asymmetries need not be complete. For a given direction of motion, any ratio of forward and backward spin is admitted. Indeed, there may be no preferred spin direction, half of the electrons may have spins directed forward and

half backward. And for leptons with their spin in a given direction, the direction of motion may be partially forward and partially backward. There may be no preferred direction of motion; half of the electrons may be moving forward and half backward, so the overall distribution is isotropic.

3. From the Uncertainty Principle, $j\hbar \cdot d\theta = \hbar$, and for spin $j = 0$, the uncertainty in angle $d\theta$ is complete. No direction is defined by the state, and no direction can be defined by decay products.

4. With respect to the direction of spin, the ratio of forward and backward intensities takes the following forms for leptons and antileptons:

$$\text{(Leptons)} \frac{F - B}{F + B} = -\beta \qquad \text{(Antileptons)} \frac{F - B}{F + B} = \beta$$

Hence

$$\text{(Leptons)} \frac{F}{B} = \frac{1 - \beta}{1 + \beta} \qquad \text{(Antileptons)} \frac{F}{B} = \frac{1 + \beta}{1 - \beta}$$

consistent with the result that the intensity (or probability) of emission of leptons (or antileptons) with "unnatural" helicity is proportional to $(1 - \beta)/2$ and proportional to $(1 + \beta)/2$ for "natural" helicity. The factors of ½ are inserted to normalize the helicity contribution to the emission probability to 1.

5. The weak quarks d_w and s_w and the strong quarks s and d are related by equations similar to those used in Chapter 3 to discuss coordinate transformations:

$$d_w = d \cos \theta + s \sin \theta \qquad \text{and} \qquad s_w = -d \sin \theta + s \sin \theta$$

and

$$d = d_w \cos \theta - s_w \sin \theta \qquad \text{and} \qquad s = d_w \sin \theta + s_w \cos \theta$$

Although the strong u-quark is the same as the weak u_w-quark, $u = u_w$, it is convenient to use the notation u_w when weak transitions are considered.

6. If an $SU(2)$ symmetry obtains such that the weak interactions do not differentiate among the three W-bosons or among associated lepton and quark pairs (such as the e^-, $\bar{\nu}$, or the d, $\bar{u}$), it follows that the probability for the W^0 to decay to any of the pairs depicted must be one-half that for the decay of the W^- or W^+ to a pair.

7. In particular,

$$d = d_w \cos \theta_c + s_w \sin \theta_c \qquad \text{and} \qquad s = -d_w \sin \theta_c + s_w \cos \theta_c$$

The decay takes place only between the d_w parts of the quark amplitudes:

$$s \to d = d_w \sin \theta_c \to d_w \cos \theta_c \to -[d_w \to d_w] \sin \theta_c \cos \theta_c$$

The effective charge strength of the interaction is equal to the UFI strength multiplied by $\sin \theta \cdot \cos \theta \approx 0.25$.

8. Breaking down the s and d to their s_w and d_w parts,

$$d \to s = [d_w \cos \theta_c + s_w \sin \theta_c] \to [-d_w \sin \theta_c + s_w \cos \theta_c]$$
$$= -[d_w \to d_w] \sin \theta_c \cos \theta_c + [s_w \to s_w] \sin \theta_c \cos \theta_c$$

Because, from the universality conditions, the basic amplitude for $d_w \to d_w$ is exactly equal to that for $s_w \to s_w$, the cancellation is exact and the decay cannot take place.

9. Using the relations between helicity and the direction of emission of leptons (and now quarks) discussed earlier, we can deduce the scattering angular distributions where the angle θ of scattering is defined by Figure 15.21. Taking the direction of the incoming particles as the forward direction so that forward scattering represents the particles continuing almost undeflected, both the antineutrino and quark angular distributions will take the form $1 + \cos \theta$ from the helicity rules for antiparticles and for particles. The probability of finding that the particles are scattered at an angle of θ is just the product of the two probabilities, or

$$P \, (\cos \theta) = (1 + \cos \theta)^2$$

16

Cosmology—
The World's Beginning
and End

By a knight of ghosts and shadows
I summoned am to a tourney
Ten leagues beyond the wide worlds end:
Methinks it is no journey.

Tom o'Bedlam

Cosmological Principles

In work published about 1650, Archbishop James Ussher concluded that the universe was created in 4004 B.C., a date that was inserted into the reference margins of the authorized version of the (King James) Bible. The mechanisms of the creation to which he refers are described metaphorically and poetically from a tradition that has been traced to the Sumerian Gilgamesh epic probably written about 3000 B.C., from an oral tradition nearly contemporary with Ussher's date. Scientific cosmology is not much older than a century but it is not less poetic, less awesome, or less mysterious than Ussher's cosmology of Genesis.

We are concerned here with cosmology because it may be necessary to understand the character of the universe in the instant of birth—in the first trillionth of a second after creation—in order to understand the particles and forces created at that time. Cosmology differs somewhat from other areas of physics in that the observations from which one draws evidence concern phenomena that usually cannot be affected by intrusive actions. One cannot do experiments, in the usual sense of the word, on the universe. Because of this constraint, there is more emphasis on axiomatic construction of theories generated *a posteriori* to fit a known body of observations. These differences between astrophysics and other physics are quantitative rather than qualitative, however. Generally, cosmological theories do have predictive power concerning observations not yet conducted, and they can be—and have been—proved to be in error.

Perhaps the first example of a rational axiomatic cosmology (and its disproof) was provided by H. W. M. Olbers in a famous paper published in 1826 in which he systematically presented ideas that go back to Kepler. Olbers investigated the simple assumption or axiom that the universe is infinite and unchanging (in gross structure) in space and time, and he demonstrated that this assumption is wildly incompatible with the simplest of observational facts: the sky is dark at night!

Olbers' reasoning can be expressed straightforwardly. If the density of stars is the same everywhere, and if the universe is infinite, a line in any direction will eventually intersect the surface of a star. In Olbers' universe, if you look in any direction at night, you will see the surface of a star that will be as the surface of the sun. The brightness of the sky at that point, and at any point, must be like the brightness of the sun; thus, not only will the sky not be dark, it will be intolerably bright.

What deviations are then required from this simplest of cosmological models in order that its conclusions fit observations? There is some very large-scale structure in the distribution of stars, inasmuch as stars are grouped in galaxies containing typically 100 billion (10^{11}) stars, and galaxies themselves are grouped in larger supergalaxies. But the supergalaxies appear to be distributed throughout space such that Olbers' paradox is not removed.[1] A severe limitation on the relation of Olbers' result to that which we know of the universe is evident when we consider that the information we have on the uniformity of space extends only to distances of a few billion light-years—and then to times extending only a few billions of years in the past. The largest telescopes, such as the 200-inch telescope at Mount Palomar, can photograph only galaxies that are not much further than 1 billion light-years away. But if the universe is homogeneous, as Olbers suggested, the average distance from the earth to the surface of a star is more than a billion times greater than a trillion light years. And light reaching the earth from such a star would be emitted more than a billion trillion years ago! Deviations from the ideal model of Olbers that are quite small locally can radically change his conclusions.

In particular, if the universe had a beginning at some particular time much less than 10^{20} years ago, Olbers' results would be affected because the light from far-away stars that must contribute most to the brightness of the sky could not yet reach the earth. Similarly, if all far-away stars were moving away from the earth with high velocities, the light from those stars would be weaker (and more red), and Olbers' conclusions would be vitiated. We know now that both of these effects occur: the universe, and time itself, had a specific beginning (perhaps 15,000,000,000 years ago), and the universe is expanding in such a manner that the far-away stars are receding from the earth (and from each other) so fast that the radiant energy reaching the earth from those stars is severely attenuated.

All serious scientific cosmologies are axiomatic, and all begin from the assumption of a Cosmological Principle to the effect that, aside from local fluctuations, observers at different parts of the universe view essentially the same picture. The local fluctuations can be on a grand scale: as I write this, I am in New Haven, Connecticut, a specific spot in the northern hemisphere of the third planet of a medium-sized star found in one of the spiral arms of an unexceptional spiral galaxy. All this detail is to be regarded as being as particular as the specific chair in which I am sitting, and as irrelevant from a cosmological view. Olbers' theory used this assumption of the Cosmological Principle.

Although scientific comologies are axiomatic, if they are to have any meaning to us they must confront observation, just as Olbers' model confronted the observation that the sky is dark at night. It is therefore necessary that the theories concern observable quantities. Such basic quantities as time and distance must be defined in an appropriately operational way in terms of real measurements. It is also necessary that any cosmological theory reduce to or agree with the description of the behavior of local objects that make up the body of physical theory on the local level. Furthermore, we must, as a matter of practicality, assume that the description of nature that accounts for local experience largely holds for very different times and very different places. And if the laws of nature we have constructed to account for local phenomena are different in regions far from our experience, those differences must be rational in the sense that a complete theory, valid for all time and space, must exist that reduces to the local description, here and now.

Surely there may be such differences; cosmologies need not follow local physical theory over the enormous extrapolations from experience involved in considering the entire universe. Laws such as the Law of Conservation of Energy or the Second Law of Thermodynamics need not be true for the universe as a whole. The observational evidence upon which such laws or summaries of experience are based simply does not extend to the vast times, distances, and masses we must consider in cosmology.

However, if we must not hold on too tenaciously to the concepts we have developed locally, we cannot afford to give them up too easily. We *must* extrapolate backward in time and far outward in distance—we cannot arrange a visit. Then the only guides we have are from the present and from our local universe, and we have to believe that those guides are sufficient to reconstruct the past and to predict the future. Our test of the efficacy of the extrapolations will be the logical consistency and completeness of the picture we construct—just as the validity of the local description we have constructed from extrapolations of our direct sensory evidence follows from its logical consistency and completeness.

The Red Shift and the Expanding Universe

About 1935, Edwin Hubble completed a series of measurements of the spectra of light emitted by distant galaxies which strongly suggested that our universe was expanding. Because the galaxies are made up of large numbers of stars, the light from a galaxy has very much the same character as the light from an average star such as our sun. As viewed through a prism or reflected by a grating, that light is seen as a nearly continuous rainbowlike spectrum marked by dark lines produced by the absorption of light by specific elements found in the outer atmospheres of the stars. Such absorption of the solar spectra by elements in the atmosphere of the sun results in the Fraunhofer absorption lines, named after the nineteenth-century Munich optician who constructed diffraction gratings to make the first accurate map of the solar spectra.

Two prominent dark absorption lines are found in the near ultraviolet and result from the absorption of light by calcium atoms in the sun's atmosphere. Hubble found that the light from distant galaxies was very much like the light from our sun (or a nearby star in our galaxy), except that the two prominent dark lines from calcium absorption were shifted towards the red end of the spectrum. Moreover, the fainter (hence farther) the galaxy, the greater the shift. If the Cosmological Principle is to hold and the far-off stars are not fundamentally different from the sun, the simplest explanation of the so-called red shift is that the other galaxies are moving away from ours, and the further the galaxy, the faster it is receding. The color change then follows from the *Doppler Shift*.

The Doppler Shift describes the shift in frequency or wavelength perceived by an observer of a wave emitted by a source moving toward or away from the observer. We hear the siren of an approaching ambulance at a higher frequency, and hear a lower note as the ambulance speeds away. The high-frequency, short-wavelength, ultraviolet light emitted by the stars of a distant galaxy seems to be shifted to lower-frequency, longer-wavelength, redder light as a consequence of the speeding away of the galaxy. Because the shift in frequency is proportional to the velocity of the source,[2] the further galaxies seemed to be receding faster.

From the reduction in frequency or increase in wavelength of lines such as the calcium absorption lines, Hubble and his collaborators, followed by many other astronomers, measured the relative velocities of a large sample of far-off galaxies and correlated the velocities with distances.[3] These correlations led them to the remarkable conclusion that all galaxies were receding from the earth (or from our local galaxy), and the farther the galaxy, the greater the velocity of recession. Indeed, the velocity was proportional to the distance. If we label the hypothetical distance such that the velocity of recession, so calculated, would be the speed of light, as R_0, the formula

$$\frac{v}{c} = \frac{R}{R_0}$$

describes the recession quantitatively[4] and $R_0 \approx 20 \cdot 10^9$ light-years. If this rate of expansion had always been in effect, we would have to conclude that 20 billion years ago, all the galaxies—all the matter we can perceive in the universe began in an instant of time in a colossal explosion, the so-called Big Bang. (The expansion seems to have been faster in the past, hence "time zero" is better estimated as about 15 billion years ago.)

This result, that all galaxies are retreating from us, does not in itself contradict the thesis of the Cosmological Principle that there is no privileged position or privileged observer. Although all galaxies are receding from the earth, to an equal degree they are all receding from each other. The earth, or our galaxy, is not a special, plagued spot from which all are fleeing.

Although this mutual recession could be explained qualitatively by classical models,[5] the application of principles from the General Theory of Relativity suggests that we should consider that space itself is expanding—after all, space is defined only by markers such as galaxies—and it is that expansion which we see by way of the red shift of the light from receding galaxies. This description of an expansion of space itself can be understood with the aid of a two-dimensional model of our three-dimensional space. Consider that the galaxies are mapped as inkspots on the (two-dimensional) surface of a balloon, as suggested by Figure 16.1, and the balloon is being blown up or expanded so that the diameter of the balloon increases at a rate of 1 inch a minute. When the balloon is 10 inches in diameter, 10 minutes after the "birth" of the universal balloon, the distance between any two points will be increasing at a rate of 10% per minute. A flatworm sitting on a spot on the balloon will see all marked points on the balloon receding, and the velocity of recession will be proportional to the distance. A point 1 inch away will be receding with a velocity of 0.1 inches per minute; a point 3 inches away will be receding three times as fast at a rate of 0.3 inches per minute. Perhaps the clever flatworm will express the recession of the inkspot galaxies of his universe quantitatively in the form

$$\frac{v}{c} = \frac{R}{R_0}$$

where v is the velocity of recession of a galaxy, c is found to have the value of 1 inch per minute, R is the distance to the galaxy, and R_0 is 10 inches. He may also deduce that a Cosmological Principle holds; his equation describing the expansion must be valid for any flatworm observer on any point of his universe, the surface of the balloon. And, if he is a worm of genius (presum-

Figure 16.1 The expansion of the balloon that is the flatworm's universe. At the left is the universe 10 minutes after the Creation; at the right, 20 minutes after.

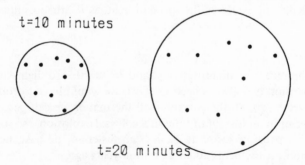

ably the flatworm of Chapter 7), he may conclude that his two-dimensional space is curved in a three-dimensional manifold he can understand only abstractly, just as we, his three-dimensional counterparts, conclude that our three-dimensional space may be curved and expanding in a four-dimensional manifold that we cannot experience directly.

Beyond the limit that v, the velocity of recession seen by the flatworm, is equal to the velocity of the signal that gave him information about the position of distant inkspots, the flatworm can have no information concerning his balloon universe. For our three-dimensional universe, in the limit that v, the velocity of recession, is equal to the velocity of light c, the wavelength received by the observer would be infinite, and no light signal—or any other signal—can ever be received. We are then not operationally connected to any existence beyond R_0, the event horizon, although our cosmological models may extend beyond this boundary.

Our universe, like the flatworm's, is the same everywhere, but it is not the same at all times. Taking the Hubble constant as 20 billion light-years, 1 billion years ago the scale of the universe was 95% of its present value, the galaxies were 5% closer than they are today, and the average density of matter in the universe was 15% greater than it is now. On the other hand, an observer 1 billion years in the future will see the galaxies 5% farther away and a mean density of matter about 85% as great as it is now.

The Curvature of Space and Time

The concepts of space and time are empty without the existence of matter: if there are no markers and no events, distance and elapsed time, defined operationally as the distance between events and the time between events, can have no meaning. We cannot then inquire into the nature of space and time without recognizing that we must deal with a space-time that contains matter.

Only gravity seems to operate between matter in the large—between stars and galaxies—and from the General Theory of Relativity, gravity is an effect of the curvature of space-time that follows from the existence of matter. Our universe, containing matter, must then be curved locally near centers of mass and may be curved in the large as a consequence of the entire mass of the universe. We can estimate the magnitude we might expect of such a curvature.

To estimate the magnitude of the curvature of space induced by the matter of the universe, we adopt a simple model that assumes that the universe is a sphere of matter with a density taken from our observations of the nearby universe. We might then calculate the radius such that a beam of light, affected by gravity according to the Equivalence Principle (Chapter 7), will travel in a circle about that sphere.[6] From such a calculation, we will find a radius near the Hubble radius of 20 billion light-years. The results of such an estimate do not establish a curvature of space, but they do tell us that we must consider possible curvatures of space in the construction of a cosmology.

How can we measure the curvature of the universe? If the Cosmological Principle is valid, the curvature can be determined by plotting the number of galaxies observed against the square of the distance from the earth. The basis of this method can be illustrated by watching our clever flatworm analyzing the curvature of his two-dimensional universe in Figure 16.2, where he plots the density of inkspot galaxies as a function of distance r. Here, he constructs concentric circles with radii that differ by 1 inch (his unit distance) about his home galaxy, and he counts the number of inkspot galaxies that fall between the circles. On a flat surface he finds that the number per inch increases linearly with distance. If the Cosmological Principle holds for inkspots and the number of inkspots per unit area is the same everywhere in his surface universe, there will be twice as many inkspots between the 9.5-inch circle and the 10.5-inch circle as between the 4.5- and 5.5-inch circles. On a sphere, the density will not increase that fast. There will be fewer than twice as many per inch at 10 inches than at 5 inches. The flatworm attributes this to a curvature of his universe and calls that curvature[7] positive. A negative curvature is also possible. If the surface is "saddle shaped" as at a mountain pass, the number of inkspots per inch will increase more than linearly; there will be more than twice as many inkspot galaxies per inch 10 inches from his position than at 5 inches. He calls this a negative curvature.

Inhabitants of a three-dimensional universe curved in a four-dimensional manifold, we proceed to measure the curvature in a manner similar to that of the flatworm. Instead of the flatworm's plot of the number of galaxies in an area between two concentric circles separated by a unit distance, as defined in his surface universe, we must determine the number of galaxies in a shell

Figure 16.2 Above, the flatworm determines the curvature of his universe by measuring the number of inkspot galaxies per unit distance from a center. Below are the graphs he constructs.

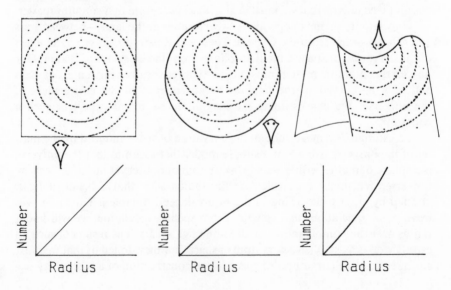

of unit thickness bounded by two concentric spheres separated by a unit distance, as a function of the distance to the shell. If the count increases quadratically with the distance of the shell, the universe is flat; if a shell twice as far has four times as many galaxies and a shell three times as far has nine times as many galaxies, and so on, the universe is flat. If the count increases more slowly, our universe exhibits a positive curvature; if the count increases faster than linearly, our universe must be curved negatively.

Just as space may be curved substantially over so large a region as the universe, we should presume that time could also be curved or distorted. Just as space is curved in a gravitational field, time is also distorted (Chapter 7). Although the curvatures of space over nearby regions caused by local distributions of matter are very small, the total curvature of the entire universe may be substantial. In the same sense, the smallness of local distortions of time should not convince us that very large distortions might not exist on universal time scales.

Measurements of space curvatures and time distortions are coupled. Any look out to great distances is also a look back into time. Galaxies observed at distances of a billion light-years are observed as they were a billion years in the past. From our elementary description of the expanding universe, we would expect that the galaxies would be about 5% closer together at a distance of a billion light-years simply as a result of the fact that we would be viewing that part of the universe as it was a billion years ago when the expanding universe was smaller than today. It is then necessary to differentiate between the *world picture,* which is the description of the universe as it can be seen by an observer on earth using a telescope that necessarily looks back into time even as it looks out into space, and the *world map,* which describes the entire universe as it might exist at a moment in time. It is usually the world map that we discuss, and always the world picture that we observe. As a result, the interpretations of observations are not completely obvious. If the galaxies seem to be denser at great distances, does this result from the greater density of the universe in the past, or is it indicative of a negative curvature of the universe? But we know from the Special and General Theories of Relativity that space and time are inherently coupled, and the behavior of the universe at great distances must be correlated with its behavior at far times. We have seen in local regions that distortions in both space and time are induced by gravity—that is, by the presence of matter. We can then expect that the character of the expansion of the entire universe in space and the history of that expansion in time must be related to the mean density of matter in the universe.

This relation among the rate of expansion, the density of matter, and the curvature of space can be suggested by a simple classical model that provides an accurate description of the expansion for small distances (distances over which the velocities are small compared to the velocity of light) and suggests the character of the relativistic expansion at higher velocities and over greater distances. In pursuit of some insight into the conceptual relations among these factors, let us consider an initial state of expanding matter in the form of a sphere packed with a constant density of galaxies such that each galaxy is moving away from the center with a velocity proportional to its distance from the center of the sphere. For such a system, the distribution of galaxies will not change with time, as the entire sphere expands against the mutual attraction of the gravitational force acting between the galaxies. Each galaxy undergoes an acceleration toward the center induced by the gravitational attraction of the other galaxies. Hence, each galaxy is slowing down very much as a particle ejected upward from the earth's surface will slow down under the pull of the earth's gravity.

If such a particle thrown upward from the earth (at an altitude just above the atmosphere, so that we need not consider air resistance) has a small velocity less than the *escape velocity*,[8] it will rise for a while, slow down, and then fall backward to the earth. If the velocity is larger than the escape velocity, it will slow down but retain enough velocity to escape the earth and go on forever. Quantitatively, the escape velocity is proportional to the square root of the mass of the earth or the square root of the mean density of the earth.

The same kind of relation holds for a sphere inscribed in a universe holding a constant density of galaxies (Figure 16.3). A galaxy at the edge of the sphere of radius r will escape from a galaxy at the center—that is, our galaxy—if it moves away with a velocity proportional to the square root of the density of matter in the sphere and proportional to the radius of the sphere—or the distance from our galaxy. Then, for any sphere and any distance between galaxies, there will be an escape velocity proportional to the square root of the density of matter in the universe and proportional to the distance between the galaxies. If the density is low and the expansion velocity exceeds the escape velocity, the expansion will slow down but still continue forever. If the density is high and the expansion velocity is less than the escape velocity, the universe will expand until it reaches a maximum size, and then the galaxies will fall back together again in a contracting universe. Figure 16.4 suggests the variation of the scale of these expanding universes as a function of time.

Figure 16.3   A sphere inscribed in a universe filled with galaxies.

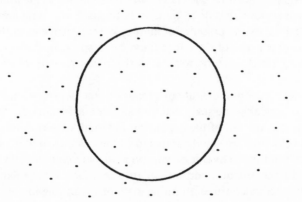

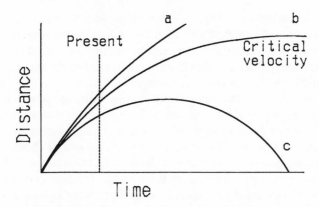

Figure 16.4 The variation with time of the distance between two galaxies for (*a*) recession velocities that are greater than the critical escape velocity, (*b*) a velocity equal to the escape velocity, and (*c*) for a velocity less than the critical velocity.

Not surprisingly, the spatial curvature of the universe is connected with the time evolution in a plausible fashion.[9] If the expansion velocity is less than the escape velocity, the universe is closed in time as it collapses (time ends at the collapse), and it is also closed in space (and finite) with a positive curvature. If the expansion velocity is greater than the escape velocity and the universe expands forever without limit, the universe must be open and infinite, and the curvature must be negative. If the expansion is exactly at the escape velocity, the curvature must then be neither positive or negative, but zero; the universe must be *flat*.

By looking out in distance and backward in time, we should be able to determine the change in expansion rate and the curvature of space. At the present time we can say only that space is very nearly flat and that the expansion velocity is very nearly the escape velocity. This is a most remarkable result; after 15 billion years of expansion, the rate is almost the escape velocity.

The unique character of our (nearly?) flat universe with its commensurate rate of expansion can be illustrated by a simple example. Assume that a rocket is shot off the earth and 1 billion years later and about 30 trillion miles away ($5.7 \cdot 10^{13}$ kilometers), an observer would see the rocket body moving slowly away from the earth at a speed of just 14.5 inches per second (37 centimeters per second), which is very close to the velocity of escape from the earth's gravitation at that great distance. For the velocity to be nearly the escape velocity after 1 billion years, the original velocity has to be equal to the escape velocity at that time to an accuracy of better than one part in a billion! With some assurance, we can trace the history of the universe to the first second of its life. For the flatness at that time to persist until now, the expansion velocity at that early time must have been equal to the escape velocity to an accuracy of better than one part in a billion billion (10^{12}). If we believe, as some do, that we can sensibly consider times as short as 10^{-35}

seconds, the original velocity must have been accurate to about one part in 10^{50}! At any rate, it can hardly be an accident that the universe is almost flat.

Knowing that the universe is nearly flat, we can check to see if the observed density of the universe is sufficient to account for the expansion rate determined from the red-shift measurements. And here we run into trouble. The mass we can see, the mass of the shining stars, appears to account for only a few percent of the needed mass. From measurements of the rotation of galaxies and of the motion of pairs of galaxies held together by gravity and rotating about their common center, we know that galaxies hold a great deal of dark matter and are many times heavier than the masses of their visible stars. However, there is still missing mass; we have reliable knowledge of little more than 10% of the mass required to account for the gravitational slowing of the expansion rate.

Where then, is the extra mass? Although there is some tenuous cold gas between the stars of a galaxy, the mass of this gas (largely hydrogen) is known and not large. Perhaps there are massive black holes left over from the Creation or from large stars that burned out in the youth of the universe. Or maybe the mass comes from many small things. Physicists have observed that the universe is filled with low-energy photons produced in the original explosion and expect that there are similar densities of neutrinos. We know that the photons, through their energy (taken as mass through the relation $E = mc^2$), contribute to the mass of the universe, but not so much as stars. Neutrinos, if they are as massless as photons, should contribute similarly. But if the neutrinos have a rest mass, which the photons do not, the mass of the neutrinos produced in the original explosion might hold the universe together. To understand the origin of photons or neutrinos, we must then study the early universe.

The First Three Minutes[10]

If the universe is expanding at a known rate, it must have been hotter and more dense in the past. The origin of this variation of temperature and density with time may be more obvious if we reverse the clock and consider a contracting universe. Surely as the universe contracts it becomes more dense. And in a contracting universe, first the galaxies, then the stars, and then the individual atoms, nuclei, and so forth fall toward each other under their mutual gravitational attractions, and thus gain energy. Because the temperature of a system is simply the mean energy of the elements that make up the system, the contraction produces a very hot, dense universe.

Setting the clock correctly again, at very early times the mean energy making up the particles of the universe must have been very high—the universe must have been hot and very dense. The variation of density and temperature with time[11] for the early universe is suggested by Figure 16.5. Although the graph extends to earlier times, the history of the universe much before the first tenth of a second is only conjecture—a remarkable conjecture to which

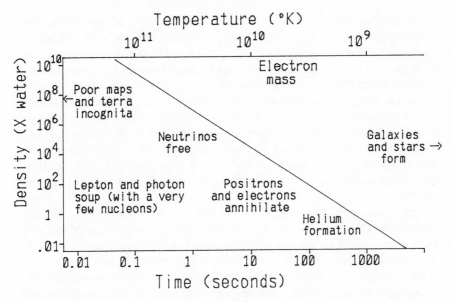

Figure 16.5 The approximate variation with time of the temperature and density of the early universe.

we will return in Chapter 18—but from the first tenth of a second through the first few minutes, we can recreate the character of the universe with remarkable precision and certainty. The evolution of the universe from that time to the present still poses puzzles concerning the condensation of matter to stars and galaxies.

One-tenth of a second after birth, "the Universe is simpler and easier to describe than it will ever be again," in the words of Nobel Laureate Steven Weinberg. All the matter and energy now within our "event horizon," consisting of a sphere of radius near 15 billion light-years, occupy at that early time a sphere with a radius of about one-half of a light-year. That sphere, and space beyond, is filled with an immensely dense gas of photons, of neutrons and protons, and of electrons, positrons, neutrinos, and antineutrinos, all colliding and interacting[12] with each other at a rate such that a thermal equilibrium obtains. The density is about 4 billion times the density of water, and the average energy of the particles is about 8 MeV—corresponding to a temperature of about 10^{11} °K—the same for each kind of particle. The density is so great that even the neutrinos interact in times small compared to the universe age of 0.1 seconds. The energy is so high that the masses of the electron and positron of about 0.5 MeV are negligible compared to the average thermal energy. (The photons at this time have energies like those emitted now as high-energy gamma rays in nuclear reactions.)

Because for the light particles the thermal energy is much greater than what rest mass they may have, there are almost equal numbers of each particle.[13] In particular, there are about $5 \cdot 10^{34}$ photons per cubic centimeter, about equal numbers of positrons and of electrons, and half as many of each of the three positrons and of electrons, and half as many of each of the three different neutrinos and antineutrinos. The masses of the neutron and proton are about 100 times as great as the energies defined by the temperature; hence, nucleons and antinucleons are not in equilibrium. Presumably at some earlier and hotter time there were as many nucleons and antinucleons as photons, but as the temperature dropped in the expansion, the nucleons and antinucleons annihilated, leaving only a small residual of about 1 nucleon for each billion photons—so few as to have a negligible effect on the universe at this "radiation-dominated" period. Because the mass difference between the neutron and proton is only about 1.2 MeV, small compared with the mean thermal energy of about 8 MeV, there are almost equal numbers of neutrons and protons.

About a second later, the density has dropped by a factor of 100, the temperature by a factor of 10, and the neutrinos and antineutrinos are decoupled from other matter as the universe becomes relatively transparent to them. After another 10 seconds, with the density about 350,000 times that of water and the temperature about 10^8 °K, the annihilation of positrons and electrons to two photons is no longer balanced by the production of positron-electron pairs by photon collisions, and all positrons and most of the electrons vanish, willing the energy they carry to the photons. At this time, the universe is still too hot for nucleons to combine to form nuclei, just as water vapor at a temperature greater than 100 °C (212 °F or 373 °K) is too hot to condense into droplets. Only when the universe is about 3 minutes and 45 seconds old, with a density about 1000 times the density of water and the temperature about 1,000,000,000 °K (about 60 times the temperature of the sun), will it be cool enough for neutrons and protons to combine to form deuterium, the first step of a set of reactions that lead to the formation of helium. From the conditions defined here, we can expect with but little uncertainty that about 26% of the nucleons will combine to form helium; no heavier nuclei will be formed from this primordial soup. Here, ancient history ends.

Medieval history, the formation of galaxies and stars from condensations of the expanding cloud of matter, and modern history, including the production of heavy nuclei in the star explosions we call novae and supernovae (which lead again to life and the contemplation of the entire process), we can ignore as not bearing critically on the considerations of particles and forces we address.

From the vantage of the present, 15 billion years after the Creation, we can speak confidently of the first few minutes of the evolution of the universe—indeed, we reach back to the first tenth of a second! How can we be so sure that we are right? Remarkably, we can look back and see the explosion. We have noted that a look out in space is also a look back in time. If we look out 15 billion light-years, we will be looking 15 billion years into the past and seeing the Creation! We have done just that. The bright ultra-ultraviolet flash of the hot universe at one-tenth of a second, radiating photons with a wavelength of about $3 \cdot 10^{-12}$ centimeters, has been shifted toward the red to form microwave radiation with a wavelength of about 0.5 centimeters, which corresponds to the emission by a cold black body at a temperature of

3°K. In 1964 Arno A. Penzias and Robert W. Wilson of Bell Labs used an advanced antenna design to look out from Crawford Hill in Holmdel, New Jersey, to detect the microwave radiation generated at the Creation.[14]

There is also related evidence for a singular creation derived from an examination of the nuclear population of the universe. From the temperature history derived from the present intensity of the 3°K microwave energy, corresponding to a photon density of about 500 photons per cubic centimeter, the helium abundance was calculated to be about 26% of the matter in the universe (with the rest of the mass largely hydrogen). And that is just what we see in the sun and stars.

Although we know so much, we know so little. The following are a few outstanding questions to be answered when physicists know much more about the evolution of the universe before the first tenth of a second and much more about the fundamental forces and particles that dominate the earliest instants.

1. The ratio of nucleons to photons is about 1 to 1 billion. At some early time there were probably about as many nucleons and antinucleons as photons; then, as the universe cooled, the nucleons and antinucleons annihilated each other, leaving the residue that makes up matter in the universe. But we live in a world of particles, not antiparticles. Why the asymmetry between particle and antiparticle that leaves us a particle residue?[15]

2. The annihilation of electrons and positrons leaves again a slight residue of electrons. Why? Moreover, there are almost exactly (to at least 1 part in 10^{30}) the same number of electrons left over as protons. The universe is very nearly neutral electrically. Why?

3. As we look out at the 3 °K microwave radiation, we find that the intensity is the same in all directions to at least a factor of 1 part in 10,000. But we are now looking at regions that were not causally connected at the time the radiation was emitted (information would have had to travel faster than light to link the regions). Hence, different parts of the universe evolved independently in almost exactly the same way. Surprising?

4. We have mentioned the remarkable flatness of the universe. Who ordered that? And where is the mass that must be there if the universe is as flat as we see it?

Notes

1. An appropriately selected infinite set of hierarchies can give us dark nights, but the constraints are severe and seem contrived.

2. Consider a galaxy emitting light at a frequency of f vibrations a second moving away from the earth at a velocity of v meters per second which, for simplicity, we take as much smaller than the velocity of light, c. The light emitted at a time $t_0 = 0$ will reach the earth at a time $t_0' = x/c$, where x is the distance from the galaxy to the earth when the light flash is emitted. Light emitted 1 second (and f vibrations) later at $t_1 = 1$ will be received at a time

$t_{1'} = 1 + (x + v)/c$, just as the galaxy traveled an extra distance of v meters during the second that elapsed and the light has that much further to go. The frequency received on earth is the number of vibrations observed, divided by the time elapsed on earth, $f' = f/(1 + v/c)$; hence, the observer on earth perceives a lower frequency. Because the wavelength is inversely proportional to the frequency $\lambda = c/f$, the observed wavelength $\lambda' = \lambda (1 + v/c)$ is increased by a factor proportional to the velocity. For very large galactic velocities, relativistic formulas lead to qualitatively similar relations.

3. These enormous distances are difficult to measure, and large errors (of factors of near 10) were made in early attempts to define distance scales. Briefly, we determine the distances of near stars by triangulation, using as a baseline points on the orbit of the earth (rather as a surveyor measures distance from a terrestrial baseline). Then, noting that stars of the same color have similar intrinsic brightnesses, we determine distances to far-off stars in our galaxy by their brightness, assuming that that brightness varies inversely with the square of the stars' distance from the earth. We estimate the distances of nearby galaxies by assuming that certain bright stars in the galaxy, which are easily identified because their brightness changes periodically with time (cepheid variables), emit the same light as such stars in our galaxy; and again we use the inverse square law to determine distances. Then, noting that similar galaxies have similar brightnesses, we use that brightness to find the distance of far-off galaxies. (More recently, the brightness of quasars is used to probe distances beyond that at which galaxies can be seen.)

4. For very high velocities such as the velocities of quasars (exploding centers of galaxies?) seen near the edge of the universe where v is a large fraction of c, relativistic formulas that are somewhat more complicated must be used.

5. For example, if gas molecules (or galaxies), all moving with different velocities, are contained in a balloon in a vacuum and the balloon bursts, the molecules (or galaxies) will move away from each other with their initial velocities. At times after the explosion marking the origin of that universe the molecules will be receding from the point of explosion (and from each other) with velocities proportional to their distance from that point (and their distances from each other). However, the Cosmological Principle will not hold; the density of molecule galaxies will vary with distance from the original position of the balloon.

6. The gravitational acceleration a_g near a sphere of radius R is expressed by the relation

$$a_g = \frac{Gm}{R^2} \quad \text{where} \quad m = Vd \quad \text{and} \quad V = \tfrac{4}{3}\pi R^3$$

where m is the mass of the sphere, V the volume, and d the density. Light traveling in a circle of radius R proceeds with a centripetal acceleration $a_c = c^2/R$. Assuming that the acceleration of the light is caused by gravity and equating the two accelerations, $a_g = a_c$,

$$\frac{c^2}{R} = \tfrac{4}{3} GRd \quad \text{and} \quad R = \frac{c}{\sqrt{4\pi Gd/3}}$$

Estimating the density of the universe as $d \approx 10^{-29}$ gm/cm^2, and $G = 6.67 \cdot 10^{-8}$ cm^3 gm^{-1} sec^{-2} from terrestrial experiments, $R \approx 1.8 \cdot 10^{28}$ centimeters, or about 20 billion light-years.

7. The curvature at a point on a surface is defined as $k = (1/R_1) \cdot (1/R_2)$, where R_1 and R_2 are the radii of curvature at the point taken in different orthogonal directions. Hence, for a sphere the curvature is simply $k = 1/R^2$. For a saddle, the two curvatures are in opposite directions and have different signs; hence, k is negative.

8. An object of mass m will escape from a spherically symmetric mass distribution M if its total energy $E = T - V$ is greater than zero, where T is the kinetic energy and $-V$ the potential energy of the body:

$$E = T - V = \frac{mv^2}{2} - \frac{GMm}{r} \geq 0; \quad \text{then} \quad v_e = \sqrt{2GM/r}$$

where G is the gravitational constant, M is the mass of the matter enclosed by a sphere of radius r, and v_e is the escape velocity. Given r as the radius of the earth and M as the mass of the earth, $v_e = 11$ kilometers per second or 6.93 miles per second.

For a galaxy at a distance r from another galaxy that can be considered the center of a sphere of galaxies of density ρ gm/cm^3, $M = 4\pi dr^3/3$ and

$$\frac{v_e}{r} = \sqrt{8\pi Gd/3} \quad \text{or} \quad \left(\frac{v_e}{r}\right)^2 = \frac{8\pi Gd}{3}$$

Note that the escape velocity is proportional to the square root of the density and to the distance from the galaxy arbitrarily chosen as the center.

9. From the General Theory of Relativity, the relation of curvature k, expansion rate v/r, and density of matter in the universe, d, can be expressed as

$$\left(\frac{v}{r}\right)^2 = \frac{8\pi Gd}{3} - kc^2$$

where the first term on the right defines the escape velocity and the second the curvature (which has dimensions R^2). Hence, if the velocity is greater than the escape velocity, the curvature must be negative; if the velocity is less than the escape velocity, the curvature must be positive. When the universe doubles in size, the density decreases by a factor of 8 and curvature decreases by a factor of 4.

10. The title of this section is that of a book by Nobel Laureate Steven Weinberg describing the beginning of the universe.

11. Consider two typical objects (say, galaxies or atoms) separated by a scale distance S. If the scale is reduced by a factor of 2, the density of matter will increase by a factor of 8; $d \propto 1/S^3$, where "$\propto$" means "proportional to." For relativistic particles such as photons, the density of particles will also increase by the factor of 8. But the wavelength λ of each photon will decrease by a factor of 2; hence, the energy $E = \hbar c/\lambda$ will be doubled. Altogether, for photons and other relativistic particles the energy density will vary as $1/S^4$. Then, from the Einstein relation connecting mass and energy ($E = mc^2$) the mass density will vary as $1/S^4$. The Stefan-Boltzmann law tells us that the energy density is proportional to the fourth power of the temperature $T^4 \propto d$; hence, $T \propto 1/S$.

For a flat universe, we have shown that the expansion velocity is proportional to the scale and the square root of the density. Hence,

$$v_e \propto \frac{dS}{dt} \propto \frac{S}{d} \propto \frac{1}{S}$$

which holds if $S \propto 1/t$. Then for a newborn universe of relativistic particles,

$$d \propto \frac{1}{t^2} \approx \left(\frac{600 \text{ sec}}{t}\right)^2 \text{ gm/cc} \quad \text{and} \quad T \propto \frac{1}{\sqrt{t}} \approx \frac{1 \text{ MeV}}{\sqrt{t}} = \frac{10^{10} \text{ °K}}{\sqrt{t}}$$

12. If the particles of a system make many interactions, the energy of the system will come to be shared among the particles (indeed, shared among each category of energy—kinetic, potential, rotational, etc.). When the distribution is complete, the system is said to

be in *thermal equilibrium*. At the densities and temperatures (mean energies) of the early universe, the particles—even the neutrinos—make very many interactions in times short compared with the lifetime of one-tenth of a second. Moreover, the interactions are all reversible; even as an electron e^- and positron e^+ annihilate to form two photons γ, two photons collide to form an electron-positron pair. Some important and typical interactions are as follows:

$$\gamma + \gamma \longleftrightarrow e^+ + e^-; \qquad e^+ + e^- \longleftrightarrow \nu + \bar{\nu}; \qquad n + \nu \longleftrightarrow p + e^-$$

where ν and $\bar{\nu}$ and n and p are the symbols used to designate neutrinos, antineutrinos, neutrons, and protons.

13. At temperatures greater than the rest mass of the particles, each quantum mechanical state of a particle will be present with equal probability. For the light particles, there will be a weight factor proportional to the number of different spin directions available and a small correction (which we neglect here) for the effects of the Pauli Exclusion Principle on fermions. The photon has two different possible spin directions (polarization directions), as do electrons and positrons. However, massless neutrinos and antineutrinos, as described by conventional models, have only one spin direction. But there are three different sets of neutrinos and antineutrinos: one set associated with electrons, one with muons, and one with tau-leptons. Hence, there will be about as many electrons as photons, the same number of positrons, and three times as many neutrinos or antineutrinos.

14. An accusation of excessive poetic license can be levied with regard to the early time cited. The universe did not become transparent to radiation until about 300,000 years had passed, and the temperature was a little hotter than the surface of the sun. The photons observed by Penzias and Wilson might best be considered to date to that time.

15. We know that galactic cosmic rays, largely injected into space from supernovae explosions in our galaxy, are made up of matter, not antimatter. Hence, it is most probable that our galaxy is wholly a matter galaxy. It is much more difficult to determine if the very high-energy cosmic rays we detect that originate in other galaxies are wholly matter. Hence, some lingering uncertainty as to the composition—matter or antimatter—of far-off galaxies remains. It seems that it is still possible to argue that there are equal numbers of nucleons and antinucleons in the universe segregated in different galaxies. However, the problem of the segregation mechanism remains.

17

Gauge Invariance—
The Unification of Fields

Our job in physics is to see things simply, to understand a great many complicated phenomena, in terms of a few simple principles.

Steven Weinberg, Nobel Prize Lectures, 1979

The confusion of the past is now replaced by a simple and elegant synthesis. [This] standard theory may survive as a part of the ultimate theory, or it may turn out to be fundamentally wrong. In either case, it will have been an important way-station, and the next theory will have to be better.

Sheldon Lee Glashow, Nobel Prize Lectures, 1979

It is worth stressing that even for the simplest grand unifying model . . . the number of presently ad hoc parameters needed by the model is still unwholesomely large—22. . . . We cannot feel proud.

Abdus Salam, Nobel Prize Lectures, 1979

Global Symmetries and Local Symmetries

The larger part of physics is concerned with the determination of connections among observations of phenomena and then connections among descriptions that encompass sets of phenomena. In the macroscopic realm we find striking similarities between electromagnetism and gravity: both gravitational forces and electrostatic forces between bodies are proportional to the products of the strengths of sources residing in the bodies (charges and masses) and fall off with the square of the distance between the elements. Conversely, the weak nuclear forces have a very short range—less than 10^{-15} centimeters—and might thus seem closely related to neither electromagnetism nor gravity.

However, we now are able to understand that electromagnetism and the weak interactions are deeply entwined, whereas we are only at the threshold of establishing a connection of gravity with either. The unifications we seek evidently lie at the roots of the descriptions of phenomena and not at the level of surface likenesses.

At this time we believe that the deep relations among forces are most evident if we consider the forces in a manner that emphasizes certain symmetries between the force fields and the de Broglie matter fields. We begin by discussing the character of global and local symmetries using a model introduced in Chapter 5 that is close to our direct experience. Figures 5.1 and 5.2 show a contour map of an island and a map of the topographical gradients on the island. The contour map defines a scalar field; the gradient map defines a vector field. At each point of the contour map, the field is defined by one number, the altitude above sea level. At each point of the gradient map, two numbers are required to define the field in this two-dimensional manifold. The vector gradient can be defined by a number proportional to the slope and a number giving the direction of the maximum slope, or numbers proportional to the magnitude of the force on a ball at rest and the angle of direction of the force—or the magnitude of the force on the ball in an x-direction and in a y-direction.

The relation between the vector and scalar fields is simple: the force on the ball at any point on the map is in the direction of the steepest slope (and then perpendicular to the direction of a contour line through the point) and proportional in magnitude to that slope (or inversely proportional to the spacing of the contour lines in the direction of the steepest slope).

To the hiker strolling along the hills, the important observable is the slope labeled ∇h and called the gradient of the altitude; the absolute value of the altitude is irrelevant and unmeasurable. Therefore, observables such as the force on the rolling balls do not change if there is an overall change in altitude h. If the entire hill were raised uniformly 100 meters, the forces on balls rolling on the hills, which are proportional to the value of the gradient ∇h, would be unchanged. The hiker would find his trail unchanged. This invariance is considered to follow from a *global transformation* symmetry, where the term "global" implies that the change must happen everywhere if there are to be no observable consequences. If the raising of the land only occurs locally over some restricted area, there would be cliffs created at the demarcation of the change, to the consternation of the hiker. The system is not invariant under a *local transformation*.

If invariance under a local transformation (a local symmetry) were achieved, some other field would have to be introduced to compensate for the physical changes imposed by the changes in the potential field. We assume that a local transformation alters the altitude $h(x, y)$ at every point x, y by the addition of a height $\alpha(x, y)$ that varies from point to point. Generally, the slopes will be changed by such a transformation, and the experience of the hiker, or the forces on a ball placed at any point, would be modified. The force on a ball at any point would remain invariant if an appropriate additional force counter to gravity were introduced to compensate for changes in alti-

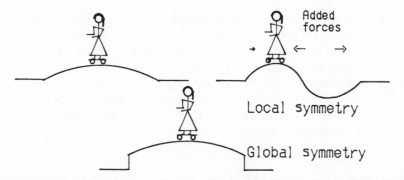

Figure 17.1 At the left, Olivia maneuvers on a hill with her skateboard. Below, she skates on the hill after it is raised uniformly. At the right, she performs on a hill that is distorted, but extra forces (shown by the arrows) act on her so that the total forces are the same as on the original hill.

tude.[1] This additional force would generally have special characteristics defined by the local transformation α.

We may better understand the character of global and local symmetries with the assistance of our researcher friend Olivia, who obliges by performing on a skateboard on a hill (shown at the left in Figure 17.1) while global and local transformations are effected. When the entire hill on which she skates is raised a few meters in a global transformation, as shown in the lower portion of the figure, Olivia observes no effect on her performance; a global symmetry is evident. Then, as illustrated at the right, the altitude of each local point on the hill is changed, but other forces (perhaps some magnets installed in the hill that act on Olivia's steel skateboard?) are cunningly arranged so that the total force acting on her at any point on the surface is the same as when she was on the undisturbed hill. Again, her skating routine is undisturbed by the changes.

Going further, assume that mechanisms are developed so that for *any* arbitrary deformation of the surface on which Olivia skates, extra forces are automatically supplied that compensate exactly for the different slope of the surface. With the total forces unchanged, she will skate just as she did on the undisturbed hill; a local symmetry is evident.

The existence of local symmetries in nature seems to be broad and important. We will see that the thread that seems to connect the disparate forces we have discussed is to be found in their descriptions in terms of local symmetries.

Gauge Invariance and the Electromagnetic Field

To consider global and local symmetries of the electromagnetic field, it is necessary to cast the description of the fields in a different form. It is useful to begin by considering only those phenomena involving the forces on test charges derived from effects induced by other stationary charges. In short, we consider initially only electrostatics.

All the information necessary to describe the effects of those stationary electric charges on a test charge, placed at arbitrary coordinates x, y, and z, are contained in the values of a scalar potential field $\phi(x, y, z)$. Excepting the expansion from two dimensions to three, the field can be considered to be similar to that illustrated by the contour map of Figure 5.1, where we would plot ϕ instead of h (or hg). Continuing in a manner similar to the procedure described in that section, we can derive a vector field $\mathbf{E}$, the electric field, in terms of the gradient of the scalar potential field

$$\mathbf{E} = -\nabla\phi$$

where again, the direction of $\mathbf{E}$ at any point (x, y, z) is in the direction that ϕ decreases most quickly—the direction of steepest slope—and the magnitude of $\mathbf{E}$ is proportional to that slope.

Proceeding with arguments parallel to those applied to Olivia on her skateboard, we see that a global increase in potential—an increase of the potential everywhere by the quantity alpha, an action described by changing $\phi(x, y, z)$ to $\phi(x, y, z) + \alpha$ where α is a constant—through an appropriate disposition of external charges will not change the values of the electric field or the forces on any test change in that field. All observations of electrostatic phenomena (or, indeed, any phenomena) are then invariant under the change in potential that defines a global symmetry. Of course, a local change in ϕ, where the value of ϕ is changed a different amount at every point, will generally change the gradient of ϕ, the value of $\mathbf{E}$, and the forces on charges. Unless some compensating field exists, there will be no local symmetry.

Time-varying magnetic fields can provide such a compensating mechanism through their (Faraday's law) contribution to the electric field. The change in electric field generated by a local change in ϕ may be compensated for by an electric field generated by a properly chosen time variation in the magnetic field at that point. Hence, electrostatics can be made to exhibit a local symmetry: the force on an electric charge will not change, although the electrostatic potential may vary from point to point if a magnetic field can be found that will vary with time in a compensatory manner.

However, a local symmetry for electric fields alone, without a similar local symmetry for magnetic fields, cannot be fundamental or very interesting, as the local symmetry would not hold for moving charges (acted upon by the magnetic field) or from the view of a moving observer. Can a local symmetry be found for the magnetic field such that the vector potential $\mathbf{A}$ can be changed locally, by a different amount at each point, without changing the magnetic field $\mathbf{B}$, derived from $\mathbf{A}$, which describes the forces on moving charges?

If the vector potential $\mathbf{A}$ is changed arbitrarily (though smoothly from point to point), the magnetic field that is the circulation or curl of $\mathbf{A}$ must also be changed in general, and such a change in the magnetic field will result in a change in the force on a moving charge. However, if the change were balanced by the magnetic field produced through a time variation of the electric field (Ampere's law), the moving charge would be unaffected by the two changes and the local symmetry would be retained. Hence, the magnetic field can be made to exhibit a local symmetry: the forces on a moving charge will not change as a consequence of the changes in the magnetic field that follow from changing the vector potential from point to point if an electric field can be found that will vary with time in a compensatory manner.

If such a symmetry of the magnetic field could be constructed, can it be made such that it does not interfere with a local symmetry for the electric field? Can a local symmetry be designed such that the vector potential A and the scalar potential ϕ change locally, but both electrostatic and magnetic fields are unchanged? This would seem to be difficult because the compensation of a change in the electric field through a variation of the magnetic field would change the force on a moving charge, and the compensation for a change in the magnetic field by a variation in the electric field would change the force on a stationary charge. It is hardly obvious that such compensations can be made to balance through a mechanism that differs from point to point—and from one time to another—but remarkably such is the case.

The necessary correlated changes in the electric and magnetic fields required for local symmetry are choreographed through properties of a new scalar field $\Lambda(x, y, z, t)$ that is allowed to vary arbitrarily (though smoothly) from point to point and from time to time. Both the electric and magnetic fields at each point and time will remain unchanged if the gradient of the field Λ at that point is added to the vector potential A, and the rate of change of Λ with time (divided by the speed of light c in scientific units) is subtracted[2] from the electrostatic potential ϕ. Electromagnetism exhibits a local *gauge invariance* under such appropriately chosen transformations. What is perhaps more important is that the requirement of invariance under a local gauge transformation *defines* the electromagnetic field.

The de Broglie Waves of Charged Particles in an Electromagnetic Field

According to well-tested recipes of quantum mechanics, the phase of the de Broglie wave describing a charged particle depends on the scalar and vector potentials explicitly—and on the electric and magnetic fields only through their relation to the potentials. If the potentials change at a place and time (x, y, z, t), the phase of the de Broglie wave will change at that point and time. Indeed, because the potentials ϕ and A can have nonzero values in regions where the fields E and B are zero, the phase of the de Broglie wave can be affected by a change in the potentials even though the electric and magnetic fields are zero at that point (the Bohm-Aharanov effect).

Figure 17.2 shows an imaginary experiment, similar in principle to more subtle experiments that have been conducted, which illustrates the kind of physical effect produced by a change in the phase of the de Broglie wave of a charged particle induced by a change in the vector potential A over the path of the particle. The shielded solenoid magnet shown in cross section in the

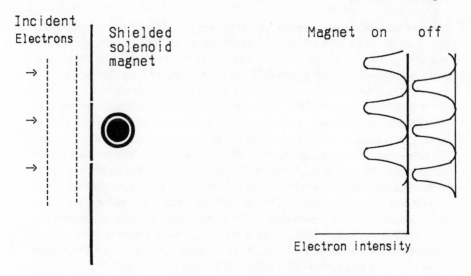

Figure 17.2 Although the shielded electromagnet produces neither a magnetic nor an electric field in the path of the electrons, the vector potential generated by the magnet changes the phase of the electrons passing through one slit by $+90°$ and through the other slit by $-90°$, thereby modifying the interference pattern. The right-hand curve shows the intensity distribution with the magnet off, and the left-hand curve shows the distribution with the magnet on.

figure will not generate a magnetic field outside the magnetic shield (which may be of soft iron), but the magnet will induce a vector potential that will produce opposite phase changes in the de Broglie waves passing from the two slits to the detector. The change in the relative phase of the de Boglie waves from the two slits in turn modifies the interference pattern.

It follows that the changes generated by a gauge transformation in the vector and scalar potentials at a point may affect the phase of a charged-particle de Broglie wave at that point, although the electric and magnetic fields are unchanged. Upon calculation we find that the phase of the de Broglie wave at the place and time (x, y, z, t) is rotated by an angle λ, where

$$\lambda(x, y, z, t) = \frac{e}{h} \Lambda(x, y, z, t)$$

From this recipe it follows that the application of two different gauge transformations, Λ and Λ', produces a resultant phase change that is the sum of the individual phase changes generated by each transformation taken alone and independent of which order the changes are made in: $\lambda + \lambda' = \lambda' + \lambda$, the transformations *commute*.

Although the phase of the de Broglie wave is changed at every point and at every time by an amount $\lambda(x, y, z, t)$, which can have quite different values at each point in space and for any time, the changes must be so correlated that no physical result is changed. The phases of the de Broglie waves in front of the two slits of Figure 9.2 are changed by the transformation λ, but the sum of the changes elsewhere conspires to give exactly the same diffraction pattern. Nothing observable is changed by the gauge transformation. The matter field, represented by the de Broglie wave, exhibits a local symmetry under the gauge transformation that changes the phase of the wave everywhere and at every time by an angle $\lambda(x, y, z, t)$. The effects of the changes in the electromagnetic potentials induced by the gauge transformation field $\Lambda = (h/e) \cdot \lambda$ compensate for the phase changes in such a manner as to hold physical observables invariant under the transformation.

The local symmetry of the de Broglie waves of a charged particle under the gauge transformation that modifies electric and magnetic potentials in a manner such that no observations are affected is even more striking when considered in reverse. Suppose that we apply an arbitrary phase change to a de Broglie wave of a particle so that the phase of the wave is changed in an arbitrary manner at each point of space and each time. Can we find a field that will change in such a way as to preserve every observable under the changes? That field must compensate for the effects of the phase changes from place to place and from moment to moment.

Surprisingly, there is a field that fulfills these stringent requirements! Indeed, the field is completely determined by the symmetries of the transformation to be a vector field, which, when quantized, leads to a field particle with spin $1\hbar$. The change in phase of a charged particle is induced by the emission (or absorption) of this field particle, which also changes the phase of a second charged particle through subsequent absorption (or preceding emission) so that the overall symmetry is retained. Moreover, because a change in the field at one point must be compensated by a change far away, even as we can observe interferences between segments of a wave that are separated by large distances, the field quantum adjusting the phases must travel at the speed of light over these distances. Therefore, the mass of the quanta must be zero, and the field must describe long-range forces. Less transparently, the commutation character of the transformations requires the field particle, the photon, to carry no charge. The field so described is the electromagnetic field!

In summary, the requirement of a local symmetry with respect to a phase change of the de Broglie wave of a charged particle defines the electromagnetic field as the requisite compensating field. The existence of the symmetry in nature logically requires the existence of the forces. This symmetry with respect to phase rotation has a place in the mathematics of group theory, where it carries the label $U(1)$. Without looking into group theory, we will use the label as a convenience.

Gauge Fields for the Weak Interactions—The Electroweak Force

The definition of the electromagnetic field as the field defined by the symmetry of nature under a local phase transformation of the de Broglie waves of charged particles suggested to C. N. Yang and Robert Mills that the prin-

ciple of invariance under local transformations might be the unifying principle that would define other forces. In 1954, with this general aim in mind
and with insight into the strong interactions as understood at that time as a
particular goal, Yang and Mills proceeded to develop a broader gauge theory
designed to establish a field that would compensate for a broader class of local
changes. Although the particular theory they developed did not describe the
strong interactions correctly, the result was sufficiently elegant that it seemed
that nature might well use such a scheme somewhere.

The first suggestion that nature did adopt a variation of the Yang-Mills
gauge theory was made in 1967 by Steven Weinberg and, independently, by
Abdus Salam and John Ward, who considered the possibility that the weak
nuclear forces might be described by such a scheme. Here the transformation
of note included the rotation of the phase of the de Broglie wave of charged
particles at every point and at every time, but added the transformation of
leptons and quarks to other leptons and other quarks.

In discussing the gauge theory that has been developed to describe the
weak and electromagnetic interactions, we will proceed by describing the theory in three "stages," where the second stage not only adds to the results of
the first stage but *modifies* those conclusions. Then the third stage supple-
ments and modifies the results of the first two stages. The separation of the
description into three parts is made for pedagogical simplicity and does not
necessarily correspond to the path of development used by God who constructed the forces in this way or by the physicists who described His works.
Indeed, the theory is whole, although we discuss it part by part.

This may also be the place to review the meaning of the verb "construct"
and the noun "theory." When we say theory, we mean a description of nature.
All descriptions of a part of nature are necessarily incomplete and cannot be
exactly correct, but "theory" here labels a description that can be used, within
certain limits, with a high degree of confidence. We use the verb "construct"
(or *"develop"*), as in the context above, to mean "reveal." Our mature theories reveal the character of nature. We should not assume that the path that
emerges is one chosen from many almost equally attractive alternatives. For
the most part, the revealed theory is the only logically consistent description
we know. And the path is narrow; the constraints of logical consistency define
the theory with great precision. When the pieces of the jigsaw puzzle finally
begin to fall into place, it is very hard to believe that those pieces could also
form quite another picture.

The First Stage—The Basic Fields

It is an underlying premise of the theory of the weak interactions that a fundamental symmetry exists to the effect that the weak interactions do not differentiate between members of certain pairs of particles that take part in the
weak interactions. Electrons and electron-neutrinos, muons and muon-neu-
trinos, tau-leptons and tau-neutrinos, as well as the three quark family pairs

$u_w - d_w$, $c_w - s_w$, and $t_w - b_w$—and their sets of antiparticles—constitute such closely related pairs. If, everywhere, the members of each of these pairs were interchanged, nature would be the same as viewed through the weak interactions: a global symmetry would exist. This $SU(2)$ symmetry under the exchange of two paired particles—for example, the electron and electron neutrino e^- and ν_e—is very much the same as the $SU(2)$ symmetry of the proton and neutron in the strong interaction world discussed in Chapter 14 and is very much like a symmetry under an arbitrary rotation in a three-dimensional abstract coordinate system that results in an appropriately defined exchange of $(e^- - \nu_e)$ identities.

The existence of a global $SU(2)$ symmetry suggests the important possibility of a local $SU(2)$ symmetry where a different rotation in the $SU(2)$ coordinate system could be admitted for each lepton or quark; each could change arbitrarily to its partner, but no property of nature, observed through the weak interactions, would be changed. Just as nature is invariant with respect to geometric rotation and then invariant with respect to the change in the values of the components of the space vector, nature is presumed to be fundamentally invariant with respect to a rotation of the weak vector in the abstract $SU(2)$ space and then invariant under changes accompanying such a rotation in the identities within the lepton or quark partnerships.

As was the case with the establishment of the local invariance under phase change of the de Broglie waves of charged particles, additional fields must be added if the requisite invariance is to be retained. Because the invariance under rotation in the internal particle space is more complicated than local phase invariance alone, more than one field is required to retain the invariance. Indeed, four photonlike fields must be added, where three of the fields are closely related and correspond to a kind of "initial," symmetric set of three chiral fields. Each of these fields is coupled to left-handed leptons and quarks and right-handed antileptons and antiquarks with the same weak interaction strength, but they carry different electric charges; one field is positively charged (the field quanta are positively charged), one is neutral, and one is negative. The electron changes to a neutrino by emitting a negatively charged quantum of the weak field that is absorbed by some other lepton or quark. The neutrino changes its phase to compensate for the phase changes of other particles through the absorption or emission of a neutral weak quanta. To these proto-weak fields is added a less closely related field corresponding to an initial electromagnetism, which couples equally to left- or right-handed charged particles. (Here we use "initial" to distinguish the fields that are introduced in the first stage of the description of the theory from the final fields observed in nature, which turn out to be mixtures of the initial fields.)

The coupling to charged particles of the primordial electromagnetic field is described by the charge strength e'. The symmetry of this field is the same $U(1)$ symmetry held by the electromagnetic field. The weak charge of the initial triplet of weak fields is g', and those fields carry the $SU(2)$ symmetry we have noted. As with electromagnetism, the local transformations adjust the

identity of a fermion, through the emission (or absorption) of a quantum of
the weak field that is absorbed (or was emitted) by a second particle, in such
a manner as to retain the symmetry. Because the field quanta are charged,
their phase is changed by their emission and absorption of photons; hence,
the weak and electromagnetic fields are inextricably mixed. The symmetry of
the coupled fields is just the combined symmetry of the basic fields that are
introduced. We label that symmetry as $SU(2) \times U(1)$.

Although this model of the weak and electromagnetic forces elegantly
describes many features of the weak interactions, there are also serious failures. In particular, calculations of some processes lead to probabilities for the
processes which are greater than 1! Such breakdowns in the calculations can
be traced to the massiveness of the W field quanta; the theory works well for
massless W' but founders if these field particles are given the mass required
to explain the observed short range of the weak interaction. Although any
model of a part of nature can be expected to fail to explain phenomena that
have important features that extend beyond that segregated part, these calculations fail more broadly. The simple model of massless quanta avoids
infinities associated with nonphysical probabilities but gives the wrong
answers to too many questions. The theory must therefore be modified to fit
the observation that the weak interactions have a short range, which requires
in turn that the weak interaction field quanta have large masses.

Even as the theory founders on the mass of the W force particles, the theory fails to handle massive lepton and quark matter particles. It describes
massless leptons, quarks, and W-bosons, but it falters when the known
masses are assigned to these particles. The elegant electroweak theory is a
theory of massless particles. Something must be added if the electroweak
model is to describe this universe made up of massive particles.

The Second Stage—Mass, Symmetry Breaking, and the Higgs Fields

The (Lagrangian) equation constructed to describe the weak interactions
using gauge invariance as a guide accounts for many things, but not for mass.
The universe described is interesting and elegant, but not ours. The simple
gauge description of the weak interactions must be incomplete; what part of
nature's design has been omitted? An examination of the simple theory tells
us something of what must be added.

Massive particles with a total angular momentum of $1\hbar$, such as the W-
bosons, have three differentiable spin directions; however, massless particles
traveling always at the speed of light can have no more than two differentiable
spin directions: along the direction of flight and opposite to the direction of
flight. Massive spin 1 particles must then be described by fields that have
three numbers at each point; massless particles require only two. The Lagrangian that had been considered provided just enough numbers (or field components) to describe massless intermediate bosons but not enough for massive bosons. If the field particles were to have mass, numbers must be found
elsewhere.

It seems that nature provides such numbers by the addition of extra scalar Higgs fields (first described in 1964 by Peter Higgs, then at the University of Edinburgh), each of which gives one set of numbers or field components. The quanta of these fields are the scalar (spin zero) Higgs bosons labeled H. With the addition of two pair (at least) of Higgs fields into the fundamental Lagrangian equation, there is freedom to give the intermediate particle mass. In the simplest model, a pair of fields with H^+ and H^0 quanta and another pair with H^- and H^0 quanta are added. With the additional Higgs fields, the fundamental Lagrangian has sufficient numbers to describe fields of massive particles; however, it is hardly evident that the primeval fields of the equation will conspire, in solution of the equations, to describe massive quanta.

One can consider a Higgs field an all-pervasive universal background field that modifies the simple electroweak gauge theory somewhat as the earth's gravitational field modifies the simple exposition of Newton's First Law that particles travel in a straight line with constant velocity. To a limited observer, the pervasive background of gravity causes bodies that should move in a straight line according to Newton's laws to follow curved trajectories. To the same observer, the pervasive background of the Higgs fields gives mass to particles that are massless according to the simple electroweak theory. The simple symmetry of the straight line predicted by Newton's laws is broken by the gravitational field. Although the fundamental electroweak Lagrangian equation deals symmetrically with the four original massless fields, the Higgs fields break that mass symmetry by contributing to the construction of four new vector fields, three mediated by massive intermediate bosons, from the old massless vector fields and the additional Higgs fields. In mathematical terms, the solutions of the equations describing fields mediated by particles of different mass have lost their symmetry although the equations themselves treat the different fields symmetrically.

Such *spontaneous symmetry breaking* is evident in nature in less exotic circumstances. Although electromagnetism, as described by Maxwell's equations, is spatially symmetric and defines no special direction, a set of magnets (such as iron atoms) tends to line up in some (arbitrary) direction—the energy of an aligned set of magnets is less that of a randomly oriented set. The set of magnets *spontaneously* breaks the symmetry of the electromagnetic field laws. At the left of Figure 17.3 is a set of small bar magnets mounted on gimbals, like compass needles. Although set originally in random directions, through their mutual attraction the needles will spontaneously line up in some arbitrary direction. Similarly, iron atoms, each acting as a small magnet, align themselves spontaneously in each of the microscopic domains that make up iron metal.

At the right of Figure 17.3 we see an even simpler spontaneous symmetry breaking. Although the wine bottle is symmetrical about its center of rotation, the ball in the bottle will not rest at the center but somewhere near the edge. The forces on the ball are symmetric about the center of the bottle, but the energy of the ball will be greater at the center than at the edge, and the ball spontaneously breaks the symmetry by finding a position of lowest energy.

From the spontaneous breaking of the symmetry of the massless fields by

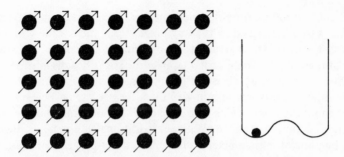

Figure 17.3 At the left, a set of magnets on pivots lines up spontaneously in an arbitrary direction as a consequence of their mutual attractions. At the right, a ball is asym-
metrically located in a symmetric wine bottle. Wine bottles have a bulge in the bottom
to segregate sediment.

effects of the Higgs fields, the two initial charged massless weak fields (W^+
and W^-) mix with their related Higgs fields (H^+ and H^-) to give the W-fields
mass. The two electrically neutral massless fields (W^0 and G^0) mix with the
neutral Higgs fields (H^0) so as to generate one massive neutral weak field, the
Z-field, one neutral field (γ) to be identified with electromagnetism, and one
residual neutral Higgs field. Upon quantization of the fields, massive spin 1
charged intermediate bosons W^+ and W^- emerge that mediate the charged
weak interaction processes; there is a massive intermediate neutral spin 1
boson, the Z^0, which is transferred in the neutral weak interactions, and a
neutral massless spin 1 photon γ that transmits electromagnetic forces. One
scalar spin zero massive Higgs particle H^0 remains.

Figure 17.4 suggests the stages of the description of the electroweak forces.
At the left of Figure 17.4 is the system of fields with the Higgs mixing mech-

Figure 17.4 The relations between the fields implied by the fundamental Lagrangian
equations describing the electroweak interactions and the fields defined by the solution
of the equations. The Higgs fields are labeled H, the weak fields W and Z, and the
protoelectromagnetic field G; the final electromagnetic field is denoted by the photon
insignia γ.

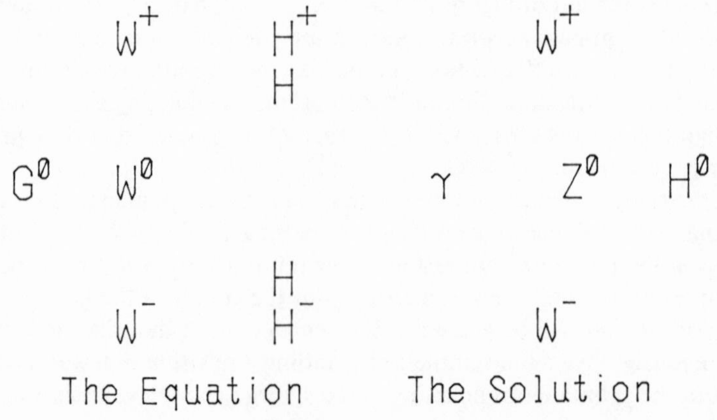

$$W^+ \qquad \begin{matrix} H^+ \\ H \end{matrix} \qquad\qquad W^+$$

$$G^0 \quad W^0 \qquad\qquad\qquad \gamma \qquad Z^0 \quad H^0$$

$$\begin{matrix} H \\ W^- \quad H^- \end{matrix} \qquad\qquad W^-$$

The Equation   The Solution

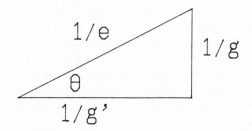

Figure 17.5   Geometric description of the relations between the coupling strengths.

anisms introduced in the fundamental Lagrangian equation; at the right is the configuration of fields defined by the solution of the equation.

After the symmetry breaking, the effective electromagnetic coupling or charge e and the weak couplings or charges are defined in terms of fundamental couplings g and g' of the two original fields by a parameter θ_W determined in some way outside the territory mapped by the electroweak gauge theory. Similarly, the masses of the charged and neutral intermediate bosons, the W^+, Z^0, and W^-, are defined in terms of that mixing angle θ_W. Figure 17.5 shows the relation between the coupling strengths.

Knowing the value of the electric charge e, knowing g_w, the strength of the weak coupling to the W from the weak interaction measurements of quantities such as the lifetime of the muon, and knowing the weak interaction coupling to the neutral Z through analysis of neutrino interactions, we may determine the value of θ_w to be about 28°. The mass of the W is then predicted to be 79 GeV/c^2, and the mass of the Z would be 90 GeV/c^2. The values of the masses determined in 1983 of 81 GeV and 93 GeV are in accord with the predictions and set the value of θ_w more accurately.[3]

We cannot derive the masses of leptons and quarks at this time from the simple Higgs field backgrounds—perhaps more Higgs fields are needed. However, we can demonstrate that the Higgs background will provide mass to the matter particles as well as to the W and Z force particles.

The Third Stage—Renormalization

After the Higgs fields were added to the original electroweak Lagrangian equation, a description of nature emerged that seemed attractive. The model accounted correctly for much that had been observed, no observation contradicted the model, and the model had predictive power; it predicted the existence of weak neutral currents that had not been seen at the time. Moreover, the model fit other natural constraints by exhibiting appropriate symmetries. Energy, momentum, and angular momentum were conserved, the model predictions were invariant under a Lorentz transformation, and many more subtle difficulties seemed to be avoided. However, Yang-Mills fields with massive quanta had never been renormalized; a large set of calculations of the probabilities of certain interactions gave unphysically infinite results. Physicists had began to consider that their basic methods of calculation were correct, and any theory that admitted processes that gave infinite results upon calculation must be wrong. Nature was believed to eschew nonrenormalizable theories that led to infinities that could not be removed by the addition of a small number of renormalization terms to the fundamental Lagrangian equation. The Weinberg and Salam-Ward gauge theory of electroweak forces could not be and was not accepted as long as the renormalizability was not established.

Then in 1971, Gerhard 't Hooft, a student at the University of Utrecht in the Netherlands, solved that last essential problem and showed that the infinities could be made to cancel through a renormalization procedure. The theory must be right! To speak in Einstein's manner, "The Old One could not be so malicious as to allow a theory so beautiful, so unique, and so accurate in describing what was known, which was not a correct description of nature."

A Summary of the Electroweak Gauge Theory

Insofar as the electroweak gauge theory described electromagnetism and the charged weak current interactions in an almost final form in 1972, the model could be considered only as an elegant and possibly valid summary and consolidation of what was known; the theory added nothing to the description of the low-energy phenomena then available for observation. The prediction that positively and negatively charged W's must exist with a very high mass near 75 GeV and that there must be an even heavier neutral boson labeled the Z could not then be tested.

However, the electroweak model did make quite specific predictions concerning the existence of a weak neutral current and the properties of that current, and these predictions were validated by a series of experiments at many laboratories, beginning about 1973. (The first important result followed from work at CERN in Geneva by a group led by A. Lagarrigue.) With the receipt of the various neutral current results, all in accord with the predictions of the electroweak model of nature, the validity of that description seemed to be assured. Moreover, with information from the neutral currents, the value of the weak mixing angle was determined, and the masses of the W and Z could be calculated accurately; the W was to have a mass of about 80 GeV/c^2, and the neutral gauge boson, the Z, was to have a mass of about 90 GeV/c^2.

Then in 1983, Carlo Rubbia, Simon van der Meer, and their (approximately 100) collaborators working at CERN attached a final seal to the theory by finding the W's and Z and determining their masses to be 81 GeV/c^2 and 93 GeV/c^2, just as predicted.

Aside from the general complexity of the electroweak description of forces, required to match the complexity exhibited by nature, this model of nature differed from the simple gauge description of electromagnetism in a more fundamental manner. The field quanta of electromagnetism, the photon, carries no charge and is thus not a source of further field. But some of the field quanta—the photons of the electroweak force—are charged; the W-bosons carry an electric charge *and* a weak charge. Some special properties of fields of charged photons become evident when the local gauge transformation is applied twice to the same particle.

For electromagnetism, where the forces are mediated by uncharged photons, the phase of the de Broglie wave will be rotated by each transformation; under two transformations, the final phase will be simply the sum of the two imposed phase changes, and the order of the phase changes is immaterial. The phase changes commute; if the two phase changes are labeled λ and λ', $\lambda + \lambda' = \lambda' + \lambda$. The phase of a de Broglie wave describing the charged particle is altered by each emission of a photon and by each absorption of a photon, but the total phase change is unaffected by the order of emission or absorption.

However, the extended gauge theory of electroweak forces includes the rotation of a vector in a three-dimensional internal space, and the results of rotations in a three-dimensional space are not, in general, independent of the order of the rotations about different axes; such rotations do not commute. Figure 17.6 shows a brick that is rotated first by 90° about the z-axis and then by 90° about the x-axis. The order of the rotations is then reversed, and we see that the final orientation of the brick depends on the order of the rotations.

In the stage-one description, considered here for simplicity, an absorption of a W^+ boson by a particle results in a rotation of the particle description in the internal three-dimensional space. A subsequent emission, for example, of a neutral W-boson will result in another rotation. Because rotations in a three-dimensional space do not commute, the final state of the particle will depend on the order in which the absorption and emission occur.

Figure 17.6 A demonstration that angular rotation in three dimensions does not commute. The result of two 90° rotations of the brick depends on the order of the rotation operations.

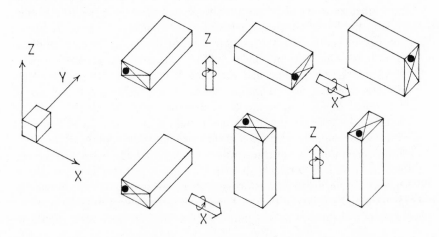

We call those gauge theories which describe fields such as electromagnetism, where the gauge transformations commute, *Abelian theories.* The electroweak model is a non-Abelian theory. (Here we use mathematical language concerned with group theory results developed by Niels Henrik Abel in the early nineteenth century.) The fields in such non-Abelian theories carry charges and are themselves sources of further fields. Physically, such non-Abelian gauge theories are richer and more complicated than Abelian models and better suited to fit the observed richness and complexity of nature.

In summary, the electroweak description of electromagnetism and the weak interactions has the following important properties:

1. The fields are required to preserve invariance under a local transformation and are thus defined by that requirement.
2. A local symmetry exists that is broken (and is then a hidden symmetry of nature) by a spontaneous symmetry-breaking mechanism.
3. The resultant gauge theory is non-Abelian; the photons derived from quantization of the fields carry charge and are the origin of further fields.
4. The theory is renormalizable.

A majority of physicists have now adopted the working hypothesis that all nature can be described by one such sufficiently complex local symmetry that leads uniquely to all the fields and forces now seen in nature—gravity, electromagnetism, and the strong and weak nuclear forces. This field will have the salient properties listed above and contains the electroweak theory discussed here as a piece of a whole.

Gauge Fields for the Strong Interactions

The strong interactions are manifestly complicated, seemingly more complicated than even the electroweak interactions. There are six different kinds (flavors) of quarks; each quark of a specific flavor will carry one of three different color charges. Antiquarks carry anticolor charges related to color charge rather as a negative electric charge is related to a positive charge. Moreover, the forces between quarks of a different color—that is, carrying different color-charges—are attractive only for the "white" octet combinations, so that all hadrons are made up of white quark-color combinations.

If all quarks everywhere had their colors changed in an appropriately coordinated manner, no observable would be changed as a consequence of such a global $SU(3)_c$ symmetry. Such a recipe would result in a kind of exchange of the colors of the three quarks making up a nucleon, so that octet combinations would be retained and the nucleon would remain white.

However, if this kind of transformation were applied to each quark, with a different "rotation" for each, the octet symmetry of the nucleon would be upset under such a local transformation; the nucleon would no longer be white, and the character of the nucleon would be changed dramatically. If the universe is to exhibit a *local* symmetry such that no observable would be affected by an arbitrary local transformation, additional fields must be provided to compensate for the changes.

Can a set of fields be found that would act to retain invariance under a change in the local color condition where each quark changes color arbitrarily? If the observations of nature are to be invariant under the changes, the field must act so as to retain the octet symmetry of the colors making up any hadron; the hadron must remain white. In close analogy with the electromagnetic gauge transformations, we find that gauge fields can be introduced that affect the quark colors in such a manner that the octet color symmetry is retained and the hadrons retain their observable character in the face of an arbitrary change in the color of any quark. Indeed, as with electromagnetism, we find that the requirements of invariance under the local guage transformation *define* the character of the force fields. Eight color fields are so defined, each of which is analogous to the electromagnetic fields defined by the gauge transformation of electromagnetism. The quanta of the fields are called *gluons*. (We use this word because the gluon fields define forces that hold or glue the quarks together to form the hadrons, rather as electromagnetic forces hold electrons and nuclei together to form atoms.) As there are eight different color fields, there are eight different gluons. Because of the similarities of the strong interaction theory of color charges to quantum electrodynamics (QED), the strong interaction gauge theory is labled quantum chromodynamics (QCD).

The gluons are massless bosons and have a spin of $1\hbar$, as do photons. Like photons, they are electrically neutral and do not carry a weak interaction charge. But unlike photons, gluons are not wholly neutral; each gluon carries a color charge and an anticolor charge. There are nine combinations of the three colors and anticolors making up an $SU(3)_c$ octet and singlet, but only the octet set of eight gluons is required to retain the invariance of the quark color octet under the local transformation.[4]

Local $SU(3)_c$ color symmetry is preserved through the absorption and emission of gluons by quarks in a manner analogous to the preservation of the simpler $U(1)$ symmetry of electromagnetism through the emission and absorption of photons by charged particles. A quark is free to change its color arbitrarily, but that change is accompanied by the emission (or absorption) of a gluon, just as a charged particle changes its de Broglie phase through the emission or absorption of a photon. The gluon, like the photon, traveling at the speed of light, is then absorbed (or was previously emitted) by a second quark that will have its color shifted in just such a manner as to compensate for the original change and retain the overall color (octet) symmetry. Assume that one of the three quarks making up a neutron changes its color from red to yellow. In the course of this change it will emit a red-antiyellow gluon that must be absorbed by a yellow quark, changing that quark to red. After the transaction, there will be a red quark and a yellow quark as before with symmetry unchanged. The gluon transfers the color from one quark to another so as to retain the $SU(3)_c$ octet symmetry.

As with the $SU(2)$ gauge transformation defining the weak part of the basic electromagnetic interaction, two "rotations" in $SU(3)_c$ space do not necessarily commute. In general, the color change of a quark under two gluon emission and absorption processes[5] depends on the order of the processes. The color gauge transformation is non-Abelian.

Notes

1. The original transverse force on the ball of mass m will be $F = -mg\nabla h$. With the local transformation, the new force would be $mg\nabla(h + \alpha)$. The additional force required so that the local transformation α does not change the total force on the ball must have the form $F' = mg\nabla\alpha$. The mathematical form of F' and the value are defined by the character of the local transformation α.

2. The magnetic field **B** is described in terms of the *vector potential* $\mathbf{A}(x, y, z)$ through the curl operation

$$\mathbf{B} = \nabla \times \mathbf{A}$$

The magnetic field is proportional to the circulation of the vector potential field. The three components of the vector potential at any point **A**, together with the electrostatic potential ϕ, transform as a Lorentz four-vector

$$[\phi, A_x, A_y, A_z]$$

in the same manner as the space-time four-vector $[ct, y, y, z]$. This description of electromagnetism leads to an especially clear picture of the local symmetry of the electromagnetic field by admitting a simple description of an additional field that induces local transformations of the potential functions that leave the field quantities **E** and **B** unchanged. If we take an arbitrary scalar field $\Lambda(x, y, z, t)$, which may have a different value at each point in space and time, and add to the vector potential the gradient of the field Λ and subtract from the scalar potential the rate of change of the field with time (divided by the velocity of light c), we leave the values of **E** and **B** unchanged. This *gauge transformation* is written formally

$$\mathbf{A} \rightarrow \mathbf{A} + \nabla\Lambda \quad \text{and} \quad \phi \rightarrow \phi - \frac{1}{c}\frac{\partial\Lambda}{\partial t}$$

Transformations of this kind change $\phi(x, y, z, t)$ and $\mathbf{A}(x, y, z, t)$ at every point in space and at every time in an almost arbitrary fashion but in such a way that the fields E and B, together with the forces on charges and currents, are unchanged. The electromagnetic field is said to exhibit a local symmetry under such a gauge transformation Λ, where $\Lambda(x, y, z, t)$ is an arbitrary scalar function of the coordinates and time.

3. The square of the electron charge is $e^2 = 1/137$ in natural or dimensionless units (units of $e^2/\hbar c$); the weak coupling strength to the W, g_w, and the coupling to the neutral Z, g_z, are

$$g_w^2 = 10^{-5}\frac{M_w^2}{M_p^2} \quad \text{and} \quad g_z^2 = 10^{-5}\frac{M_z^2}{M_p^2}$$

where M_w, M_z, and M_p are the masses of the W, Z, and proton. From this relation among the weak charge, the strength of the low-energy weak interactions, and the W mass, together with the equations defined by Figure 17.5,

$$g_w = \frac{e}{\sin\theta} \quad \text{and} \quad M_w = \frac{37.3 \text{ GeV}}{\sin\theta}$$

also

$$g_z = \frac{e}{\sin\theta\cos\theta} \quad \text{and} \quad M_z = \frac{37.3 \text{ GeV}}{\sin\theta\cos\theta}$$

4. The six nonneutral gluons carry the colors $r\bar{b}$, $y\bar{b}$, $y\bar{r}$, $b\bar{r}$, $b\bar{y}$, and $r\bar{y}$. The neutral combinations are not so simple or unique. The two states $\sqrt{\frac{1}{2}}\,(r\bar{r} + y\bar{y})$ and $\sqrt{\frac{2}{3}}\,(b\bar{b}) - \sqrt{\frac{1}{6}}\,(r\bar{r} - y\bar{y})$, or any appropriate (orthonormal) linear combination of the two, complete the octet.

5. Consider a simple example where a red quark emits q red-antiyellow gluon to change to a yellow quark (λ). The the quark emits a yellow-antiblue gluon and changes to blue (λ'). The two changes ($\lambda + \lambda'$) are the emission of the red-antiyellow gluon and the emission of the yellow-antiblue gluon. What happens if the two processes are reversed? The red quark should first emit a yellow-antiblue gluon (λ'). But to conserve color charge, the quark must then carry red, blue, and antiyellow charges. However, there is no such quark state. Hence, the two transformations cannot be reversed ($\lambda + \lambda' \neq \lambda' + \lambda$), and the set of transformations is non-Abelian.

18

To the Ultimate Theory—
Through a Glass Darkly

Future Paths and Wild Conjectures

We have now drawn a sketch of a map of reality as far as that reality has been explored. The concepts marked on the map are rather well tied to observation. Most, but not all, physicists believe that the map is unlikely to be seriously flawed—but it is certainly limited. There are important areas within the strong and electroweak interactions that are untouched. And a vast, unexplored region surrounds the area we have surveyed. In the electroweak theory, what is the origin of the weak angle θ_w? Why are there three sets of leptons? What sets the values of the lepton masses? How is the value of the electric charge assigned? What is the origin of chirality—why is God left-handed? And what is the basis of CP invariance breakdown? Physicists believe that most of these questions cannot be answered within the electroweak theory but must be found as part of unifield theory of all interactions.

There are as many unanswered questions concerning quantum chromodynamics, the theory of the strong interactions. Some of our problems with the strong interactions follow simply from our inability to solve the equations that we believe describe QCD, but more fundamental problems also exist that may be intractable except as consequences of a unified theory. What sets the strength of the color force? Why do the quarks have the masses they do? Why are there three families of quarks? Or are there more families with very high masses?

Outside the area we have mapped, albeit without the detail we might like and with important features glossed over, there is an unknown territory such as that marked *terra incognita* on ancient charts. It is probably going to be very difficult to explore that unknown territory. It is likely that the next level of knowledge of fundamental matter lies at very small distances, hence at very high energies—energies so high that we will not be able to make measurements at such levels in the forseeable future, if ever. We may never be able to set foot on this unknown land, and we may thus be forced to infer its character by indirect means. The history of physics (which is virtually the history of twentieth-century physics) shows that for the most part, we have understood nature through the examination of observations and not through conjecture per se. With little certainty of help from observation, can we expect to proceed further? Can we construct a good map of terra incognita if we can never set foot there? We have some comfort that we may.

The General Theory of Relativity may constitute a critical exception to the rule that theory must be based on large sets of detailed observations. Einstein's way of thought, built on philosophy, aesthetics, and intuition, may dominate physical theory in the future. The elegance of the General Theory was immediately compelling (to those who understood the ideas), although at the time of its publication, there was no observational evidence supporting the theory;[1] Einstein himself seemed to have little doubt concerning the validity of his description of gravity. Along with others who studied and understood his description of gravity, he was convinced of its validity because of the simplicity of the fundamental ideas and the breadth of experience covered by the theory. All the pieces fell together so nicely that it seemed most improbable that the fit was accidental. And the theory was falsifiable; the theory generated predictions (albeit few) that did not follow from the more primitive ideas that it displaced or enhanced.

The search for a unified theory follows a similar path to that trod by Einstein. Physicists are searching for a simple idea that fits the complexity of experience so well that the fit cannot reasonably be accidental. Such an idea has not yet emerged clearly, although there are candidates. Perhaps we are close to God's Equation; perhaps we are far away. But most physicists believe that we are at a point in scientific history when a search for that Equation can be sensibly conducted. The rest of this chapter represents an attempt to describe the probable direction of future searches as seen from the middle of the ninth decade of the twentieth century.

Grand Unification Theories and the Standard Model

Practically all extrapolations of our present knowledge presume that the four forces we know now are fundamentally one force. From this view, at very small distance and at complementarily high energies, all forces—color forces, electromagnetism, weak forces, and gravity—are identical in strength and character. There is but one force at such small distances. Moreover, the matter particles, the leptons and quarks, are indistinguishable, and the force particles—W's and photons—are the same; there is but one force particle and one field particle. Even force and matter may be joined, leaving only one particle and one force.

Because we do not observe such symmetry in our environment, the symmetry must be unstable and thus broken at larger distances and lower energies by variants of the symmetry-breaking mechanisms discussed previously; the background of Higgs-type scalar fields may break the basic symmetry. Because the full symmetry, if it obtains at all, holds only at energies on the order of the Planck energy equal to about 10^{19} proton masses, or distances on the order[2] of the Planck distance of 10^{-33} centimeters, the masses of the symmetry-breaking particles must also be extremely large and almost inaccessible. The broad set of conjectures concerning such unifications are brought together under the general label of Grand Unification Theories (GUTS). The electroweak unification of the electromagnetic and weak forces, together with the gauge theory of the strong color forces, constitutes the *Standard Model,* which in turn is a base of the more inclusive, and more speculative, GUTS models. The Standard Model is a long-distance, low-energy subset of any GUTS.

An especially simple Standard Model theory or set of theories is called $SU(5)$. Although the most straightforward version of $SU(5)$—minimal $SU(5)$—does not appear to have been adopted by nature, a discussion of this model will be useful since Her design is probably similar to $SU(5)$ in many ways.

This $SU(5)$ conjecture, which is modest inasmuch as only the electroweak and color forces are considered (leaving gravity to a further unification), assumes a consolidation of the $SU(2) \times U(1)$ electroweak theory and the $SU(3)_c$ color description of the strong interaction. According to this model, at the unification energy the three colors of the $SU(3)_c$ color symmetry and the two lepton varieties (such as e^- and ν_e) of $SU(2)_w$ weak symmetry combine so that the five varieties are equivalent under an interchange, thereby creating an $SU(5)$ symmetry.

We have noted that the effective electric charge increases slowly as the energy increases or the distance of interaction becomes smaller. Conversely, the effective color charge becomes weaker at small distances or high energies. The weak charge, slightly greater than the electric charge, also gets weaker at high energies. Figure 18.1 shows the variation of the value of the square of the charges—in natural units of $1/\hbar c$—as a function of energy and distance. From this extrapolation, it appears that the couplings become equal at the very high energy of 10^{14} proton masses, the energy scale of the $SU(3) \times SU(2) \times U(1)$ unification.

An $SU(n)$ gauge theory has $n^2 - 1$ vector bosons as field quanta. Hence, there are three bosons W^+, W^-, and Z^0 for the $SU(2)$ weak interaction and eight gluons for the $SU(3)$ color interactions. Then there must be 24 intermediate $SU(5)$ vector bosons. At the unification energy where the $SU(5)$ symmetry holds, the 24 gauge bosons must be coupled equally to the 24 known basic fermions such that no feature of the universe would differ upon a

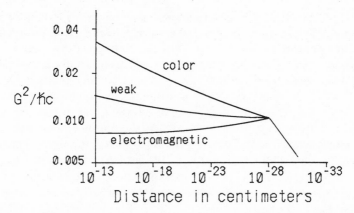

Figure 18.1 The square of electromagnetic, weak, and color charges is plotted as function of distance and energy in units of the square of the electric charge at large distances. The unit of distance is the distance complementary to a momentum of 1 GeV/c: $\hbar/(\text{GeV}/c)$.

change of label of any basic fermion. The small observer who failed to differentiate between a neutron and proton using a strong interaction meter in Chapter 14 will fail to differentiate among the basic fermions under these conditions using strong or weak field meters. Nor will the observer be able to differentiate among the 24 vector bosons required to enforce the local symmetry.

Valid at very small distances, at larger distances the symmetry is broken spontaneously by interaction with the background of scalar Higgs fields, and the 24 bosons and the 24 fermions take on different identities. We can count the bosons that are important at the larger distances and lower energies accessible to our measuring instruments. Physicists believe they have seen 12 fundamental vector bosons: the 3 weak bosons, the 8 gluons, and the photon. There must then be 12 more. At the same time, we know of 24 fundamental fermions—and 24 antifermions—that must be interchangeable at the unification energies. There are 18 quarks and 6 leptons: 6 flavors of quarks, each with 3 colors, and 3 families of a charged lepton and a neutrino. At low energies, we know that the 3 kinds of leptons, the 3 sets of quarks, and the 3 colors are separately conserved. At the unification energy only charge will be conserved, as the gauge symmetry is defined by the invariance under free change of the fermions. Hence, the 12 missing gauge bosons must change quarks to leptons, violating the low-energy rules that separate the overall $SU(5)$ symmetry into $SU(2)$, $SU(3)$, and $U(1)$ subsets. Since the separation holds to high energies—indeed, until near the unification energy—we know that the 12 vector bosons that must mediate transitions that violate the low-energy, long-distance conservation rules must be very heavy; they must have masses near the unification mass of 10^{14} proton masses. According to this model, very massive Higgs bosons break the symmetry of electroweak and color forces at energies below the unification energy. Then, lighter Higgs bosons break the

$SU(2) \times U(1)$ symmetry of electromagnetism and the weak forces to generate the very low-energy universe in which we live. Feynman diagrams representing characteristic transitions mediated by the various classes of fundamental $SU(5)$ vector bosons after symmetry breaking are shown in Figure 18.2.

Although quarks and leptons are known to be conserved separately at low energies and for interactions at large distances, if a unified description of nature such as $SU(5)$, or any GUTS model, is valid, transitions from quarks to leptons must occur at some level. Hence, the proton and neutron cannot be absolutely conserved and must decay, albeit slowly, to leptons. Figure 18.3 suggests the dominant decay mode of the proton according to the $SU(5)$ model, where the proton decays to a positron and a neutral pion (which in turn decays to two high-energy photons through known low-energy processes.)

Presumably the very large masses of the X and Y intermediate vector bosons cause the basic interaction to take place only over the very small distances complementary to these masses. Because the probability of the interacting particles (such as the two u-quarks combining to initiate proton decay, as shown in Figure 18.3) being that close together is very small, the decay rate will be very low. For specific models the decay rate can be calculated. Such a calculation, based on the minimal $SU(5)$ model, leads to a proton lifetime of 10^{29} years with an uncertainty of perhaps a factor of 50. The median value of the lifetime corresponds to a decay of one proton per year for 300 kilograms of matter. Enormous detectors—such as a pool containing 8000 tons of water viewed by phototubes sensitive to light emitted by the passage of fast particles through the water, all placed in a deep salt mine near Cleveland, Ohio, to

Figure 18.2 Typical fermion transitions mediated by the different (after symmetry breaking) vector bosons of the $SU(5)$ model. The six different heavy X-bosons with a charge of 4/3 and the six Y-bosons with a charge of 1/3 mediate different quark color combinations, which are not defined here.

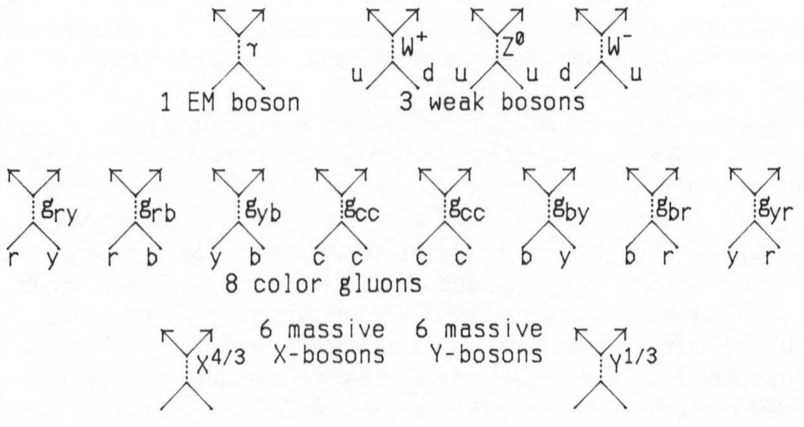

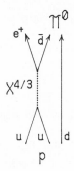

Figure 18.3 The character of the proton decay to a positron plus a π^0. The proton is formed of $u + u + d$ quarks, and the two u-quarks interact to form an X-boson that decays to a positron and a $\bar{d}$ quark. The $\bar{d}$ quark and the leftover d-quark form a neutral pion, which later decays to two photons.

reduce cosmic ray backgrounds—have been used to search for proton decay. The negative results of such experiments have led to limits on the decay rate that appear to be less than one-hundredth of that expected from the minimal $SU(5)$ model. Hence, although the uncertainties are large in both experiment and calculation, it seems that God has chosen to do things in another (and to us, more complicated) way. However, some form of this symmetry must apparently hold, and many extensions and variations of $SU(5)$ are now being considered.

Another characteristic of $SU(5)$, shared by almost all other GUTS models, is the prediction of the existence of magnetic monopoles with the enormous masses of the unification energy, masses near 10^{16} proton masses. Such masses cannot be created in the conditions that now hold in the universe, but they would be expected to arise in the very early universe (the first 10^{-40} seconds), when the temperature of the universe corresponded to energies greater than the electroweak-QCD unification mass, as a consequence of phase differences between adjacent Higgs fields produced in an uncorrelated manner at the Creation. However, if such monopoles were produced at the predicted rates, the mass of the monopoles would soon dominate the mass of the universe, and the resultant mass density would be so great as to produce a very large positive curvature in the universe; but we know the universe is quite flat. Our knowledge of the early universe is hardly definitive, and it is possible that the monopole production may have been very much reduced by mechanisms (such as "inflation") that we will take up later.

Whatever their origin, we know that heavy monopoles must be rare. No one has observed a monopole, and if monopoles of any large mass exists now in quantities such that more than one monopole per month would pass through an area the size of a football field, the magnetic monopole current in the galaxy would short-circuit the galactic magnetic field, which is a few millionths the size of the earth's field.

Subquarks

One hundred years ago, Mendeleev's set of the 90 or so different atoms (or nuclei) found in nature constituted the elementary particles. Then this list was simplified by the understanding that all atomic nuclei could be constructed of only two elements, the neutron and proton. As of 1950 it seemed that nature was made up of only a few "elementary" hadrons—the neutron, proton, and pions—and leptons—the electron, muon, and neutrino. Later, the hadron list increased to hundreds, to be reduced only when it was noted that all known hadrons could be constructed from sets of three quarks. But then the number of quarks increased to 18 and the number of leptons to 6. Add to these 24 antifermions, the 12 known vector bosons, and then 12 as yet unseen vector bosons required for the simplest unified theory, as well as ancillary Higgs scalar bosons, and our simplest description of nature requires more than 75 fundamental particles.

We have then gone from atoms to nuclei to "elementary particles" and on to quarks, leptons, and vector bosons. Would the Old One have set up so complicated a system? Jonathan Swift notes,

> So, naturalists observe, a flea
> Hath smaller fleas that on him prey;
> And these have smaller still to bite 'em;
> And so proceed ad infinitum.

Are there then subquarks making up quarks, and sub-subquarks making up subquarks—ad infinitum? There are interesting and elegant simplicities that seem to emerge from further (but not infinite!) generations of subquarks or preons, and this is one path of exploration. Again, it has not yet been possible to link such models with observational evidence.

Supersymmetry

Driven by aesthetic considerations—and the hope of solving the manifestly difficult problem of bringing gravity into the unification fold—physicists have devoted considerable energy to a bold hypothesis (or set of hypotheses) that for each boson, nature has produced a sibling fermion, and vice versa. The electron, a fermion with spin ½, must then have a sister, the selectron, a boson with spin zero. Similarly, the photon, a vector boson with spin 1, is expected to have a fermion brother, the photino,[3] with spin ½. Moreover, in the simple theory a local symmetry called *supersymmetry* must hold such that nothing observable will be changed if any fermion changes to its associated boson— or any boson changes to its associated fermion.

If such a symmetry were to hold, the sibling particles must have the same masses and interact symmetrically with other particles. According to this hypothesis, for each fermion F_a there exists a related boson B_a, and for every boson B_b, there is a related fermion F_b. Moreover, the couping of F_a to B_b is equal to the coupling of F_b to B_a. Hence, the two transitions of Figure 18.4 will occur with the same intrinsic strength. Because no one has seen evidence for the existence of a selectron, a photino, or any other member of the set of associated particles postulated by supersymmetry in measurements that probe energies as high as 20 proton masses, supersymmetry must be a very high-energy, very short-distance symmetry, badly broken at low energies—if it exists at all.

With no supporting experimental evidence, why is supersymmetry interesting? There are several reasons. Supersymmetry leads to attractive formal or mathematical simplicities that provide a high degree of cancellation of infinities that plague calculations in GUTS theories. It appears that the inclusion of supersymmetric particles at very short distances results in cancellations of forces that otherwise generate infinite cross sections for some processes. Perhaps more important, the idea is grand. Supersymmetry connects internal symmetries such as color with the space-time Poincaré symmetries of nature that describe the invariance of nature with respect to displacement in space or time or direction of orientation; moreover, supersymmetry connects matter (fermions) and forces (bosons)—particles and fields. Perhaps the most important feature of supersymmetry, however, is that the field required or defined by the local symmetry has the form of gravity! This theory of gravity is commonly called *supergravity.*

To many physicists it seems that the only attractive way to incorporate gravity into the unification of all forces is through some kind of supersymmetry. We can gain some insight into the relation of supersymmetry to gravity by considering the gauge field that must mediate the transformation of a fermion to a boson. Because the intrinsic angular momentum of the particle must change and angular momentum must be conserved, the angular momentum of motion of the particle must be changed, and upon a local transformation the particle must be displaced in space and momentum as an acceleration—the gauge field provides an acceleration of the particle just as gravity does. Hence, supersymmetry acts as a gauge theory of gravity that

Figure 18.4 Two transitions connected by supersymmetry that have equal intrinsic strength: F_a and B_a are related as supersymmetric fermion-boson pairs, as are B_b and F_b.

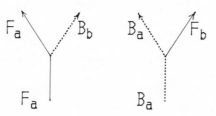

may serve as a quantum generalization of Einstein's General Theory of Gravitation. Because the particle that changes from a fermion to a boson may also change internal symmetry values such as color or flavor, the gauge field connects the internal symmetries with the dynamic space-time Poincaré symmetries that lead to the conservation of energy and momentum.

More Dimensions Than Four—Kaluza-Klein Theories

In 1921 a young Polish physicist, Theodor Kaluza, wrote Einstein concerning his views on the possibility of extending the General Theory of Relativity to include electromagnetism by writing a General Theory in five dimensions—in a space with four space dimensions plus time. Such a model was and is of great interest, inasmuch as a change in gauge for the familiar four-dimensional (including time) electromagnetism is simply equivalent to a displacement in the fifth dimension. Invariance under a gauge transformation reduces to simple invariance under translation in the fifth dimension—to be added to the invariances under displacement in the other three space dimensions and one time dimension already known. But if this model is to describe our universe, where is the fourth spatial dimension? Can God's universe contain more than three space dimensions though man and beast see but three?

In 1926 Swedish physicist Oscar Klein solved that problem by showing that the fourth dimension could be rolled up into a tight circle too small to perceive. Such a state of affairs is suggested by the experience of the one-dimensional lineworm shown in Figure 18.5 moving along a two-dimensional tubular universe where the second spatial dimension, normal to the axis of the tube, is rolled up into a tight circle—the radius of the tube. If the radius of the tube is much smaller than any distance that the worm can sense, he will describe his universe in terms of only one dimension—the dimension along the axis of the tube. Theories based on the five-dimensional space introduced by Kaluza with one dimension rolled up according to the prescription of Klein became known as Kaluza-Klein theories, and this name is now applied to models where there are more than five dimensions and more than one dimension is rolled up.

Aside from the simple description of gauge invariances that derive from the consideration of a universe with more dimensions, other relations that

Figure 18.5 A lineworm moving in a space that has two dimensions with one tightly rolled up. The worm thinks he lives in a one-dimensional space.

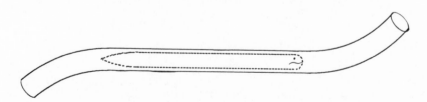

seem rather arcane in four dimensions take simpler forms in higher-dimensional spaces. We illustrate such simplifications by considering a fable concerning the flatworm's study of nature in his two-dimensional universe. According to this story, early in the history of flatland science, flatworms studied a cubic crystal of table salt and concluded that a salt crystal sometimes has the shape of a triangle and sometimes that of a rectangle. From our exalted three-dimensional position, we know that the salt cube sometimes is cut across a corner by the two-dimensional space of the flatworm, resulting in a triangular projection in flatland; it is also sometimes cut so as to form a square or rectangular projection. Salt crystals are less regular in flatland than in our three-space. Later, after flatland science evolved greatly, flatworms studied the properties of the Z (neutral intermediate vector boson) and found that there are three such particles with spins of -1, 0, and $+1$ in units of $\hbar$. In three-space we claim that there is but one Z with a total angular momentum of $\frac{1}{2}\hbar$, but the Z can be oriented so as to have angular momentum components of -1, 0, and $+1$ normal to the flatland space. For salt crystals or elementary particles, the description is simpler in three-space than in two-space. Similarly, our four-dimensional description of nature may find simpler forms in higher-dimensional universes.

According to this multidimensional view of nature, we are tricked the same way as the worm in Figure 18.5. We presume that we may live in a manifold of more than three space dimensions where we do not notice the extra dimensions that are rolled up into multidimensional generalizations of a very small circle.

Superstrings in Ten Dimensions

Although the attitudes supporting the intense interest by physicists in highly speculative theories are related to those which supplied Einstein's motivation for the construction of the General Theory of Relativity, the differences are such as to admit the possibility that we may be at the threshold of a new way of considering physics, hence a new way of considering nature. The construction of the universal law of gravitation represents an elegant paradigm of classical procedure in physics. Newton established his general laws on the basis of Kepler's compact, phenomenological description of the orbits of the planets, derived in turn from the observations of the positions of the planets (observations of nature) made by Tycho Brahe. Einstein, relying on observations of nature only to the extent of the equivalence of gravitational and inertial mass, relied upon the unique character of the description of nature he derived as his guarantee that he must have found nature's design. In his searches for that design, he found and discarded mechanisms that were unnatural, ugly, and contrived, and he found and discarded processes that did not fit simple observations. Einstein did not see the General Theory as one of many almost equally plausible relativities, but as the only plausible theory, the only design worthy of the Old One.

Today, armed with the Lagrangian field theory methods that have provided an accurate description of the electromagnetic field at large distances and low energies, physicists are hunting for a Grand Unified Theory of all interactions, using as a criteria for the examination of candidate models not the contact of the GUTS model with observation but the internal consistency of the model. And the criteria of internal consistency seems to have an astonishing power.

The calculation of low-energy electromagnetic phenomena through quantum electrodynamics has been complicated by the appearance of anomalies and infinities. Straightforward calculations of simple processes can lead to results that give anomalous violations of conservation laws such as the conservation of charge or the conservation of energy. Similarly, calculations of some probabilities lead to values greater than 1 and to infinite values for quantities that are manifestly finite. Both the anomalies and the infinities in the calculation, which arise from the nonphysically large probabilities, can be traced to effects at very high energies and very small distances and attributed to the implicit assumption that electromagnetism can be considered apart from other forces and that the weak interactions and gravity can be completely disregarded. Although the results of the calculations can be recovered by techniques such as renormalization that segregate the infinities at very small distances—presumably generated by the neglect of other forces—into sets of processes for which the measured values of mass and charge can be substituted, the breakdown at small distances remains a fundamental limitation of the theory. But any anomalies and infinities that occur in simple calculations based on complete, unified general theories cannot be excused as a consequence of the neglect of certain forces, as all interactions are considered.

Hence, if a unified theory is valid, calculations of simple processes, such as the scattering of one primordial particle by another in the high-energy, short-distance limit where the underlying symmetry of the theory is not yet distorted by symmetry-breaking mechanisms, must lead to plausible results. Although the scattering cannot be observed and the results of a calculation of the scattering cannot then be compared to measured values, we can expect that the result will not exhibit anomalies and violate simple conservation laws, and the result should not lead to nonphysical infinities and probabilities that exceed 1. If many theories could be constructed that would have these weak virtues of not predicting nonsense, the constraint to sense would be a necessary criterion but would not have much discriminatory power. However, it seems that it is extremely difficult to construct a sensible theory—so difficult that the emergence of such a model seems almost a miracle, a miracle worthy of the Old One.

The insight that many physicists believe may lead to the derivation of a theory that is not nonsense, hence to the unraveling of nature's grandest puzzle, is the realization that the fundamental particles may not be points but lines or *strings*. As so often happens in science, the concept of particles as strings began as an effort to describe phenomena now understood in rather different ways. These original models described hadrons as taut, vibrating strings of mass rather as lines drawn in a three-dimensional space. The frequencies of vibration of a classical taut string, such as a guitar string, vary as the square root of the tension of the string;[4] to tune a guitar string that is flat (registering too low a frequency), one tightens the string. In the original hadron string theory, the tension was adjusted so that the frequencies f multiplied by Planck's constant h corresponded to known masses m of the hadrons: $hf = mc^2$. Although some serious logical difficulties with the string ideas emerged, the spectrum of vibration and rotation energies of the extended string seemed to fit the distribution of masses of hadrons rather well, a result that we now understand more deeply as following from the quark structure of the hadrons. The quark-antiquark systems we call mesons are known to rotate and vibrate rather as stringlike structures (see Figure 17.7), such that the quantum spectrum of these vibrations and rotations is manifest as particles with discrete masses.

Although descended intellectually from the simple hadron string ancestor, the Grand Unified Superstring Theories are far grander and more powerful constructs. Described in ten space-time dimensions (nine space dimensions and one time dimension) rather than four, and incorporating supersymmetry in most forms of the theory, the string is shorter and the tension is very much greater; hence, the characteristic energy is nearly the Planck mass 10^{19} M_P rather than the nucleon mass. The string is probably in the form of a closed loop (though open superstring theories are also interesting). Moreover, instead of describing excited states of hadrons, the superstring models are theories of all particles.

These string loops are very small, on an order of magnitude of the Grand Unification characteristic distance of about 10^{-33} centimeters, which is the scale of the rolled-up extra dimensions. Hence, probes at lower energies with poorer spatial resolution still see the elementary particles as points. Like the simple one-dimensional guitar strings considered in Chapter 10 (Figure 10.12), the 10-dimensional superstrings can be considered to support standing waves, and, for the closed loop models, running waves that travel about the loop. Here, the sets of waves constitute the base of primordial fundamental particles; each particle corresponds to a different wave. Different models of the string support different sets of waves where these sets display special symmetries. Grandly, these symmetries are much broader than the $U(1) \times SU(2)_w \times SU(3)_c$ symmetries evident at the much larger distances and much lower energies accessible to us; hence, they require many more intermediate gauge particles to mediate a local symmetry. If the theory is viable, the large symmetry must contain the lesser symmetries we observe as a subset.

An especially interesting model displays an arcane $E_8 \times E_8$ symmetry, where E_8 is the largest "exceptional" finite Lie group formerly of interest only to mathematicians. These two E_8 groups are linked only by gravity; in this model, one of the E_8 groups describes our universe and the other a parallel shadow universe detectable by us only through mutual gravitational interactions. An E_8 group generates 496 intermediate force particles compared with the 12—the photon, two W's, the Z, and the 8 gluons—which describe low-energy forces, or the 24 bosons postulated for an $SU(5)$ symmetry. Remarkably, almost miraculously, the anomalies and infinities that plague unification theories all seem to cancel neatly for this superstring model. Moreover, the cancellation requires the contributions of all the 496 gauge particles.

Although this E_8 universe—with 1 force, 10 dimensions, and 496 gauge particles— is surely elegant, what does it have to do with that universe—of 3 forces and gravity, 4 dimensions, and 12 gauge particles—constructed for us by the Old One? First, we must believe that 6 of the 9 space dimensions are curled up tightly with radii of curvature of about 10^{-33} centimeters breaking the original symmetry. Moreover, the 6 curved dimensions curl up to form a rather special curved space with parts like cones coming to a point and with holes like a doughnut (in a rough three-dimensional analogy), and the character of the subspace determines some of the characteristics of our world. For example, it seems that the three quark and lepton families we see may correspond to six doughnut holes in the rolled-up six-space. But we have no natural understanding as to why six dimensions curl up so tightly and three remain extremely flat; the radius of curvature of the flat dimensions is known to be larger than the Hubble radius of the universe. The ratio of the radii of curvature of the curved and flat dimensions is greater than 10^{60}, the ratio of the smallest distance defined by the universe and the largest.

There are both internal and external difficulties with superstring theories. We do not yet understand superstrings; the superstring concept is not known to flow from some unifying principle. And we cannot yet relate the theories satisfactorily to the world we know. If one of the extant candidate superstring models finally wins out as the only model that seems to escape nonsense in describing the properties of nature at distances of 10^{-33} centimeters, how can we be sure that this is the model actually chosen by Nature as the basis of Her universe? To establish the validity and relevance of the model beyond reasonable doubt, we must understand how spontaneous symmetry breaking can deliver the low-energy, large-distance $U(1) \times SU(2)_w$ electroweak forces and the $SU(3)_c$ color forces that we know act over distances of the order of 10^{-13} centimeters. The theory is separated from observations by about 20 factors of 10. Can we ever bridge that gap, which is greater than the gap between the classical physics of Newton and the quantum physics of the atom and nucleus? We do not now have any indication that we can.

What are the odds? Are we really on the brink of understanding God's Master Plan, or are we merely engaging in recreational mathematics? At this time, most physicists hope for the former and fear the latter. We can say with moderate assurance that by the turn of the century, superstrings will be either accepted as the basic blueprint of the universe, or a half-forgotten item on the junkpile of the history of science.

Conjectures Concerning the Very Early Universe

Chapter 16 covered the history of the early universe from a period about one-hundredth of a second after the Creation until the present. That history (albeit with some gaps) is quite reliable. The temperature at that beginning of that history corresponds to average particle energies of about 12 MeV, which lie in regions where physicists have conducted extensive laboratory studies. If we extrapolate to much earlier times, the temperature and average energy per particle are higher, beyond those accessible to us in the laboratory, and we lose contact with that which we know well. We must then rely on conjectures concerning the interactions of the elementary particles at very high energies. Consequently, conjectures concerning the very early universe are inextricably linked with conjectures concerning fundamental particles and fields at very small distances—and very high energies. We cannot understand the very early universe, extremely hot and highly compressed, without understanding particles and fields at extremely short distances and high energies. Conversely, the very early universe is the only laboratory we have, or are likely ever to have, in which we can examine the consequences of models of particles and fields as they interact at such short distances and high energies.

Although the true story of the very early universe is not known, there are elements of this prehistory that can be identified with some reliability. Figure 18.6 shows a conventional view of temperature and density as a function of time in the very early universe, along with historic notes. In order to provide a coherent scenario for the discussion of ideas concerning this initial epoch, we assume that we have unearthed an ancient diary written by an ancestor of the small observer introduced in Chapter 14, which recounts one version of the very early history of the universe. This saga begins a few instants after the Creation and reads as follows:

> My watch reads 10^{-42} seconds—though I don't really know how it was set. [Neither do we; we do not know what time zero means.] Gravity seems to exist already in a conventional fashion. The density of the universe [mainly

Figure 18.6 Approximate density and temperature as a function of time in the very early universe, according to a noninflationary extrapolation.

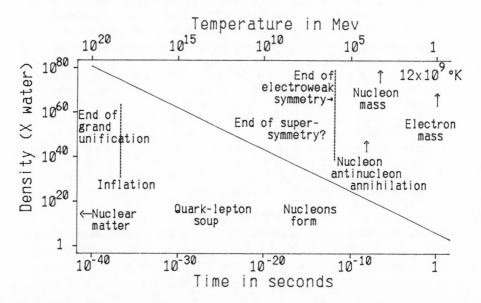

in the form of energy] is about 10^{90} times the density of water,[5] and the average energy of the particles is about 10^{18} proton masses, corresponding to a temperature as read by my thermometer of about 10^{31} °K. [Here we pass over some complaints about hot weather.] The universe seems quite symmetric inasmuch as even with all my meters, I cannot distinguish one particle from another. Also, the universe is electrically neutral—I can't get any reading on my electric field meter at all. . . .

My watch now reads 10^{-34} seconds, and the density and temperature have dropped a bit. The density is now only (!) about 10^{77} times the density of water, and the temperature has dropped so that the average energy of the particles I see moving by is about 10^{14} proton masses. A phase change[6] is occurring as the unified force symmetry I mentioned is beginning to break down into separate strong QCD color forces and electroweak forces. As a consequence of this loss of symmetry, the particles are assuming different identities in a kind of internal crystallization. Quarks and leptons now seem different and tend to keep their identities upon collision. I see that this phase transition [which acts somewhat as a crystallization] is not completely uniform, and at some of the more nonuniform points in space [which are like defects in crystals], pieces of field are torn off that carry magnetic monopole charge. These monopoles have a mass of about 10^{16} proton masses. But when the quarks and leptons do change into each other, I see that the antiquarks are changing to leptons slightly more easily than the quarks to antileptons (violating CP symmetry). Hence, there is a slight (perhaps 1 part in a billion) excess of quarks over antiquarks and an equal excess of leptons over antileptons. The entire system remains electrically neutral as it has been since I came on the scene. . . .

I have a lot of empty pages in my diary, as nothing very dramatic has happened for a long time. It is as if I have been traveling (in time) through a desert. My watch reads 10^{-10} seconds, and the density and temperature are now much lower: the average energy per particle is only about 100 GeV, the density is only about 10^{25} times the density of water, and the quarks have largely condensed to baryons and antibaryons. I believe that I am seeing the beginning of another phase change where the electroweak force splits into a weak force and an electromagnetic force. The gas of massless vector bosons and scalar Higgs bosons is changing to massive charged W's and a neutral massive Z and a massless photon. The undifferentiated leptons are also taking on specific identities as electrons, muons, and tau-leptons.

My watch now reads $\frac{1}{100}$ of a second, and prehistory has come to an end. The average energy per particle is now about 10 MeV, 1% of the proton mass and 20 times the mass of the electron; the corresponding temperature is about 10^{11} °K, and the density is about 5 billion times the density of water. Since the last entry in my diary, the baryons and antibaryons annihilated (to form particles that eventually decayed to leptons and photons), leaving only the very small excess of baryons (about 1 baryon to a billion photons). Prehistory is now over. Ancient history begins [reasonably well recorded and understood as discussed in Chapter 16], and my diary ends.

As we expect in archeology, as many questions are asked as answered by the ancient text we have assembled—and there are certainly errors in the report. The account of the breakdown of the original force, demonstrating a full symmetry, to the subset of forces we have now is probably qualitatively

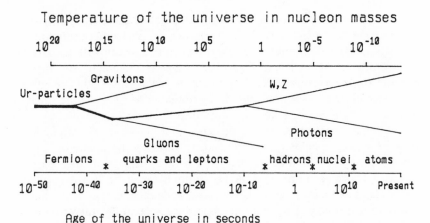

Figure 18.7 The history of the symmetry breakdowns (phase changes) that resulted in the breakup of the original one-force to the present set of four different forces.

correct: that history is shown in Figure 18.7. But where are the monopoles the diarist mentions? We have never seen one. And why is the universe so flat? If it is so flat now, it must have been flat to about 1 part in 10^{50} at the beginning of the diary and what happened before the diary began?

We suspect that this old document contains a mixture of truth and myth, and the true history of the very early universe might differ from the account of the diary in some interesting and significant ways.

The Inflationary Universe

Some answers to the questions raised by the conventional expanding universe model described in the diary are suggested by a view, largely initiated by Alan Guth, that at the very early time of the phase change that takes place when the unified force breaks down into separate color (QCD) forces and electro-weak forces, the universe expanded explosively, increasing its size by an enormous factor in a very short time. This explosion is driven by the mass of the vacuum.

Most physicists are reasonably confident that, at this time in the history of the universe, the vacuum carries no mass. By a vacuum, we mean a state of a field such that the field has least energy. Usually this is the case when there are no real particles in the field. If there is no photon in a cavity, the cavity will hold no electromagnetic energy. If there are no W or Z bosons in a region, the region will hold no weak interaction energy and less energy than if such particles inhabited the region. In either case, the vacuum is that state with no particles. Curiously, the situation can be different for the scalar Higgs particles. One can consider that in effect, there is an attractive force between the Higgs particles. The energy in a region may then be less with scalar Higgs particles present—and holding together with negative potential energy—than if there were no particles at all. The particles hold together so tightly that energy must be added to remove them and create a region with no particles.

The diagram at the left in Figure 18.8 suggests a characteristic variation of energy (or mass) density as a function of the scalar Higgs field strength that we believe plays an essential role in describing unified fields—the square of the field strength can be taken as a kind of particle density. At the right, the conventional variation of energy density w with electric field strength E is shown: for electric fields the energy density varies as the square of the field, $w = E^2/8\pi$. The point at zero scalar field, with no density of Higgs particles, is called a "false vacuum," and the minimum at larger fields, where the density of Higgs particles is greater than zero, is defined as the "vacuum," the state of lowest energy. The inflation model suggests that the universe was created with no Higgs field at all such that it was initially in such a false vacuum state. Typically, the shapes of the curves of energy density versus field strength vary with temperature, and the curve of Figure 18.8 is similar to that expected for Higgs-type particles at the very high temperature of the very early universe. At later times and at lower temperatures, the false vacuum fades away and the energy varies with field strength rather as the variation of energy density with electric field shown to the right in Figure 18.8.

According to one inflationary concept, at the time (listed as 10^{-34} seconds in the ancient observer's diary) when the unified force broke down into separate electroweak and QCD forces, the universe was caught in the false vacuum shown in Figure 18.8 generated by the forces between the scalar Higgs particles taking part in the unified force. Hence, the vacuum had a mass.

According to the "escape velocity relation" discussed in Chapter 16, the velocity with which one point in the universe recedes from another (the expansion velocity) is proportional to the distance between the two points multiplied by the square root of the mass (or energy) density of the universe. For a universe filled with radiation or a universe filled with particles, the density decreases sharply as the distance between points (or particles) grows and the rate of expansion slows with time. However, if the vacuum has a definite

Figure 18.8 At the left, the energy density versus the strength of the scalar field is plotted for a field of scalar particles that attract one another. A similar graph is shown for the electric field at the right.

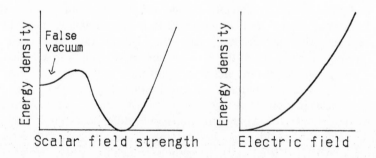

mass density, that density will be constant, and the velocity at which one point recedes from another will be simply proportional to the distance between the points;[7] when the distance between two points doubles, the velocity of recession doubles. The expansion then proceeds explosively (or exponentially), as suggested by Figure 18.9. At the time (as read by the observer's watch) of 10^{-34} seconds, the inflationary universe expands a factor of about 10^{50} in a time of perhaps 10^{-33} seconds.

During this very large and sudden expansion, the temperature of the universe drops precipitously from about 10^{28} °K (or equally, degrees Celsius) to a value near zero. The expansion stops when the false vacuum decays to the true vacuum in a process quite similar to the alpha-particle decay of heavy nuclei. In the course of this decay, the energy stored in the vacuum as a dense gas of Higgs scalar particles is discharged in a flash of radiant energy as the Higgs particles decay to lighter particles heating the universe up to a temperature of about 10^{25} °K; the universe then continues to expand in the normal, noninflationary manner that continues to this day.

Three special problems of the "regular" expansion scenario recorded in the diary find solutions in this inflationary projection.

1. Because the universe expands enormously after the monopoles are created, the density of original monopoles is then very small—consistent with our failure to find any—and the reheating does not bring the temperature to the monopole-creation level; therefore, we conclude that no more monopoles were made after inflation.

2. The 3 °K microwave radiation coming today from opposite directions has been measured to have the same intensity and frequency to within at least

Figure 18.9 The solid curve shows the distance between characteristic points of a universe during the inflationary period. The dashed curve suggests the same distance under the standard expansion hypothesis.

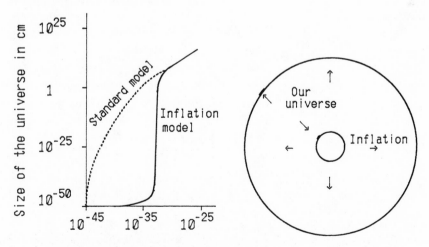

Age of the universe in seconds

1 part in 10,000. As we have noted before, light that comes from a great distance also comes from the distant past. Figure 18.10 shows the space-time path of two (microwave) light rays coming to us from opposite directions, one from the north (the direction of Polaris), the other from the opposite direction, labeled south. The ray, shown as a dotted line, defines the position of the signal as a function of time back to 300,000 years after the Creation when the rays emerged as the universe first became transparent. If we were to plot this pattern for light coming from all directions, we would have a four-dimensional cone; hence, the light rays are said to define a "light cone." Events that take place outside the light cone cannot have been affected by any event generated at the time and place described by the apex of the cone. We see that the two origins of the north and south light cones cannot be encompassed by any cone generated by an event that took place at any earlier time back to that of origin, or time zero. At the time when the light was emitted, 300,000 years after the Creation, the two regions of emission were more than 100,000 light-years apart. Hence, according to the standard model of the expansion of the universe, there was absolutely no communication, ever, between the regions from which the 3-degree radiation were emitted. Then why is the intensity and frequency of the radiation from opposite directions the same to a factor of 1 part in 10,000? If the regions evolved separately, how could they reach the same temperature with such precision?

The inflationary scenario solves that problem inasmuch as before inflation, when the region now occupied by the universe within our event horizon was extraordinarily small, those regions were connected causally. Hence, the

Figure 18.10 A light-cone diagram plotting the space-time position of light rays reaching an observer on earth at the present time from opposite directions, north and south. No action taken at time zero can affect both the light sources at *N* and *S*. All slanted lines are light-ray paths. (The diagram is not to scale.)

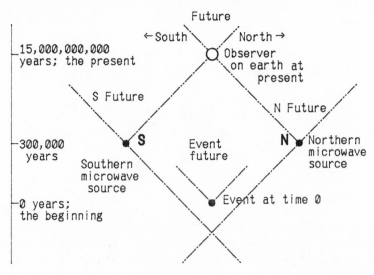

regions were connected at some very early time and must have been at the same temperature. The view that this equivalence of temperature persisted through inflation and for the next 15 billion years is quite plausible when examined in detail.

3. The standard expansion theory does not explain the extraordinary flatness of the universe. But this flatness is a natural result of inflation. We can understand this on a metaphoric level with the help of the model introduced in Chapter 16, which compares the expanding three-dimensional universe curved in a fourth spatial dimension to the two-dimensional surface of a balloon curved in a third dimension. Consider that before inflation the balloon was very small, perhaps the size of a pea, and that the sector now our region of the universe (the region within our event horizon with a radius of about 15 billion light-years) is but a tiny speck on the pea-size balloon. When the balloon is then inflated very quickly to a size greater than our entire observable universe, the small surface area representing our space would seem very flat. And the expansion we are considering is more than 10^{20} times greater than that from a pea to the present observable universe!

There are severe technical and conceptual difficulties accompanying the impressive qualitative successes of the simple inflationary ideas just sketched. The false vacuum, stable classically, is not stable when quantum mechanical fluctuations and barrier penetration are considered. A small region of the massive space can decay to the true vacuum by tunneling through the potential barrier along the dashed line in Figure 18.8. When this happens, a bubble of real vacuum forms that expands at a velocity near the velocity of light. From the Uncertainty Principle, we know that different regions of space will form bubbles at different times, so we might expect that space would be full of different size bubbles—and be much more nonhomogeneous than the space we know.

This difficulty with nonhomogeneity has been surmounted by the postulation of a more complicated and somewhat special interaction of various scalar Higgs particles so as to slow down the decay of the inflation. With this model, large domains are formed by the crystallization-like phase change, where the symmetry may differ somewhat from domain to domain. Again, the universe is nonhomogeneous but on so large a scale that we would not notice it. Presumably we are occupying a domain about 10^{35} light-years across— a size enormously larger than our visible universe of but $15 \cdot 10^9$ light-years. For either inflation model we know little about the Higgs interactions supposedly responsible for inflation from other sources—such as the study of elementary particles and fields.

The Creation?

The history of the universe from the first one-thousandth of a second after creation until the present is now a well-documented part of knowledge. We have been able to understand something about the very early universe and are able to make attractive conjectures concerning times as early as 10^{-40} seconds. But what have we to say about what happened before that? To consider such early times, we must have a useful model of quantum gravity to extend the General Theory of Relativity. Perhaps a 10-dimensional superstring theory will help here. At temperatures corresponding to the Planck energy expected at such time, time itself will be tied into knots, and then what is the meaning of "before"? What is meant by "after"? And by "beginning"? I doubt that we are close to an answer.

And then where did everything come from? There is about

$$1,000,000,000,000,000,000,000,000,000,$$
$$000,000,000,000,000,000,000,000,000$$

tons of matter (10^{78} nucleons) within the 15 billion light-year radius of our event horizon. Who ordered that? We really have no answer, and we are not even sure that we know how to ask the question, since even the basic conservation laws are suspect. Emmy Noether's deduction of the conservation of energy from the invariance of the Lagrangian with time is not secure when everything is changing rapidly with time. However the uncertainty, an interesting relation exsits that may not be wholly an accident.

How much energy is required (naively following the rules we use locally in our small corner) to add one stationary particle of mass m to an empty universe? From special relativity, we would say mc^2. With no particles and no forces in that empty universe, the particle has no potential energy to add or subtract to the mass-energy, so the total energy is also mc^2. Now assume that there is a sphere of great mass in the universe that generates an attractive force on an additional particle. Figure 18.11 shows the potential energy of a mass m in the field of the central mass as a function of distance from that

Figure 18.11   The potential energy of a mass m as a function of distance from a large mass. At the distance shown, the potential energy $-mc^2$ is equal in magnitude to the rest mass, and the total energy is zero.

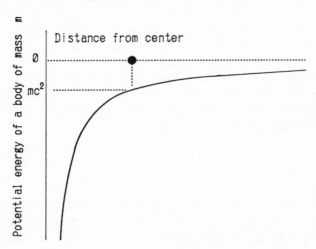

mass. Notice that a point exists such that the potential energy is $-mc^2$. Surely if we were to add this mass m to the universe very far from the central mass, where the gravitational potential energy is negligible, an additional energy of mc^2 would still be required to add the particle to that universe. But what if we were to add the particle to the universe at the indicated position so that the potential energy was $-mc^2$? Then the total energy would be zero! No energy would be required to add the particle to the universe. With care, we could fill up the universe with properly distributed matter so that the total energy of the universe (as we have calculated it) would be zero. And this is our universe. Calculated in this manner, the total energy of our universe is zero—and the total energy has always been zero. No work (or energy) at all is required to create a universe! Is this complete nonsense? We don't know.

Philosophy, Theology, and the Anthropic Principle

If we separate the concerns of man into that which has to do with man and the welfare of men and that which has to do with God, or Nature, or things of the spirit, and if we call the first set the province of humanism and the second, that of religion, then the inquiry into the character of nature described here is religious; and those who pursue that inquiry follow a religious vocation though they be clothed in secular garb. Einstein's references to God ("Der Alte") was, if sometimes light-hearted, not jocular. If his pantheistic "God of Spinoza" was not the God of the Jews or Christians or Moslems, his confidence in the rationality of nature was religious in kind. Then what lessons on the nature of God can we draw from the results of our "religious" studies of physical reality?

I respond to my personal question with a personal answer. Excepting those who base their religious convictions upon literal interpretations of ancient metaphors, I find no compelling reason for anyone to emerge from the study of nature with a faith different from that he or she may have possessed upon entrance. Although the world is not flat and was not constructed 6000 years ago, physicists know nothing that contradicts the cores of various religious beliefs held by most people today, and some have found a deeper faith as a result of their inquiry. (The eminent English theoretical physicist J. C. Polkinghorne said in the epilogue of his fine book on elementary particles, *The Particle Play*, "The pursuit of science is an aspect of the *imago dei*. Therefore it does not seem strange that these words written while Professor of Mathematical Physics in the University of Cambridge will be published while I am an ordinand studying for the Anglican priesthood at Wescott House.")

Although it is difficult to consider the intricate pattern divulged by our studies of nature without being touched by the awesome mystery of existence itself, should the miracle of the subtleties of elementary particles and the universe of space and time that they inhabit and define touch us more, or differ-

ently, than the miracle of the grain of sand or the wildflower described by
William Blake?

> To see a world in a grain of sand
> And a heaven in a wild flower,
> Hold infinity in the palm of your hand
> And eternity in an hour.

I think not. Writ small or large, particle or grain of sand, flower or galaxy, the
miracle is the same.

But what is man's place in this universe of galaxy and flower? Do we hold
a special position? *Was the universe created for us?* The long-held view of an
anthropocentric universe was shaken by Copernicus and largely dispelled in
the nineteenth century. "Melancholy, long, withdrawing roar of the sea of
faith," Matthew Arnold mourns in his poem "Dover Beach." Curiously, it is
now not unthinkable scientifically to muse over the possibility that an
Anthropic Principle acts, and the universe might really be especially designed
for man. It appears that the universe must be exquisitely fine-tuned to accom-
modate us—and we are here. If the universe were only a little different, life—
and the contemplation of that universe by intelligent beings—might not be
possible.

Starting with the small, it seems probable that only the chemistry of carbon compounds can support the complexity required for the astonishing
organization of life, and no other kind of life is possible. Moreover, that
unique complexity can develop only within a very small range of physical
conditions, especially within a very small temperature range. The argument
that with so many stars in our galaxy, and so many planets, intelligent life
must abound, is interesting and appealing. But upon analysis, it seems that
few planets can be expected to be fit for life. And if conditions obtain that are
adequate for the formation of life, that formulation may be intrinsically
highly improbable. Hence, the argument, also plausible within the bounds of
our ignorance, that we may be alone is, to some, even more striking.

If this universe as we know it is so constructed that life may be found only
under rare and especially fortuitous circumstances, how precarious is the
dependence of these circumstances on the precise form of the laws governing
the universe? Considering for the moment only the strengths of the fundamental forces, how much could the value of the electric charge vary without
changing chemistry so as to limit the complexity of organic compounds and
change the improbability of life to an impossibility? The nuclei of carbon,
nitrogen, oxygen, phosphorus, and other elements necessary for life were
formed in supernovae explosions that took place billions of years ago. If the
weak interaction force important in the generation of these explosions were
a little different, would this seeding of space with the heavy nuclei necessary
for life have occurred? If that weak force were different, if the strong color
force were a little stronger or weaker, would the intricate set of nuclear reactions sensitively dependent on these forces still produce the energy of the sun
and stars? Would the stars shine in a universe only slightly different from

ours? Would the heat of the sun warm planets to the precise temperature required for the generation of life? It seems likely that the span of values of the strengths of forces that lead to this universe, hospitable to life as we know it, is small indeed.

If life may depend singularly on the detailed properties of the universe, how much more sensitive it must be to structural differences. We have discussed the attractive view that the original universe held ten dimensions, and six of these space dimensions "rolled" up, leaving the three space and one time dimension we know. If there were only two space dimensions the net of connections of neurons could not be sufficiently complex to power a brain able to contemplate the universe about it. If the universe were to hold more than three space dimensions, the gravitational force would not be such as to allow stable orbits of planets about a sun and the constant temperature required for life. Hence, aside from more subtle arguments (of which there are many) such structurally different universes could not support those who might dare inquire into the origin of their universe.

If intelligent life can be expected to inhabit only a universe almost exactly like ours—and no truly firm answer to so complex a question has been established—does this indicate a motivation of nature? Or merely chance? Are we here as a consequence of a lucky cut of the cards? Or is there another, more singular explanation?

The models of the universe in which we now place considerable confidence suggest that in the first instant of time, the universe was wholly symmetric with indistinguishable particles and indistinguishable forces. And then, upon the elapse of time, spontaneous symmetry-breaking mechanisms operated to produce the universe we know, with particles of different masses and forces of different strengths. We have emphasized the logical parallel of the formation of the universe to the formation of tiny crystallike domains of iron through the cooling and condensation of a gas of very hot iron vapor. The iron gas exhibits complete symmetry—there is no preferred direction. However, in each domain of the cold iron condensed from the gas, the spin and magnetic moment of each atom points in one specific direction, generating a magnetic field in that direction, and the directional symmetry of the original gas is lost. But different domains point in different directions determined by chance in the symmetry-breaking process of the condensation of the gas to a solid. In some interesting models of the formation of the universe, an infinite number of adjacent subuniverses are formed as the proto-universe cools, each with the primordial symmetry broken in a different way defining a different "direction." Even if only a very narrow range of these "directions" can lead to a universe that supports intelligent life, such subuniverses must exist with intelligent inhabitants who attempt to deduce the character of the primordial symmetry and its breaking. Is it then astonishing that we are chosen to inhabit one of the rare universes that can sustain us? No! At least no more astonishing than that we inhabit the only planet in the solar system that can sustain life and that, in the entire history of that universe, it is only at this particular time that we ask fundamental questions. The Anthropic Principle that served to select a special time on a special planet for such questions may as well have selected a special universe.

The Universe: Created for us? yea or nay? But what or who, is or was, the Creator? Is there a Primal cause? What was there before the beginning of space and time? What will be there after the end? We do not know and—exercising a rare humility—we are not even confident that we can know. If the universe was born in a quantum fluctuation, the inherent randomness revealed in quantum mechanics may eliminate the possibility of extrapolation before that incident. Before the beginning of the universe and after the end may be beyond the reach of rationality. With dimensions so distorted that even time may be tied into knots, can we even know what is meant by "before" or what is meant by "Cause"? Perhaps physicists must leave the Cause with theologians and philosophers.

Notes

1. Of the two specific fits of the General Theory to observation, the precession of the orbit of Mercury had not been calculated, and the (anomalous) curvature of light by the sun had not been measured at the time. The equivalence of gravitational and inertial mass was an input to the theory rather than a special consequence.

2. An appreciation of the magnitude of very large or very small numbers is difficult for scientists as well as laymen. The ratio of the Planck distance and the span of the smallest mote the human eye can perceive is about the same as the ratio of the size of the mote and the size of the universe taken as 15 billion light-years. The ratio of the Planck distance and the smallest distance yet probed ($\approx 10^{-17}$ centimeters) is about equal to the ratio of the thickness of a sheet of paper to the distance from the sun to the farthest planet, Pluto. The highest energies presently on accelerator designers' drawing boards are less than 10,000 proton masses per particle. With present techniques, an accelerator the size of the earth's orbit about the sun would still be about 1 million times too small to reach the Planck energy.

3. An expansion of particles leads to an expansion of nomenclature. The supersymmetric (SUSY) boson sisters of the basic fermions take their brother's names with an "s" suffix, the SUSY fermion brothers of the basic bosons form their names by adding the Italian diminutive "ino" to the boson names.

4. The velocity of a wave in a classical string is equal to $\sqrt{T/\rho}$ where T is the tension force and ρ the mass per unit length of the string. The fundamental frequency f of a string of length L secured at both ends is such that $f = v/2L$, and the frequency varies as the square root of the tension.

5. Numbers so large are almost without meaning. If the entire mass of the universe as far as our event horizon of about 15 billion light-years were compressed into a region the size of an atom, the density would be about 10^{80} times that of water. The density to which our diarist refers is 10 billion times greater.

6. We use the term "phase change" to describe a change in a system involving a loss of symmetry, such as the condensation of steam to water or the freezing of water to ice. Consider a small observer (again) of molecular size in a water system. There is no preferred direction, as clumps of H_2O molecules are oriented in all different directions. But then the

water freezes and the observer, now in an ice crystal, finds the H_2O molecules all aligned in specific directions in a lattice. The original symmetry, where all directions are equivalent, has been lost in the phase change from water to ice. This phase change is typical in releasing energy (80 calories per gram) as the increased order corresponds to a reduced entropy. The diarist discusses the phase change from the unification force to the separate electroweak and QCD (strong color) force.

7. From Chapter 16, footnote 11, if the distance between two points is R, the expansion rate is proportional to R/d, $dR/dt = \alpha R/d$, where $\alpha = \sqrt{8\pi/3)G}$, and G is the gravitational constant. Integrating,

$$R = R_0 e^{\alpha t}$$

where R_0 is the distance between the points at the time $t = 0$ the inflation began.

Index

Abbot, Edwin, A., 127n
Abelian, 220
Abelian, non-Abelian fields, 339–342
Aberration, 79
Absolute zero (of temperature), 44
Acceleration, absolute character of, 103–104
Action, 29n, 164
Addition of velocities, Galilean Relativity, 73
  Special Relativity, 93–94
Alpha particle, 195
Ambler, Eric, 232, 282
Ampere, Andre, 128
Ampere's Law, 129–131, 133
Angular Momentum, 16–22, 26–28, 140
  in quantum mechanics, 176–180
Anthropic Principle, 365–369
Antimatter, 236–238
Aristotle, 141
Arnold, Mathew, 366
Aston, F. W., 249
Asymptotic freedom, of quarks, 276–278
Atoms, as standing waves, 198–200
  chemical bases, 39, 41
  described by quantum mechanics, Bohr Model, 194–207
  electromagnetic radiation from, 200–202
Avogadro, Amedeo, 40
Avogadro's number, 45

Barrier penetration, 171–173
Baryon's, 214, 228n, 248
Becquerel, Henri, 280
Bell, J. S., 183
Bernoulli, Daniel, 42
Big Bang, 311
Binding energy, 197
Black Holes, 108–109, 318
Blake, William, 366

Bohm-Aharanov effect, 64, 329–331
Bohm, David, 183
Bohr Model of atom. *See* Atoms
Bohr, Neils, 6, 149, 150, 196
Boltzmann, Ludwig, 140
Boltzmann statistics, 181
Born, Max, 149
Bose-Einstein statistics, 180–182
Bosons, 181–182, 347, 351
Bottom quark. *See* Flavor
Boundary Conditions, 14
Boyle, Robert, 41
Boyle's Law, 41–43
Bradley, James, 79
Brahe, Tycho, 353
Brownian movement, 45–47
Brown, Robert, 46
Butler, C. C., 256

Cabibbo angle, 293–296
  model, 290–294
Cabibbo, Nicolo, 291
Carroll, Lewis, 230
Causality, in Special Relativity, 93
Celsius (temperature scale), 44
Chadwick, James, 249
Charge, 211–213
  color, 264–266
  conjugation, 236–238
  conservation, 211, 245–247
  independence, 249, 250, 255–256, 254–255, 306n; *see also* SU(2)
  quantization, 243–245
  of the quark, 269–271
  strong interaction, 213
Charm quark. *See* Flavor
Charmonium, 271–274
Chirality, 282, 344
Clausius, Rudolf, 44, 146
Collisions, 19